D0151272

This book surveys both the history of the earth and the nature of the processes that have controlled the earth's history. It integrates information from many fields to provide a comprehensive summary of this interdisciplinary subject. Fundamental processes such as convection, thermal evolution of the earth, evolution of the crust and mantle, orogeny and rifting are explained. Historical topics such as the origin of life, paleontologic extinction events, differences between the Archean and younger time periods, deposition of Precambrian sediments, and evolution of the atmosphere and oceans are discussed. The book then focuses on key topics in the earth sciences such as the development of modern ocean basins, the history of Phanerozoic orogenic belts, and the nature of cratonic sedimentary cover sequences.

A history of the earth can be used as a broad introduction to this vast subject for all interested in the study of the earth and will provide a concise and accessible overview for readers in other research areas.

A HISTORY OF THE EARTH

A HISTORY OF THE EARTH

John J. W. Rogers

W. R. Kenan Jr. Professor of Geology
University of North Carolina at Chapel Hill

CAMBRIDGE
UNIVERSITY PRESS

Published by the Press Syndicate of the University of Cambridge
The Pitt Building, Trumpington Street, Cambridge CB2 1RP
40 West 20th Street, New York, NY 10011-4211, USA
10 Stamford Road, Oakleigh, Melbourne 3166, Australia

First published 1993
Reprinted 1994

Printed in Great Britain at the University Press, Cambridge

A catalogue record for this book is available from the British Library

Library of Congress cataloguing in publication data

Rogers, John J. W. (John James William), 1930–
A history of the earth/John J. W. Rogers.
p. cm.
Includes bibliographical references and index.
ISBN 0-521-39480-5. – ISBN 0-521-39782-0 (pbk.)
1. Historical geology. I. Title.
QE28.3.R64 1994
551.7–dc20   93-1661 CIP

ISBN 0 521 39480 5 hardback
ISBN 0 521 39782 0 paperback

CONTENTS

PREFACE

From the standpoint of the human mind, the earth is very large and very complicated. How, therefore, can we describe its history in any book smaller than a multi-volume compendium? This book is largely an exercise in scale. I have attempted to discover, in each of the many aspects of geologic history, some level of information that is important to all geologists. What information do specialists in different fields need to convey to each other about their various disciplines in order to promote an understanding of the development of the earth? What have we agreed upon – actually, not very much – and what are the major controversies?

The book is mainly intended for senior and first-year graduate students in geological sciences, but the extensive maps and capsule discussions may make it useful as a reference for anyone in the field. Hopefully, it can provide an overview of the earth for students who will then go directly into the study of a speciality within the earth sciences, and perhaps this broad introduction will serve to promote communication between investigators in their separate fields. Although the book is directed toward an audience that has some knowledge of the earth sciences, readers with an interest in the earth but little formal background could profit from much of what is said.

The book has been fun to write. Much of my enjoyment has come from the willingness of colleagues and friends to review – in many cases shred – what I have written. I acknowledge a profound sense of gratitude towards, but shift no responsibility to, the following people. The first is Thomas W. (Nick) Donnelly, who reviewed the entire manuscript with a mixture of enthusiasm and rigor that greatly improved the final product. Others who reviewed major parts of the work include Tim Bralower, Marlene Braun, Joe Carter, Pat Gensel, Edith Macrae and Charlie Paull. I also received great help from Drew Coleman, Geoff Feiss, Allen Glazner, Chris Powell, Bob Parr, Jose Rial and Tom Rossbach. Bob Butler, Michael Follo and Kevin Stewart contributed photographs and assistance. Loren Babcock, Ed Landing, Mark Leckie and Bill Schopf sent photographs and suggestions. Laurel Kaczor assisted in preparation of some of the illustrations. My wife, Barbara, was patient, for which I remain very grateful.

Rock isolated in forest. This outcrop exhibits an unconformity between Archean and early-Proterozoic rocks in northern Australia (discussion in Needham, Crick and Stuart-Smith, 1980, and Stuart-Smith *et al.*, 1980).

1

GEOLOGIC TIME

1.0 Introduction

WE START with a rock. Perhaps it is in a forest, surrounded by soil and trees. We want to know the relationship between the rock and other rocks – is it part of a sedimentary sequence? – has it been offset on a fault? – how old is it? These relationships may be difficult to establish, particularly if the rock is in the forest and we cannot see its neighbors.

When we know as much about the rock's field relationships as can be ascertained, we take a sample into the laboratory and ask further questions of it. We might discover that it was originally a sediment, derived from two different source terranes, that was later metamorphosed and deformed at three separate times, the last of which was sufficiently intense to extract part of the original rock as a partial melt. (Alright, 'discover' may be too strong a word for a history this complicated – let us use 'infer'.) Outfitted with this information, and similar data from other rocks, we construct a history of the area in which the rock occurs. If we are very brave, we may construct a history of the earth. That is what this book attempts to do.

The earth does not divulge its history easily. Nowhere is our difficulty more obvious than in the field, where we might find that the rock that we need to study is pro-

tected from our efforts to investigate it. Perhaps it is on land owned by someone who does not welcome people with sledgehammers. Possibly it is guarded by the farmer's bull, or a cobra. It may be covered by 3000 meters of ice, 5000 meters of sea water, 10 meters of glacial till, or 60 acres of parking lot. Perhaps it is on the inaccessible face of a high mountain. As geologists, we face the fact that we cannot always obtain the information that we regard as crucial for the solution of a problem.

Assuming that we obtain adequate field and laboratory information about the rock, we embark on a process mostly restricted to the geological sciences, the task of assembling an historical record from objects whose development was never affected by human activity. Scientists in other disciplines commonly do not understand the process. How can we sift through the mass of conflicting data that the earth provides us? Wouldn't it be possible to interpret the gravity surveys in a different fashion? Why do we think that the variety of dates for the same rock obtained from different isotopic systems can be resolved into only two different events in the rock's history? How can we correlate strata between two mountain ranges separated by an alluvium-filled valley? 'Are you not,' ask the chemists and the physicists, 'drawing conclusions from inadequate and inconsistent data interpreted on the basis of unwarranted assumptions?'

Well, yes. We lack some of the 'advantages' that investigators in other disciplines have. A chemist who does not like the results of an experiment can run it again. The experiment can be repeated under more-uniform conditions of temperature and pressure, starting with a simpler mix of materials, and carefully monitored during its progress. Chemists, and scientists in similar fields, have control over their experiments.

Geologists do not have control when they study the history of the earth. The only experiment in which we are involved was run for us. It began about 4500 million years ago with a chemical mixture of extraordinary and, largely unknown, complexity and proceeded with no external regulation of the reaction. Furthermore, nobody watched it. Our task is to take the results of this experiment – this mess that has been left for us – and infer everything that happened during its progress. Because of this complexity, our explanations (models?, inferences?) are seldom complete.

Although we lack control over the experiment that has constructed the present earth, we have some advantages that members of other disciplines do not have. We have the whole earth to ponder. We make some progress. The history of the earth is better understood now than it was twenty years ago and will be even better understood twenty years from now. We have a lot of fun. Let us begin with a discussion of geologic time.

1.1 Measurement of geologic time

THE GEOLOGIC time scale is constructed from numerous different types of observations. They can be divided into those that measure relative time and those that measure absolute time. We use Ma (or MA) to signify millions of years ago.

Relative time scales

Within a local area, the relative sequence of events is obtained from traditional observations. Sedimentary suites become younger upward unless they have been deformed, and many sediments contain top and bottom criteria that permit overturning to be recognized. Similarly, faults are younger than rocks that they displace, dikes are younger than rocks that they intrude, etc. Within the limits placed by adequacy of exposure

and uncertainty of interpretation of observations, relative histories can be obtained in local areas with reasonable accuracy.

Expansion of a relative time scale from one area to others at great distance is very difficult, particularly if the areas are on different continents. Relative relationships are virtually impossible to infer for the Precambrian, which occupies approximately the first 85% of geologic history. In the past 550 to 600 Ma (the Phanerozoic), correlations can be based on fossils, with all of the attendant problems of uncertainties in the age ranges of individual organisms, variability of biotic suites from one depositional environment to another, and provinciality of organism distribution (Section 6.7). Based on dates provided by radiometric ages, fossil ages may be in error by ten million years or more in the early Paleozoic but become more precise toward the present.

The paleontologic time scale is relative in two senses. The obvious one is that skeletal remains and their enclosing sediments do not yield radioactive dates except possibly within approximately the past hundred thousand years (in which ages can be obtained from measurements of ^{14}C and from U and Th daughters not in equilibrium within their decay series). The paleontologic scale also is relative in the sense that it depends on the accuracy of stratigraphic information used to construct it. Relative ages of fossils are established originally by sequence within sedimentary columns and correlation of those columns with each other. Thus, the apparent ages of fossils can be affected by unrecognized unconformities, repetition of strata not detected by mapping, and inadequacy of exposure (among other problems).

Radioactive methods (discussed below) can be used to obtain a quantitative history of events in a sedimentary sequence, but the errors involved may be large. The only sedimentary rock or mineral that is customarily dated radioactively (by K–Ar and Rb–Sr) is glauconite, a green, K-bearing clay formed during or shortly after deposition; the results are highly controversial because of the possibility of post-depositional alteration. With that exception, all sediments must be dated by constraining their position between datable igneous (rarely metamorphic) rocks. The best ages are obtained from lava flows or pyroclastic rocks within the sedimentary sequence. Some sediments, however, can be constrained only as younger than igneous rocks underlying them and older than igneous rocks that

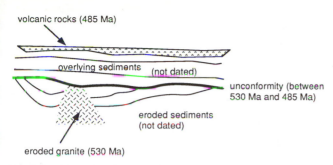

volcanic rocks (485 Ma)

overlying sediments   (not dated)

unconformity (between
530 Ma and 485 Ma)

eroded sediments
(not dated)

eroded granite (530 Ma)

Fig. 1.1. Example of dating of unconformity. Even assuming that ages of eroded rocks and overlying volcanic rocks are accurate, the age of the unconformity can be specified only as occurring between 530 Ma and 485 Ma, a range of 45 m.y. □

intrude them; the age range from these observations may be large.

Most of the major subdivisions of the Phanerozoic (fossil-bearing) part of the geologic time scale are based on unconformities or rapid biotic changes in sedimentary sections exposed on continents. Ages of unconformities can be constrained only as younger than rocks eroded below them and older than some intrusive or overlying extrusive suite (Fig. 1.1). The accuracy of dates assigned to various subdivisions of the Phanerozoic is discussed in Section 1.2.

Although stratigraphic/paleontologic correlation is the major method of relative dating, we must mention one other technique. For reasons that are very obscure, the earth's magnetic field has undergone numerous reversals ('flip-flops') of its north and south magnetic poles. Suitable rocks (Fe-bearing) retain a 'memory' of the magnetic field at the time when they formed; thus, sequences of rocks can retain the history of field changes over various intervals of time. The best-known sequences of field reversals are preserved in oceanic basalts as they form at, and move away from, spreading ridges. This process creates magnetic stripes in the Jurassic to Recent crust of modern ocean basins and presumably had the same effect on the now-destroyed crust of older oceans.

Because magnetic reversals are irregular through time rather than periodic (Section 1.3), the magnetic-reversal pattern is a 'fingerprint' unique to each interval of time and each suite of rocks formed in that interval (Fig. 1.2). This uniqueness has been used for relative dating of magnetic stripes in modern oceans and for some suites of terrestrial rocks, such as Tertiary

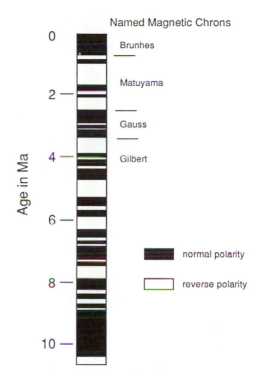

Named Magnetic Chrons

Age in Ma

Brunhes

Matuyama

Gauss

Gilbert

■ normal polarity

□ reverse polarity

Fig. 1.2. Magnetic reversal pattern for the past 10 million years (from Berggren, Kent and van Couvering, 1985). □

redbeds. Magnetic fingerprinting has also been used with far less confidence for pre-Jurassic oceanic crust preserved on land and for older terrestrial volcanic and sedimentary suites.

Absolute time scales

The most significant method for determination of the absolute age of a geologic event is measurement of the abundances of a radioactive parent isotope and its daughter decay product. This section outlines the general procedures for the following systems: Rb–Sr, Sm–Nd, U–Th–Pb, and K–Ar. Further discussion of the various isotopic systems is provided in Section 2.3. Non-radiometric dating techniques are generally less satisfactory than radiometric ones; they include fission-track analyses, rates of racemization of amino acids (Section 3.6), rates of hydration of obsidian, etc.

The various isotopic dating techniques are useful for different types of materials. Because Rb–Sr systems can be modified by metamorphism, Rb–Sr dating is most important for determination of the ages of

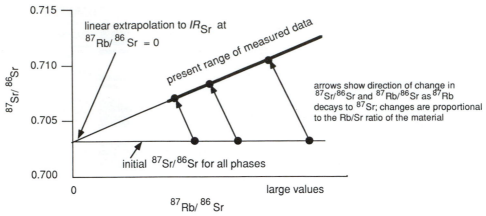

Fig. 1.3. Explanation of dating by the Rb–Sr method using a diagram that plots $^{87}Rb/^{86}Sr$ vs. $^{87}Sr/^{86}Sr$. A set of rocks or minerals that formed at the same time, such as the minerals in a quickly crystallized igneous rock, start with the same (initial) $^{87}Sr/^{86}Sr$ ratio (referred to as IR_{Sr}). Decay of ^{87}Rb to ^{87}Sr increases the $^{87}Sr/^{86}Sr$ ratio and decreases the $^{87}Rb/^{86}Sr$ ratio in each rock or mineral. Thus, the decay produces $^{87}Rb/^{86}Sr$ and $^{87}Sr/^{86}Sr$ ratios that lie on a line radiating from the IR_{Sr} at $^{87}Rb/^{86}Sr=0$. The slope of the line is proportional to the age at which the isotopic system was established; that is, it is an 'isochron' that indicates the age of initial formation of the rock or suite of rocks. The IR_{Sr} is useful in determining the source of the magma (Section 2.3).   □

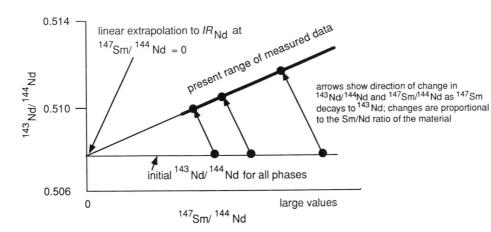

Fig. 1.4. Explanation of dating by the Sm–Nd method using a diagram that plots $^{147}Sm/^{144}Nd$ vs. $^{143}Nd/^{144}Nd$. Decay of ^{147}Sm to ^{143}Nd produces a rotated isochron from an initial $^{143}Nd/^{144}Nd$ ratio in exactly the same fashion as the decay of ^{87}Rb to ^{87}Sr (Fig. 1.3). Initial $^{143}Nd/^{144}Nd$ ratios are discussed in Section 2.3.   □

igneous rocks that have not been disturbed after crystallization and for dating of metamorphic events. Rb–Sr geochronology is particularly important for upper crustal rocks, in which the Rb/Sr ratio is high. Conversely, Sm/Nd ratios are highest in the mantle (Section 2.3), and the Sm–Nd system is particularly useful for mafic rocks. The U–Th–Pb system is affected by numerous processes that may complicate determination of the age of the material being investigated; one major problem is incorporation of old minerals such as zircon into younger igneous or sedimentary rocks, thus leading to mixing of isotopic systems established at different times. The U–Th–Pb system is very widely used because it provides information on this large variety of processes. The K–Ar system is most

useful in determining sequences of heating events in complex terranes.

All radiometric dating is based on the observation that radioactive isotopes decay at a constant rate characteristic of the individual isotope. That is, let P_0 = the number of initial parent atoms, P = the number of atoms remaining after a decay time = t, D_d = the number of daughter atoms formed by decay of P in the material being measured, and λ = the decay constant of the isotope; the decay constant is related to the half life of the parent by the relationship $\lambda = \ln2/(\text{half life})$. Then

$$-dP/dt = \lambda P$$
thus, $$P/P_0 = e^{-\lambda t}$$
also $$D_d = P_0 - P.$$

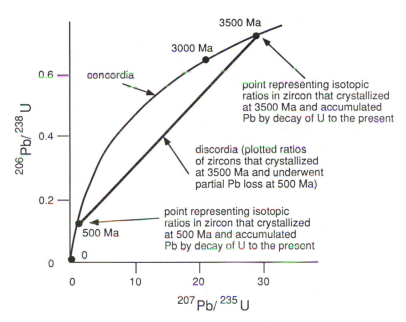

Fig. 1.5. Explanation of U–Pb dating of minerals that originally have a very high U/Pb ratio. Decay of both U isotopes from the formation of the mineral or rock to the present produces $^{207}Pb/^{235}U$ and $^{206}Pb/^{238}U$ ratios uniquely determined by the age of crystallization. The curve produced by this decay in materials of different age is referred to as 'concordia' and represents the variation through time of $^{207}Pb/^{235}U$ and $^{206}Pb/^{238}U$ ratios caused by pure radioactive decay if the materials were not modified by other chemical processes.

Unfortunately, most minerals do not contain simply a parent and all of the daughter product formed by its decay. Two major problems are 'inheritance' and Pb loss. Inheritance is most easily shown by the observation that many zircons in igneous or metamorphic rocks contain cores of older grains over which the younger grains grew. Thus, these grains contain isotopic systems that are a mixture of two or more periods of growth and decay. The problems of mixing are beyond the scope of this discussion.

Pb loss is exemplified by data from zircons. Zircons are the most commonly dated mineral, but they do not readily retain Pb in their lattices. The consequent loss of Pb after crystallization may occur over a period of time but is more commonly related to a single later event, perhaps reheating during low-grade metamorphism. In the diagram, zircons originally crystallize in a rock at 3500 Ma, and the U in the zircons begins to decay. If the decay proceeded to the present without chemical processes, the $^{207}Pb/^{235}U$ and $^{206}Pb/^{238}U$ ratios would now have the values shown on the concordia at 3500 Ma. The zircons shown in this diagram, however, exhibit some Pb loss at 500 Ma; (this loss normally is partial loss from all grains but could also be described as total loss from some grains and no loss from others). At 500 Ma, unaltered zircons would have the isotopic ratios shown on the concordia at 3000 Ma (3500 – 500), and for the next 500 m.y. their U would decay until isotopic ratios reached the concordia point at 3500 Ma. Zircons experiencing total Pb loss at 500 Ma, however, would begin again to accumulate Pb decay products from the point for time=0; in 500 m.y. these reset zircons would achieve isotopic ratios shown on the concordia at 500 Ma.

The development of U–Pb isotopic systems in the zircons at two different times causes the total zircon suite in the rock to have isotopic ratios that are a linear mixture of the concordant points at 3500 Ma and 500 Ma. This line is a 'discordia', and its intercepts with concordia show both the time of initial crystallization and the time of Pb loss. If the Pb loss was small, the zircon would be nearly concordant at the higher age and yield a good value for initial crystallization. If the Pb loss was large, an estimate of the time of Pb loss could be made with some accuracy, but the initial age of the zircon might be estimated only within large errors. □

Thus, measurement of present values of P and D_d in a rock or mineral theoretically yields its age through the ratio P/P_0, which equals $P/(P + D_d)$.

If some initial daughter (D_i) was present in the material measured, then the total amount of daughter present is $D_t = D_d + D_i$, and an estimate must be obtained of D_i in order to determine D_d. This determination is made by measuring a series of materials that have different P/D_t ratios. These measurements yield a general relationship of

$$D_t = D_i + D_d$$
or $$D_t = D_i + (P_0 - P)$$
thus, $$D_t = D_i + P(e^{\lambda t} - 1).$$

This relationship is linear between D_t and P for a constant D_i and t. Therefore, measurement of D_t and P and extrapolation to $P = 0$ yields both t and the initial D_i. The process is shown graphically for the Rb–Sr (Fig. 1.3) and Sm–Nd (Fig. 1.4) systems and further described below in the discussion of Sr isotopes.

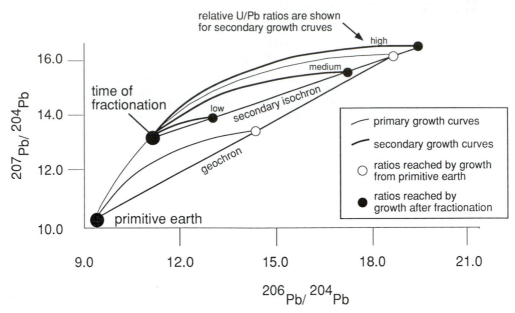

Fig. 1.6. Explanation of U–Pb dating of materials that contain significant amounts of initial Pb (see also Section 2.3). Both ^{207}Pb and ^{206}Pb are plotted as ratios against ^{204}Pb, which is not produced radioactively by any parent isotope now existing in the earth. In the figure, 'growth curves' characterized by different U/Pb ratios emanate from the ^{207}Pb/^{204}Pb and ^{206}Pb/^{204}Pb ratios in the original earth (inferred largely from studies of Pb in U-free meteorites). Without discussing the mathematics of these curves, we can say that evolution along them to the present leads to ^{207}Pb/^{204}Pb and ^{206}Pb/^{204}Pb ratios that lie on a straight line, referred to as the 'geochron'.

A rock or mineral that evolves along a growth curve for the primitive earth may undergo some event that changes its U/Pb ratio (for example, metasomatic removal of U). The altered material, then, continues to evolve its ^{207}Pb/^{204}Pb and ^{206}Pb/^{204}Pb ratios along a new growth curve emanating from the point on the original growth curve at which the event occurred. Any event that establishes different phases with different U/Pb ratios also establishes a series of growth curves that result in a linear set of points in the same fashion as the development of the geochron. The line formed is referred to as a 'secondary isochron', and its slope defines the age of the resetting event. One example is the crystallization of a magma into a rock containing minerals with different U/Pb ratios, and the resulting isochron is the age of crystallization. The isochron formed during magmatic crystallization radiates from the 'initial' ^{207}Pb/^{204}Pb and ^{206}Pb/^{204}Pb values in the magma at the time of crystallization; these initial values lie on the growth curve for the magmatic source.   □

The *Sr isotopic system* contains both ^{87}Sr and ^{86}Sr. The ^{86}Sr is not radiogenically produced in the present earth, whereas some of the naturally occurring ^{87}Sr is formed by the decay of ^{87}Rb, with a half life of 4.88×10^{10} years. Radiogenic ^{87}Sr is added to the original Sr in a rock or mineral and increases the ^{87}Sr/^{86}Sr ratio as decay continues. Thus, any rock or mineral that contains both Rb and Sr has an initial ^{87}Sr/^{86}Sr ratio and a higher present ratio. The initial ratio of ^{87}Sr/^{86}Sr is $(^{87}\text{Sr}/^{86}\text{Sr})_i$, commonly designated as IR_{Sr}. Measurement of present ^{87}Sr/^{86}Sr and ^{87}Rb/^{86}Sr ratios produces an isochron with the equation

$$(^{87}\text{Sr}/^{86}\text{Sr})_{present} = (^{87}\text{Sr}/^{86}\text{Sr})_i + (^{87}\text{Rb}/^{86}\text{Sr})_{present}(e^{\lambda t} - 1).$$

This equation permits determination of the age of the material and the initial ratio (further explanation in Fig. 1.3).

The *Nd isotopic system* produces initial ratios and isochrons similar to those produced by the Rb–Sr system

(Fig. 1.4). The Nd isotopes of interest are ^{143}Nd and ^{144}Nd, and radiogenic production of ^{143}Nd by decay of ^{147}Sm occurs with a half life of 1.06×10^{11} years (Section 2.3).

The *Pb isotopic system* contains: ^{206}Pb, formed by the decay of ^{238}U with a half life of 4.468×10^9 years; ^{207}Pb, formed from ^{235}U with a half life of 0.7038×10^9 years; and ^{208}Pb, formed from ^{232}Th with a half life of 14.01×10^9 years (also see Section 2.3). Because the isotopes ^{235}U and ^{238}U have identical chemical properties and masses that differ by only ~1%, they cannot be fractionated by chemical processes in the earth. Thus, the ^{235}U/^{238}U ratio varies only because of the more rapid decay of ^{235}U. Consequently, a rock or mineral crystallizing in the earth inherits a ^{235}U/^{238}U ratio determined solely by its age, with the ratio decreasing from an original value of 0.22 at 4.5×10^9 years ago to a present value of 0.007. Fractionation of Th from U in the earth adds complexities to the U–Th–Pb system that are beyond the scope of

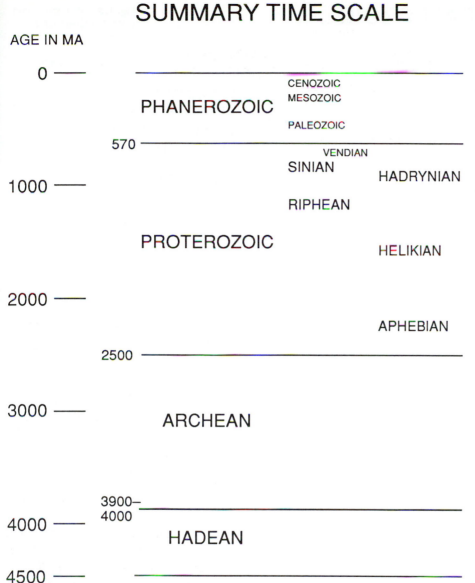

SUMMARY TIME SCALE

AGE IN MA

Fig. 1.7. General time scale. No consistent terminology exists for subdivision of any part of the Precambrian, and we include only those terms that are needed in this book or that have some general, worldwide use. The terms 'Aphebian', 'Helikian' and 'Hadrynian' are based on work in the Canadian shield (Chapters 3 and 5). 'Vendian' and 'Riphean' are originally Russian terms, and 'Sinian' is Chinese. Approximate ages are from Harland *et al.* (1990).   □

this book, and hence we discuss only the decay of U. U–Pb dating can be accomplished by two methods: 1) dating of zircons and other minerals that did not contain Pb when they crystallized (explanation in Fig. 1.5); and 2) dating of rocks and minerals that did contain initial Pb (explanation in Fig. 1.6).

The *K–Ar isotopic system* is based on the production of both ^{40}Ar and ^{40}Ca by the decay of ^{40}K with a half life of 1.25×10^9 years. The production of two daughter isotopes is accomplished by 88.8% of the ^{40}K undergoing beta decay (electron loss) to ^{40}Ca and 11.2% undergoing capture of the inner (K) electron to form ^{40}Ar. Thus, the ^{40}K/^{40}Ar ratio of a mineral or rock is theoretically capable of yielding the age of formation of the material.

Because Ar is an inert gas, however, it does not bond into any mineral lattice and is easily lost during metamorphism, weathering, or any other later event. For this reason, ^{40}K/^{40}Ar ratios in most whole rocks merely indicate the age of the last heating event, at which time all Ar was lost and the K–Ar 'isotopic clock' was reset to zero. This technique has been most useful for very young rocks, such as Cenozoic volcanics, and also as an estimate of the time at which old terranes became tectonically stable (for example, the 'cratonization ages' of many Precambrian terranes were originally determined by whole-rock K–Ar dates; Section 3.0).

The problem of Ar loss (Ar gain in some situations) can be overcome partly by study of ^{40}Ar/^{39}Ar ratios.

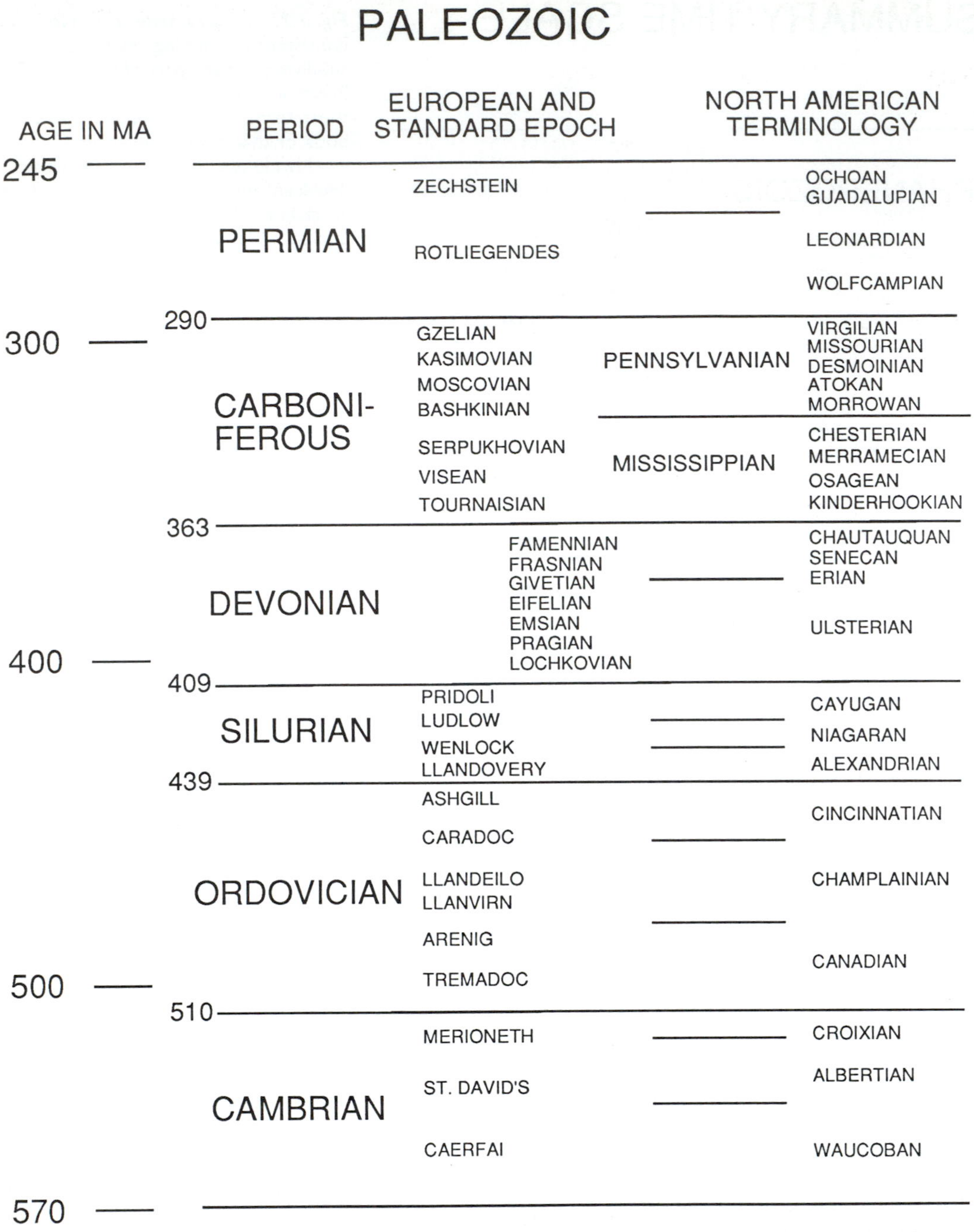

PALEOZOIC

AGE IN MA	PERIOD	EUROPEAN AND STANDARD EPOCH		NORTH AMERICAN TERMINOLOGY
245	PERMIAN	ZECHSTEIN		OCHOAN / GUADALUPIAN
		ROTLIEGENDES		LEONARDIAN
				WOLFCAMPIAN
290 / 300	CARBONI-FEROUS	GZELIAN, KASIMOVIAN, MOSCOVIAN, BASHKINIAN	PENNSYLVANIAN	VIRGILIAN, MISSOURIAN, DESMOINIAN, ATOKAN, MORROWAN
		SERPUKHOVIAN, VISEAN, TOURNAISIAN	MISSISSIPPIAN	CHESTERIAN, MERRAMECIAN, OSAGEAN, KINDERHOOKIAN
363	DEVONIAN	FAMENNIAN, FRASNIAN, GIVETIAN, EIFELIAN, EMSIAN, PRAGIAN, LOCHKOVIAN		CHAUTAUQUAN, SENECAN, ERIAN, ULSTERIAN
400 / 409	SILURIAN	PRIDOLI, LUDLOW, WENLOCK, LLANDOVERY		CAYUGAN, NIAGARAN, ALEXANDRIAN
439	ORDOVICIAN	ASHGILL, CARADOC, LLANDEILO, LLANVIRN, ARENIG, TREMADOC		CINCINNATIAN, CHAMPLAINIAN, CANADIAN
500 / 510	CAMBRIAN	MERIONETH, ST. DAVID'S, CAERFAI		CROIXIAN, ALBERTIAN, WAUCOBAN
570				

Fig. 1.8. Paleozoic time scale. European (and standard) terminology is correlated with terminology used in North America. Approximate ages are from Harland et al. (1990). □

The technique consists of irradiating the sample to be measured, thereby converting ^{39}K into ^{39}Ar. Because the ^{39}K has a constant ratio with ^{40}K, the resultant ^{40}Ar/^{39}Ar ratio is a measure of the age of the material. One major advantage of the ^{40}Ar/^{39}Ar technique is that the release of the two Ar isotopes from the sample can be measured at different temperatures. Thus, for example, a rock that was crystallized and later reheated may contain some minerals from which no Ar was lost in the later event; because of continuous production of ^{40}Ar, these minerals would have high ^{40}Ar/^{39}Ar ratios that indicate the initial age of crystallization. Conversely, minerals from which Ar was lost in the later event would have low total ^{40}Ar and low ^{40}Ar/^{39}Ar ratios that indicate the time of reheating.

[**References** – Cowie and Brasier (1989) discuss the age of the Precambrian/Cambrian boundary. Summaries of dating techniques are: Rb–Sr, Faure and Powell (1972); Sm–Nd, DePaolo (1988); U–Pb,

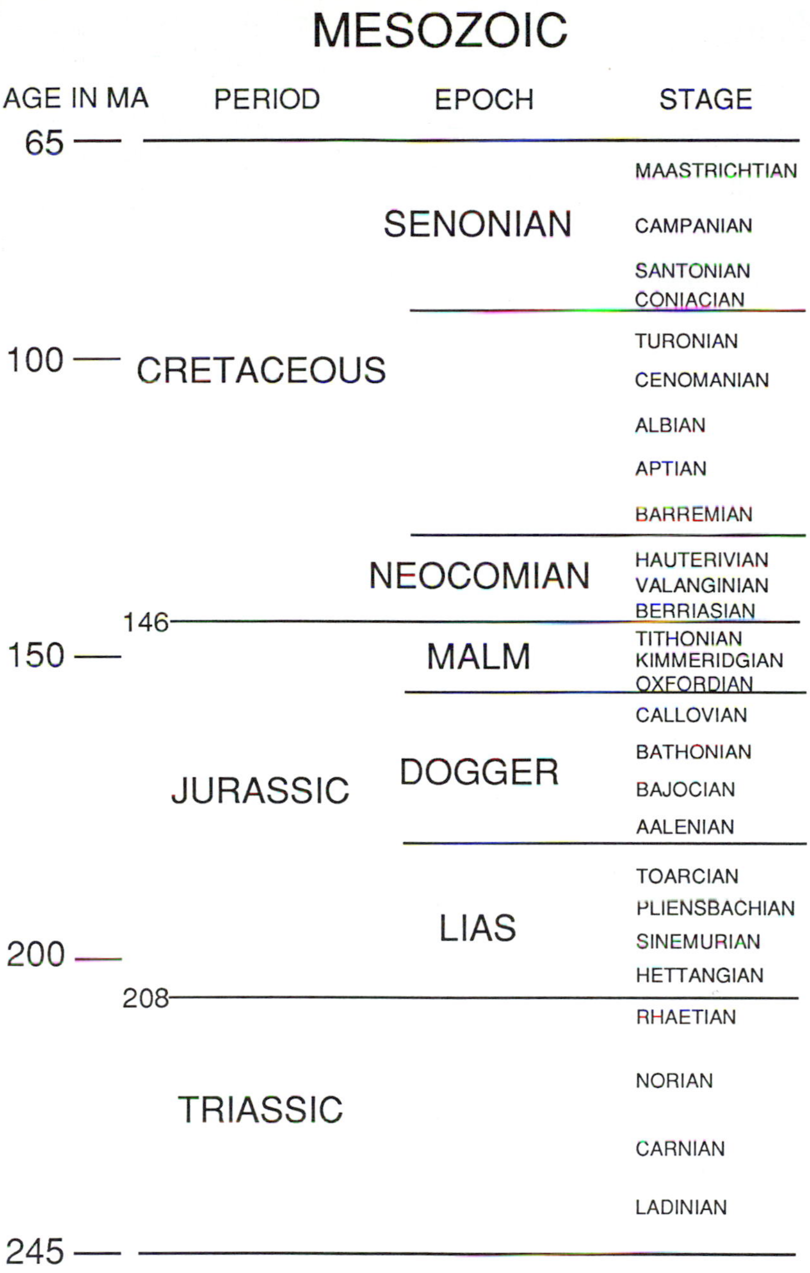

MESOZOIC

AGE IN MA	PERIOD	EPOCH	STAGE
65 —			
		SENONIAN	MAASTRICHTIAN
			CAMPANIAN
			SANTONIAN
			CONIACIAN
100 —	CRETACEOUS		TURONIAN
			CENOMANIAN
			ALBIAN
			APTIAN
			BARREMIAN
		NEOCOMIAN	HAUTERIVIAN
			VALANGINIAN
146			BERRIASIAN
150 —		MALM	TITHONIAN
			KIMMERIDGIAN
			OXFORDIAN
			CALLOVIAN
	JURASSIC	DOGGER	BATHONIAN
			BAJOCIAN
			AALENIAN
			TOARCIAN
		LIAS	PLIENSBACHIAN
			SINEMURIAN
200 —			HETTANGIAN
208			RHAETIAN
			NORIAN
	TRIASSIC		CARNIAN
			LADINIAN
245 —			

Fig. 1.9. Mesozoic time scale. Approximate ages are from Harland *et al.* (1990). ☐

Doe (1970); K–Ar, McDougall and Harrison (1988); and general geochronology, Faure (1986) and Geyh and Schleicher (1990).]

1.2 The geologic time scale

THE GEOLOGIC time scale is a somewhat flexible set of terms used to identify various time intervals in the history of the earth. The basic subdivision (Fig. 1.7) is into eras, conventionally recognized as Archean, Proterozoic, Paleozoic, Mesozoic, and Cenozoic. The beginning of the 'geological' history of the earth, represented by the formation of the oldest preserved materials, is generally regarded as the start of the Archean (about 4000 Ma to 3900 Ma). A Hadean era can be identified prior to the Archean as the earliest stages of the earth's existence. The Phanerozoic is not normally regarded as an era but a collection of eras characterized by the presence of skeletal remains of organisms. The term 'Precambrian' encompasses both the Proterozoic and the Archean.

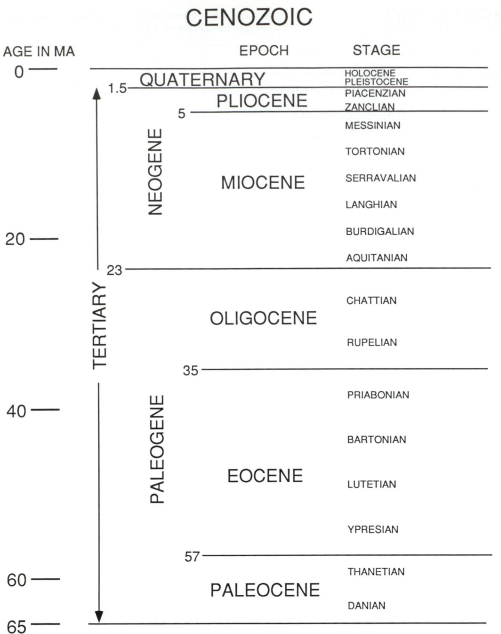

CENOZOIC

AGE IN MA		EPOCH	STAGE
0 —		QUATERNARY	HOLOCENE
	1.5		PLEISTOCENE
		PLIOCENE	PIACENZIAN
	5		ZANCLIAN
			MESSINIAN
			TORTONIAN
		MIOCENE	SERRAVALIAN
			LANGHIAN
20 —			BURDIGALIAN
	23		AQUITANIAN
		OLIGOCENE	CHATTIAN
			RUPELIAN
	35		
40 —			PRIABONIAN
			BARTONIAN
		EOCENE	LUTETIAN
			YPRESIAN
	57		
60 —		PALEOCENE	THANETIAN
65 —			DANIAN

NEOGENE · PALEOGENE · TERTIARY

Fig. 1.10. Cenozoic time scale. The spacing of names is an inadequate representation of relative durations of the Pleistocene and Holocene in the Quaternary. The Pleistocene is largely the age of glaciation in the northern hemisphere. The Holocene is the time since the retreat of the last major glaciers and occupies approximately the past 10 000 years. Some geologists use the term 'Recent' instead of Holocene. Approximate ages are from Harland *et al.* (1990).

□

Chronologic divisions in the Phanerozoic (Figs 1.8 to 1.10) are based primarily on paleontologic and stratigraphic relationships. Most of the boundaries between intervals represent an unconformity in at least some part of the world. Most chronologic/stratigraphic terms have a history (commonly quite a fascinating one) rooted in the early development of the geological sciences, and we cannot review them here. Absolute dates provided by isotopic methods are coordinated with the stratigraphic scale where possible, but many of the ages are controversial (Section 1.1). Uncertainties in the dates shown in Figs 1.8 to 1.10 may be considerably greater than ten million years in older parts of the scale and several million years toward the younger end.

The Paleozoic, Mesozoic and Cenozoic eras are divided into a number of periods. The periods, in turn, are divided into epochs and the epochs into stages. Virtually complete agreement exists worldwide on the names of periods, with the exception that North

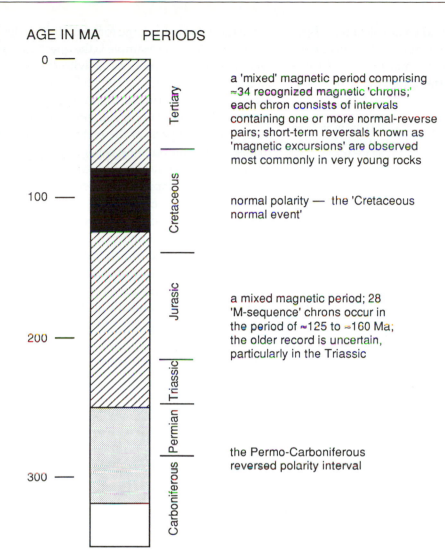

AGE IN MA   PERIODS

Tertiary — a 'mixed' magnetic period comprising ≈34 recognized magnetic 'chrons;' each chron consists of intervals containing one or more normal-reverse pairs; short-term reversals known as 'magnetic excursions' are observed most commonly in very young rocks

Cretaceous — normal polarity — the 'Cretaceous normal event'

Jurasic — a mixed magnetic period; 28 'M-sequence' chrons occur in the period of ≈125 to ≈160 Ma; the older record is uncertain, particularly in the Triassic

Triassic | Permian | Carboniferous — the Permo-Carboniferous reversed polarity interval

Fig. 1.11. Magnetic time scale. This figure is highly generalized, with some indication of the detailed magnetic record shown in Fig. 1.2. Approximate ages are from Harland *et al.* (1990).

□

American geologists tend to subdivide the Carboniferous into the Mississippian and Pennsylvanian. Names used for the epochs and stages within periods, however, are widely different in different parts of the world. Geologists in some countries use classifications that are unique to their country, a process that does not improve international correlation. The names of epochs and stages given in Figs 1.8 to 1.10 are generally European and are more widely used around the world than other classifications. In the Paleozoic, however, the North American terminology is both well established and very different, and it is also shown in Fig. 1.8.

Uncertainties in the dates of their boundaries lead to uncertainties in the duration of the various chronologic/stratigraphic intervals shown on Figs 1.8 to 1.10. Some approximation of the duration of each interval is provided by the spacing of epoch and stage names.

Most of the chronologic/stratigraphic epochs in the Paleozoic have durations of approximately 6 to 15 m.y. Similar ranges are 5 to 10 m.y. for stages in the Mesozoic, ~5 m.y. for the early part of the Cenozoic (Paleogene), and <1 m.y. to 2 m.y. for the later part of the Cenozoic (Neogene).

Calibration of the times of magnetic reversals with absolute radiometric dates has permitted the establishment of a 'magnetic-reversal' time scale (Figs 1.2 and 1.11; Sections 1.1 and 4.2). The relatively recent magnetic record indicates that small-scale reversals ('excursions') occur every few hundred thousand years (or less), but this type of detail cannot be recognized in the ancient record. Intervals in which one polarity (either normal or reversed) is dominant are referred to as 'magnetic chrons' and have average durations of a few million years. Fig. 1.11 is very generalized, showing Mesozoic and Cenozoic intervals in which rapid alternation of the magnetic field occurred plus one period of consistent normal polarity during the Cretaceous and a less-certain interval of reversed polarity in the late Paleozoic.

[**References** – Discussions of the geologic time scale are in Snelling (1985) and Harland *et al.* (1990). Information concerning the magnetic time scale is from Berggren *et al.* (1985) and Harland *et al.* (1990).]

1.3 Periodic and repetitive events

GEOLOGISTS CONTINUALLY search for periodicities in natural processes, and numerous ones have been proposed. Examples include: 100 000-year cycles of Pleistocene glaciation and deglaciation; ~30-m.y. cycles of biotic extinction and re-radiation; ~500-m.y. cycles of accretion and dispersal of supercontinents. The attempt to find periodicity is one method of looking for factors that control the processes. For example, periodicity in climate change or biotic extinction may imply extraterrestrial controls over these processes. Periodicity in accretion of supercontinents may show the time necessary for the mantle to cycle through various modes of convection.

Periodicities are difficult to prove. Repetitions may be periodic (cyclical) or simply episodic (randomly repeated). Demonstration that a process is cyclical requires sufficiently accurate data to show that the event recurs after the same time interval at each time that it occurs. For example, at any point on a coastline, tides are represented by periodic rise and fall in sealevel whose movements can be analyzed with mathematical precision. Conversely, an event may be repetitive, but the time intervals between repetitions may be so variable that we would not want to propose periodicity. An example of aperiodic behavior is reversals of the magnetic poles, which do not show any uniform recurrence time although they occur frequently. Recognition of periodicity is commonly based on the presence of modes in the Fourier spectra of the data, a process that we cannot develop here (see Sections 2.1 and 8.4 for brief discussions of Croll–Milankovitch cycles).

Most of the arguments about periodic and aperiodic processes are based on different interpretations of the observed data. The issue is generally whether uncertain-

ties in the data are too large for periodicity to be extracted from them. A prime example is the question of the occurrence of worldwide orogenic activity. For example, opinions of different geologists about the periodicity of orogeny occupy virtually the entire spectrum between two endmembers: 1) orogenic pulses can be correlated among different parts of the earth, showing episodes of intense mountain building separated by times in which compressive orogeny virtually ceased; and 2) the amount of worldwide compressive deformation is nearly constant at all times in earth history, and the recognition of separate 'orogenies' can be done only locally.

As a short digression into the philosophy of geologic research, we should finish this discussion with the observation that arguments about periodicity are themselves repetitive (and possibly periodic). About forty years ago, the predominant geologic thinking was that orogeny occurred mostly during limited intervals of time (endmember 1 above). About twenty years ago, expansion of information on a worldwide basis showed tectonism occurring during many of the presumably quiet periods, leading to the general conclusion that orogeny was relatively continuous through time (endmember 2). Further information now has suggested some fluctuation in the intensity of orogenic activity during geologic time. Twenty years from now…?

[**References** – Among the numerous discussions of periodicities in the geologic record, the ones mentioned here are: glacial climates, Broecker and Denton (1989); extinction and radiation, Fischer and Arthur (1977); and supercontinent assembly and dispersal, Gurnis (1988).]

1.4 Major events in earth history

WE CONCLUDE this introduction to the history of the earth with a brief mention of those particular events that have been major milestones in its evolution. Much of the remainder of this book is devoted to documentation of these events and their consequences. For that reason, we will merely list them and not discuss them further in this section. In a sense, we are having dessert first but will require the rest of the meal in order to know whether it is any good.

Fig. 1.12 displays a chronology of earth history on a distinctly non-linear scale; the Precambrian history of the earth is compressed, and the Phanerozoic history is

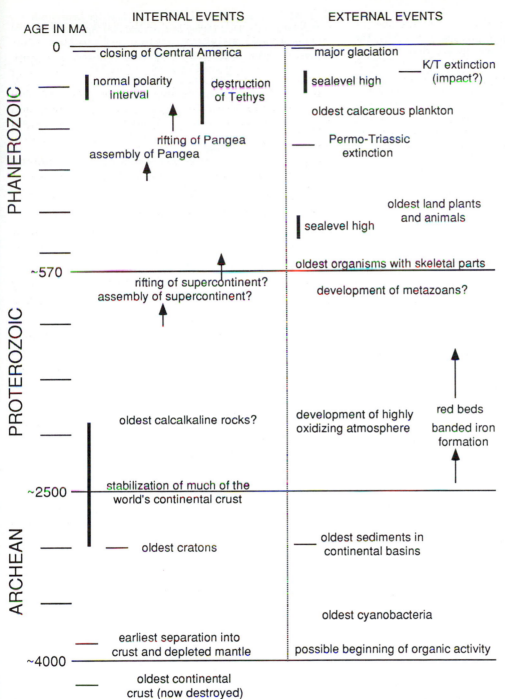

INTERNAL EVENTS

EXTERNAL EVENTS

AGE IN MA

0

PHANEROZOIC

— closing of Central America

normal polarity interval

destruction of Tethys

rifting of Pangea

assembly of Pangea

major glaciation

sealevel high

K/T extinction (impact?)

oldest calcareous plankton

Permo-Triassic extinction

oldest land plants and animals

sealevel high

oldest organisms with skeletal parts

~570

PROTEROZOIC

rifting of supercontinent?
assembly of supercontinent?

development of metazoans?

oldest calcalkaline rocks?

development of highly oxidizing atmosphere

red beds

banded iron formation

~2500 — stabilization of much of the world's continental crust

ARCHEAN

— oldest cratons

oldest sediments in continental basins

oldest cyanobacteria

earliest separation into crust and depleted mantle

possible beginning of organic activity

~4000

oldest continental crust (now destroyed)

Fig. 1.12. Summary of major events in the history of the earth. The diagram distinguishes 'internal events', which are controlled by mantle and/or crustal processes, and 'external events', which are controlled only by activities on the earth's surface. Horizontal time marks (on the left) in the Phanerozoic are at 100-m.y. intervals; marks in the Proterozoic and Archean are at 500-m.y. intervals. Horizontal lines within the diagram indicate comparatively precise times of events. Thick vertical bars indicate approximate ranges of events. Arrows indicate periods of time with uncertain limits.

□

expanded. The major events (according to one geologist!) are separated into those that occurred within the solid part of the earth and those that occurred on or above the earth's surface. Throughout this book, we investigate the relationships between these internal and external processes and inquire into possible controls of one on the other. For example, does the volume of ocean water change through time because of crustal evolution? Does the development of an oxidizing atmosphere change the composition of sediments subducted back into the mantle and, thereby, affect compositions of later magmatic rocks?

Fig. 1.12 is merely a guidepost. All of the information on it is discussed in later chapters.

References

Berggren, W. A., Kent, D. V. & van Couvering, J. A. (1985). The Neogene: Part II. Neogene geochronology and chronostratigraphy. In *The Chronology of the Geological Record*, ed. N. J. Snelling, pp. 211–60. Geological Society of London Memoir 10.

Broecker, W. S. & Denton, G. H. (1989). The role of ocean-atmosphere reorganization in glacial cycles. *Geochimica et Cosmochimica Acta*, **53**, 2465–501.

Cowie, J. W. & Brasier, M. D., eds. (1989). *The Precambrian–Cambrian Boundary*. Oxford: Clarendon Press, 213 pp.

DePaolo, D. J. (1988). *Neodymium Isotope Geochemistry*. Berlin: Springer-Verlag, 187 pp.

Doe, B. R. (1970). *Lead Isotopes*. Berlin: Springer-Verlag, 137 pp.

Faure, G. (1986). *Principles of Isotope Geology*, 2nd edn. New York: John Wiley, 589 pp.

Faure, G. & Powell, J. L. (1972). *Strontium Isotope Geology*. Berlin: Springer-Verlag, 188 pp.

Fischer, A. G. & Arthur, M. A., (1977). Secular variations in the pelagic realm. In *Deep-Water Carbonate Environments*, ed. H. E. Cook & P. Enos, pp. 19–50. Society of Economic Paleontologists and Mineralogists Special Publication 25.

Geyh, M. A. & Schleicher, H. (1990). *Absolute Age Determination – Physical and Chemical Dating Methods and their Application*. Berlin: Springer-Verlag, 503 pp.

Gurnis, M. (1988). Large-scale mantle convection and the aggregation and dispersal of supercontinents. *Nature*, **332**, 695–9.

Harland, W. B., Armstrong, R. L., Cox, A. V., Craig, L. E., Smith, A. G. & Smith, D. G. (1990). *A Geologic Time Scale 1989*. Cambridge: Cambridge University Press, 261 pp.

McDougall, I. & Harrison, T. M. (1988). *Geochronology and Thermochronology by the $^{40}Ar/^{39}Ar$ Method*. New York: Oxford University Press, 212 pp.

Needham, R. S., Crick, I. H. & Stuart-Smith, P. G. (1980). Regional geology of the Pine Creek geosyncline. In *Uranium in the Pine Creek Geosyncline*, ed. J. A. Ferguson & A. B. Goleby, pp. 1–22. Vienna: International Atomic Energy Agency.

Snelling, N. J., ed. (1985). *The Chronology of the Geological Record*. Geological Society of London Memoir 10, 343 pp.

Stuart-Smith, P. G., Willis, K., Crick, I. H. & Needham, R. S. (1980). Evolution of the Pine Creek geosyncline. In *Uranium in the Pine Creek Geosyncline*, ed. J. A. Ferguson & A. B. Goleby, pp. 23–37. Vienna: International Atomic Energy Agency.

2

Comet West (photo by U. S. National Optical Astronomy Observatories). □

2

PRINCIPAL CONTROLS ON EARTH HISTORY

2.0 Introduction

THE EARTH that we observe is so complex that we can understand it only by finding some simplifying principles that have governed its development to its present condition. This chapter concentrates on global processes that have affected the entire history of the earth. First, we discuss two fundamental processes that have controlled the development of the earth's interior – mantle convection and radial compositional segregation. Other topics are: the relationships of the earth to its surroundings, including its orbital parameters and the rate of acquisition of meteorites (Section 2.1); the thermal evolution of the earth (Section 2.2); the segregation of the outer part of the earth into various 'domains' in the upper mantle and crust (Section 2.3); and the variation in sealevel relative to continents through time (Section 2.4). We review these concepts in Section 2.5.

Convection

Transfer of the earth's internal heat to the surface can occur in two ways. One is conduction, in which the kinetic energy (heat) of molecules is transferred to lower-temperature molecules by collision without transfer of mass. Conduction is the principal method of heat transfer in the outer (rigid) part of the earth, including the crust and part of the upper mantle. The second method of heat transfer is convection, in which a mass of hot material moves upward into lower-temperature regions. Convection occurs only in non-rigid materials, but it is considerably more efficient than conduction and is responsible for most of the heat transferred upward in the mantle.

Convection occurs because the earth is too hot to be stable. The condition of mechanical stability in a gravity field in which temperature varies with depth is shown by the equation

$$(\partial T/\partial P)_s = \alpha T/\rho c_p$$

where T and P are temperature and pressure, s refers to constant entropy, α = the coefficient of thermal expansion, ρ = density, and c_p = heat capacity at constant pressure. This equation specifies the variation of T with P for a body of material moving vertically, without loss of heat (i.e. 'adiabatically'), and at equilibrium (e.g., infinitely slowly). Thus, the curve defines an 'equilibrium adiabat' (see caption for Fig. 2.1).

Convection can be demonstrated by comparing adiabatic and actual thermal gradients in the mantle. Consider a body of material (e.g., a mantle peridotite) equilibrated at the starting P and T shown in Fig. 2.1.

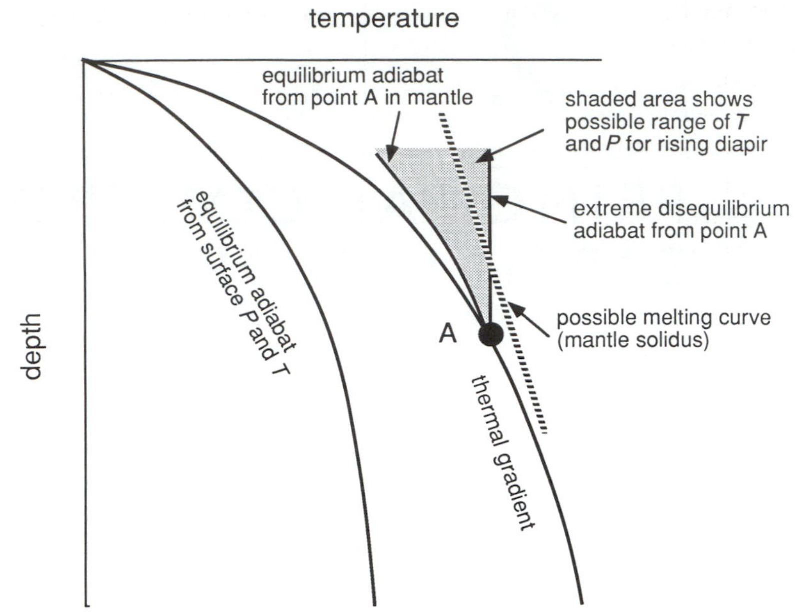

Fig. 2.1. Explanation of thermal properties of the mantle. The 'thermal gradient' is the actual change of T with depth; (depth is proportional to P). 'Adiabats' show the change in T and P for material moving vertically without exchange of heat. The 'solidus' is the P–T line along which melting begins; that is, it separates material that is completely solid from material that is partly molten.

Because of the time needed for thermal equilibration in the mantle, any spontaneous movement of a body probably will be adiabatic. Along an adiabatic path, expansion of the body converts internal energy into work, thus reducing the remaining energy and reducing the temperature of the body. Similarly, increasing pressures cause heating. Infinitely slow movement forms an 'equilibrium adiabat', with maximum change in T; more rapid movement forms a series of 'disequilibrium adiabats' with slopes as high as vertical (no change in T). Adiabats may originate from any point on the P–T grid.

Partial melting, presumably to form basaltic magma, occurs where the P–T paths of rising mantle bodies intersect the solidus. This process is commonly termed 'decompression melting' to distinguish it from melting caused simply by increase in thermal gradients.

Because adiabats are steeper on this diagram than actual thermal gradients in the earth, vertical adiabatic movement of material promotes mechanical instability that results in mantle convection (further discussed in text).   □

This starting point is at a temperature higher than that of the equilibrium adiabat at that depth. The P–T paths followed by a rising body from this starting point may lie anywhere between that of an equilibrium adiabat (maximum cooling) and a disequilibrium adiabat without temperature change (because of instantaneous movement). Regardless of the path followed, the body moving upward cannot cool off as much as the surrounding mantle and maintains a lower density than that of its surroundings. Thus, the body is unstable and continues upward. The reverse process operates for initially falling bodies.

The preceding discussion shows that the earth's mantle is unstable if its actual temperatures are higher than those along an equilibrium adiabat passing through $T=P=0$ (i.e. if it is 'superadiabatic'). In the present earth, actual temperature gradients in the mantle are estimated at ~1°C/km, approximately ten times

greater than the gradient of the equilibrium adiabat inferred from the thermodynamic properties of mantle materials. For this reason, the earth must be undergoing mantle convection.

Radial compositional differentiation

Radial compositional variation in the earth results from the tendency of elements or ionic groups with low densities to rise, and denser elements to sink, in the earth's gravity field. This redistribution reduces the potential energy of the earth, thus increasing its stability.

The principal effect of gravitative fractionation is the development of a core consisting predominantly of dense Fe–Ni. The core is separated from a mantle of predominantly silicate rock by the core-mantle boundary ('Gutenberg discontinuity') at a depth of ~2900 km.

Both the core and the mantle are internally stratified. The major transition in the mantle is at a depth of ~650 km, where velocities of all seismic waves increase sharply downward. The nature of this ~650-km discontinuity is a matter of intense debate (see Section 4.4).

Within the silicate part of the earth, the principal radial movement of elements has been upward segregation of large-ion-lithophile (LIL) elements. These elements include ions of comparatively low valence and large radii (K, Rb, Ba) plus some heavy elements that form low-density radicals with oxygen (e.g. Th and U). The earth's only significant heat-producing (radioactive) elements are all LIL (K, Th, U).

[**References** – Physical properties of the earth are discussed quantitatively by Turcotte and Schubert (1982). A recent summary of radial differentiation is provided by papers in Newsom and Jones (1990).]

2.1   The earth in space

THE EARTH is not a placidly rotating body revolving around the sun by itself. It interacts with the moon, causing the familiar tides and other, less-obvious, phenomena. It wobbles. It is struck by extraterrestrial bodies. In this section, we discuss these interactions in the order mentioned.

Orbital characteristics

The earth's orbit around the sun and its rotational characteristics are affected both by internal instabilities in the earth and by its interaction with extraterrestrial bodies. We discuss first the change in the rate of rotation of the earth and then the periodicities (Croll–Milankovitch cycles) in the earth's orbit that have been proposed to exert major control on climate.

The principal change in the *rate of rotation of the earth* through geologic time is an apparent change in the length of the day. Observational evidence for such a change comes from a study of growth rings in shallow-water calcareous organisms, particularly those living in intertidal environments. Patterns of growth rings are affected by various cycles with daily to yearly periodicities, including the day–night cycle, flooding and removal of water by twice-daily tides, and seasonal variations (such as winter and summer). The first measurements to show a change in length of the day were diurnal growth rings in Middle Devonian corals, which suggested that a year contained approximately 400 days at that time (~375 Ma). Additional measurements on a variety of organisms of different ages tend to confirm a slowing of the earth's rotation by approximately 2% per hundred million years during the Phanerozoic.

Because we cannot conceive of a practical event that could increase the earth's rate of revolution around the sun, we presume that a decrease in the number of days in the year resulted from a decrease in the rate of rotation of the earth on its axis. Changes in the earth's rotation rate could have been caused by some astronomical event, by changes in the radius and density distribution in the earth, and/or by the relationships between the moon and the earth. A reduction in rotational velocity by nearly 10% during the Phanerozoic, however, would require an increase in radius that is inconsistent with the possible changes in thermal gradients and with the small compressibilities and thermal expansion coefficients of earth materials. Similarly, reduction in rotational velocity by change in density distribution would require an increase in mass toward the earth's surface, inconsistent with the long-term trend toward segregation of heavy materials downward. By elimination, we conclude that the change in velocity of rotation of the earth results from tidal friction caused by the moon. This slowing of the earth is coupled to an increase in the distance of the moon from the earth as rotational momentum in the earth–moon system is transferred to the moon.

Combined variations in the *orbit of the earth around the sun and the orientation of the earth's rotational axis* yield cyclical variations that can be observed now and inferred in the past (Fig. 2.2). The earth does not follow a circular path around the sun but an ellipse whose shape changes through time, becoming nearly circular at some times and more 'flattened' at others. Changes also occur in the orientation of the earth's rotational axis. These interactions result in three types of periodicities, discussed below, that were first described comprehensively by A. Croll, in the 1860s, and greatly amplified by M. Milankovitch in the 1920s and 1930s. They are referred to as 'Milankovitch cycles' or, more properly, as 'Croll–Milankovitch cycles'.

The periodicities are shown by variations in: 1) eccentricity, which is the shape of the elliptical orbit of

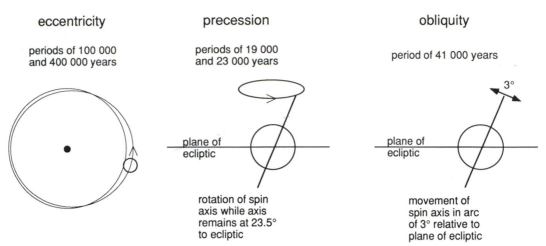

Fig. 2.2. Perturbations that produce Croll–Milankovitch cycles. *Eccentricity* is essentially the ratio of the major and minor radii of the elliptical orbit of the earth; it varies by ~3% in a cyclical pattern. *Precession* is analogous to the wobble of a spinning top; the spin axis of the earth describes a cone with an average angle of 23.5°, thus causing the precession of the equinoxes (the time of year at which the earth is closest to the sun). *Obliquity* is the angle between the spin axis and the ecliptic; currently it is 23.5°, but it varies between ~21.5° and ~24.5°. The periodicities of these various orbital variations have been proposed to exert a major effect on the earth's climate and related processes such as glaciation.   □

the earth around the sun; 2) precession, which is the orientation of the earth's spin axis without change in the angle between the spin axis and the plane of the earth's rotation around the sun (plane of the ecliptic); and 3) obliquity, which is the angle between the spin axis and the plane of the ecliptic (Fig. 2.2). At present, and possibly different in the past, the principal periodicities are: eccentricity, ~100 000 years and ~400 000 years; precession, ~23 000 years and ~19 000 years; and obliquity, ~41 000 years.

The Croll–Milankovitch periodicities exert some control over the earth's climate, but no agreement exists on the exact nature of the effects. The effects of variations in eccentricity, precession, and obliquity are related to each other in complex ways. Briefly, these are as follows.

- Greater eccentricity increases the difference in solar radiation that the earth receives between its closest and farthest distance from the sun. At maximum eccentricity, the radiation difference is ~25% and is represented primarily by high temperatures at the closest approach.
- The effects of precession are related to orbital eccentricity and to the concentration of continents (which reflect heat) in the northern hemisphere and oceans (which absorb heat) in the southern hemisphere. If the earth is in a nearly circular orbit, the orientation of the spin axis has no effect on the distribution of solar energy received. If the earth is in an orbit with maximum eccentricity, however, then the orientation of the spin axis exerts major control over the amount of radiation received by

each hemisphere. At present, the earth is closest to the sun in January and farthest away in July, which causes warm summers in the southern hemisphere, cold summers in the northern hemisphere, and high total absorption of heat by southern oceans. If the closest and farthest distances occurred in March and September, then the hemispheres would receive equal radiation during their summers (and also during their winters).

- The range of obliquity is sufficiently small that it has little effect on the amount of radiation received at the equator. At the poles, however, interaction between obliquity and precession can have a significant effect on absorbed radiation. In particular, high obliquity exposes polar regions to large amounts of sunlight during the summer.

The principal use of Croll–Milankovitch periodicities has been the interpretation of the Pleistocene climatic record (Section 8.4). The term 'Croll–Milankovitch forcing' is used by investigators who regard the orbital cycles as largely responsible for climatic variations.

Croll–Milankovitch cycles become more difficult to interpret as we attempt to apply them to pre-Pleistocene rock suites. One problem is that the periodicities measured at present cannot be presumed to have been the same in the past. These periodicities are affected by long-term changes in the earth's rate of rotation (see above) and also by the earth's gravitational interaction with other planets (primarily Jupiter). Another problem is that recognition of the cycles requires two types of evidence that commonly are highly controversial.

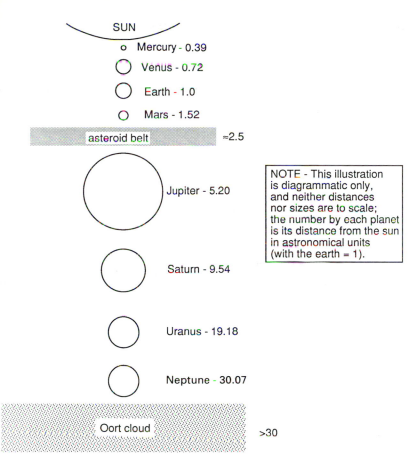

SUN

o Mercury - 0.39

◯ Venus - 0.72

◯ Earth - 1.0

◯ Mars - 1.52

asteroid belt   ≈2.5

Jupiter - 5.20

Saturn - 9.54

Uranus - 19.18

Neptune - 30.07

Oort cloud   >30

NOTE - This illustration
is diagrammatic only,
and neither distances
nor sizes are to scale;
the number by each planet
is its distance from the sun
in astronomical units
(with the earth = 1).

Fig. 2.3. Location of asteroid belt and Oort cloud. The Oort cloud is highly diffuse particulate material that extends from ~30 A.U. from the sun to extreme distances; (A.U. = astronomical unit = the distance from the earth to the sun). Most meteorites probably originate from collisions in the asteroid belt and most comets from perturbations in the Oort cloud.   ☐

One requirement for detection of the effect of orbital cycles is a repetitious (rhythmic?) sequence of features in a stratigraphic section. Examples include: 1) cycles of different rock types in Carboniferous coal-bearing cyclothems (Section 7.1); 2) thicknesses of layers in bedded cherts; and 3) alternations of limestone lithologies. The rock suites must be measured in some way that produces numerical data, and the data must be subjected to Fourier spectral analysis to extract meaningful frequencies.

The existence of frequencies does not, by itself, demonstrate control by Croll–Milankovitch forcing. The second requirement for this demonstration is sufficient information about ages to permit calculation of precise times for the duration of each cycle in the spectrum. For example, a suite of sedimentary rocks that contains 25 repeated sequences and that can be constrained to have been deposited within an interval of 10 m.y. can be inferred to exhibit a periodicity of 400 000 years. Because this time period corresponds to the longest of the eccentricity periodicities at the present time, many investigators would conclude that the sedimentary sequence was affected by Croll–Milankovitch forcing. Determination of shorter periodicities with appropriate frequencies in the same sedimentary column strengthens the conclusion of orbital forcing.

Meteorites

Investigations of the bombardment of the earth by meteorites and comets have acquired particular significance in the past decade because of the proposal that the Cretaceous/Tertiary extinction event was caused by the catastrophic environmental results of impact (Section 8.7). The earth is almost continually receiving very minor amounts of meteoritic debris, mostly dust from small meteorites that burn up in the atmosphere. The major uncertainties involve the frequencies with which large objects collide with the earth and the distribution of their sizes.

The term 'meteorite impact' is a generalization, for at least two types of bodies strike the earth – true meteorites and comets. Meteorites are composed of silicate and/or metallic phases (Fe–Ni). Most of them originate in the asteroid belt, between Mars and Jupiter, where they occasionally are rerouted into earth-crossing orbits by some combination of external gravity forces (Fig. 2.3). Some meteorites, however, apparently were ejected from the moon or Mars by impacts on their surfaces. Comets are low-density bodies, possibly mostly 'ice' around a meteoritic nucleus, that originate in the 'Oort cloud' (Fig. 2.3) of particulate matter that begins at the orbit of Neptune and extends out as far as 100 000 astronomical units (the distance between the earth and the sun). Perturbations in the Oort cloud apparently can coalesce the dust into comets, which then may be propelled into earth-crossing orbits. Meteorites are destroyed only by collisions, whereas comets gradually lose all of their mass by evaporation during passage close to the sun. Because of uncertainties about the nature of cometary nuclei, distinction between comet and meteorite impacts is problematic. For this reason, most investigators simply refer to all collisions as 'meteorite impacts'.

Evidence for meteorite impact in the geologic record consists of several unquestioned, and many more questionable, types of information. Direct evidence includes preserved shock-metamorphic rocks and minerals (glass, coesite, etc.), shock structures, and pieces of meteorite. Indirect and controversial evidence includes circular structures, commonly floored by igneous rocks and filled by sediments, that have no obvious relationship to regional structures. Many of these features, however, have been explained by processes within the earth that cause rapid and explosive release of igneous materials (diatremes, crypto-explosion structures); one example is diamond-bearing kimberlites (Section 2.2).

Impacts generally form craters approximately 5 to 10 times the diameter of the meteorite. The craters have been recognized only on land, and they are easily destroyed, or at least rendered unrecognizable, in geologically short periods of time. For example, craters with diameters of less than ~20 km would probably be destroyed within 100 m.y. unless they are covered by younger sediments. More than one half of proposed meteorite impact sites with diameters >10 km have ages <100 Ma. No impacts of Archean age are known, and fewer than ten Proterozoic impacts have been recognized.

The absence of a long-term geologic record complicates the problem of estimating past meteorite activity. Because the earth formed by accretion, the rate was obviously extremely high during the early stages of its development. Present impact rates can be estimated from the ages of the limited (approximately 100) known impact sites and the calculated probabilities of being struck by known asteroids and comets that cross the earth's orbit. As a very rough approximation, the earth is probably struck by meteorites (or comets) that create crater diameters >10 km once every 0.1 m.y., >50 km every 5 m.y., and >100 km every 50 m.y.

Because meteorites are samples of the solar system, they can be used with other data to infer the composition of the solar system at the time the earth accreted. One group of meteorites, the CI ('C-one') carbonaceous chondrites, have abundance ratios of most elements virtually identical to the ratios measured spectrometrically in the sun for all elements except H and He. The inferred average composition of the solar system is probably similar to that of the original earth except for loss of volatile elements from the earth during accretion. Thus, the earth has much lower concentrations of light and inert gases, and is presumed to have lower abundances of elements such as K and Rb, than are measured in the CI chondrites.

[**References** – The basic principles of Croll–Milankovitch cycles were (not surprisingly) first described by Croll (1875) and Milankovitch (1941). More recently, general principles of orbital variations are discussed by Goudie (1977) and Lambeck (1980). Possible geologic effects of orbital forcing are in Rosenberg and Runcorn (1975), Berger *et al.* (1984), Algeo and Wilkinson (1988), and a recent special issue edited by Fischer and Bottjer (1991). Basic principles of meteoritics are summarized from Dodd (1981). Information about impact structures is from Grieve (1987) and the two compendia edited by Silver and Schultz (1982) and Sharpton and Ward (1990). Comments about the composition of the solar system are from Anders and Ebihara (1982).]

2.2 Thermal evolution

THE HEAT originally accumulated in the earth, plus the amount produced by radioactive decay, is the energy that causes mobility in the earth's interior and its surface plates. We begin our discussion of the earth's thermal evolution with an investigation of the availability of heat in the initial earth.

The most obvious source of heat in the primitive earth was the potential energy lost as accretion occurred. The temperature generated by that heat is somewhere between two extremes. At one extreme, the maximum temperature would have resulted from instantaneous accretion of the earth from a solar-system cloud of dispersed material originally at effective 'infinite' distance from the earth's gravity well. In this event, all of the potential energy lost would have been converted into heat without dissipation, which would have generated approximately 10 000 cal/g to 15 000 cal/g. This heat would have created temperatures of several tens of thousands of degrees (Celsius), resulting not only in complete melting of the earth but also vaporization of large parts of it back into the space from which it had come. The other extreme was an 'infinitely slow' accretion, which permitted dissipation of the heat back into space as rapidly as potential energy was lost and kept the original earth cold. The actual amount of accretionary heat preserved in the earth was clearly between these two extremes, but we do not know exactly where.

In the early stages of earth history, the heat of accretion was augmented by other sources of heat generation and possibly diminished by heat sinks. Sources include: a 'giant impact' with another planet-sized body that may have created the moon; exothermic (heat-releasing) chemical reactions in the new earth; further loss of potential energy by gravitational fractionation, largely the segregation of the core; radioactivity produced by very short-lived isotopes present in the cosmos that completely decayed almost immediately after accretion; and fission reactions where the ^{235}U-rich uranium accumulated in 'ore' deposits of suficient size to reach critical mass. Heat sinks include endothermic (heat-absorbing) reactions and escape of gases with high kinetic energy back into space.

The relative strengths and effects of the preceding processes probably were important only during the first few tens, perhaps hundreds, of millions of years of earth history. By the time the earth reached a condition that permitted preservation of continental crust, about 4000 million years ago, the segregation of the core and most other effects listed above had probably been completed. We discuss the history of the earth from 4000 Ma to the present based on the assumption that some of the early heat of accretion and other sources remained in the planet and contributed to its thermal gradients, although the amount of that heat still being

Table 2.1. *Characteristics of heat-producing isotopes*

isotope	half life (in years)	heat produced (in cal/g of parent/yr)
^{40}K	1.25×10^9	0.22
^{232}Th	14.01×10^9	0.20
^{235}U	0.7038×10^9	4.3
^{238}U	4.468×10^9	0.71

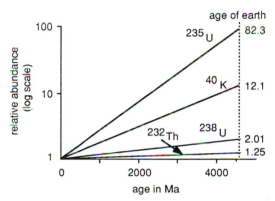

Fig. 2.4. Variation in abundances of heat-producing isotopes during earth history. The abundance–time relationships shown here represent solely the variations caused by radioactive decay of each isotope. □

dissipated from the modern earth is greatly in dispute.

We can make more reasonable estimates of the heat generated by radioactive decay at any time in earth history than we can of the 'original' heat remaining in the earth at that time. For any radioactive element, calculations of the ratios of heat produced now to heat produced at any time in the past require only a knowledge of its half life. Calculations of total amount of heat generated at any time, however, require information on the terrestrial abundance of the isotope. Four isotopes are important in terrestrial heat production (Table 2.1 and Fig. 2.4).

Relative abundances of the various isotopes to each other can be estimated within reasonable limits. The $^{235}U/^{238}U$ ratio is well known because of the inability of the two isotopes to fractionate during normal chemical reactions. At present, $^{235}U/^{238}U = 0.007$ – that is, 99.3% of natural U is ^{238}U. The $^{235}U/^{238}U$ ratio at the time of accretion of the earth was 0.22. The ratio of Th/U is also fairly well known, with a variety of evidence indi-

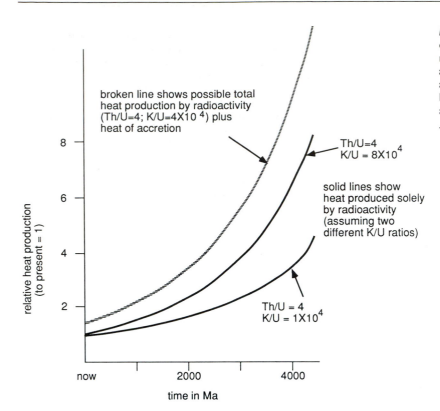

broken line shows possible total heat production by radioactivity (Th/U=4; K/U=4X10^4) plus heat of accretion

Th/U=4
K/U = 8X10^4

solid lines show heat produced solely by radioactivity (assuming two different K/U ratios)

Th/U = 4
K/U = 1X10^4

relative heat production (to present = 1)

now   2000   4000

time in Ma

Fig. 2.5. Possible variation in heat production during earth history. Curves representing only radioactive heat production can be calculated accurately if the Th/U/K ratios are known or assumed. The curve for total heat production is highly dependent on assumed K/U ratios and the amount of original heat retained. ☐

cating that terrestrial ratios are similar to the cosmic ratio of Th/U≈4. Because of its long half life, heat production from Th changes only slightly with time, and errors in the Th/U ratio are not very important.

The major difficulty in estimating relative radioactive heat production from all isotopes is the uncertainty in the terrestrial ratio of K to either U or Th. Cosmic ratios (as in CI carbonaceous chondrites; Section 2.1) are as high as 8X10^4 but are probably invalid because of loss of the volatile K during earth accretion. Many surface rocks, mantle xenoliths, and magmatic rocks generated by plumes (possibly from the deep mantle), have present-day K/U ratios of approximately 1X10^4, although estimates by different investigators differ significantly. The average terrestrial ratio is probably between 1X10^4 and 8X10^4.

Fig. 2.5 shows the probable range of variations in the generation of radiogenic heat in the earth through geologic time. Because we have estimated only relative, not absolute, abundances of the isotopes, the heat production is shown with a present-day value of 1.0. Fig. 2.5 demonstrates that present radioactive heat production is some 10% to 50% of the amount produced in the early earth, with the uncertainty caused largely by the assumed ratio of K to other isotopes.

The curve shown in Fig. 2.5 for the variation of total heat production through time contains even more uncertainties than the variations in radioactive heat alone. The problem is our lack of clear information on the amount of initial heat preserved through time. Some investigators have assumed that the earth underwent very rapid loss of original heat by convection in the absence of any blanketing crust before ~4000 Ma. If this assumption is correct, then most of the heat that has powered the internal history of the earth since ~4000 Ma has been produced by radioactivity. Other investigators propose that the earth is still living on its 'capital' of original heat and will gradually cool down so much that convection becomes impossible despite the heat generated by radioactivity.

The rate of heat loss from the earth through time is the primary control on changes in the earth's radius. Some shrinkage must have occurred as the earth cooled, but we cannot predict accurately how much that shrinkage has been. Because the materials that constitute the earth are relatively incompressible, however, changes in radius of more than ~10% are highly unlikely, and the actual variation may have been much smaller.

Efforts to construct thermal gradients for the earth at various times in the past are hindered not only by

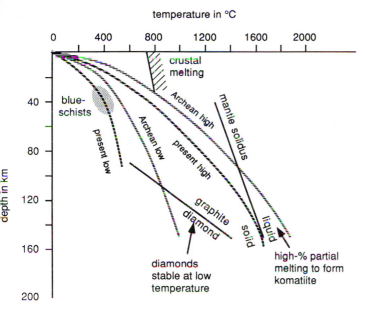

temperature in °C

blue-
schists

depth in km

present low

Archean low

Archean high

present high

mantle solidus

crustal
melting

graphite
diamond

solid

liquid

diamonds
stable at low
temperature

high-% partial
melting to form
komatiite

Fig. 2.6. Possible variations in thermal gradients through time. Phase relationships most useful in estimating these gradients are the region of crustal melting, the mantle solidus, the field of equilibration of blueschists, and the phase boundary between graphite and diamond. The 'Archean high' gradient must be sufficiently high to produce komatiites but not too high to prevent the formation of continental crust by ~4000 Ma. Similarly, the 'Archean low' gradient is fixed as higher than the field of blueschists and lower than the field of production of diamonds at depth. The 'present high' gradient is below the temperatures needed for the formation of komatiites, and the 'present low' gradient passes through the field of blueschist equilibration. ☐

questions about radioactive heat production and residual heat but also by uncertainties in the rate of heat dissipation back into space. If the entire mantle undergoes convection as a single layer, then heat release is probably rapid. If the earth is layered, with coupled convection above and below various discontinuities, then heat release is slower; one possible layer boundary is the ~650-km discontinuity in the mantle (Section 4.4).

With these various uncertainties, we must look to the geologic record for evidence concerning ancient thermal gradients. Five major types of information are available: 1) $P–T–t$ (pressure–temperature–time) paths in metamorphic rocks preserved in continents; 2) equilibration conditions of crustal and mantle xenoliths; 3) the solidus of continental crust; 4) the abundance of Archean komatiites; and 5) the presence of Archean diamonds. The use of these criteria is summarized in Fig. 2.6.

- Pressures and temperatures of equilibration in metamorphic rocks, and their changes with time, are assembled from mineral phase relationships and a variety of geothermometers and geobarometers (which are too numerous to describe). Equilibrium relationships in synchronous rock suites formed at various depths permit moderately accurate assessment of $P–T$ gradients at any one time, and disequilibrium relationships yield conclusions about the changes in $P–T$ conditions with time in one rock suite. This type of investigation generally produces inferred gradients in continental areas of any

age, including Archean, that are not significantly different from modern values of 15 °C to 25 °C per km. The major difference through time is that Phanerozoic thermal conditions in the earth have permitted subduction zones to have T/P ratios sufficiently low that blueschists can develop in them. With that exception, continental thermal gradients have been approximately constant through time, and the extra heat produced during the Archean must have been dissipated through oceanic crust.

- Determination of equilibration P and T in xenoliths in volcanic rocks provides estimates of thermal gradients in continental crust and in the present upper mantle. Continental crustal xenoliths generally confirm the 15 °C/km to 25 °C/km gradients proposed above for the upper part of the crust, with some diminution at depth. Mantle xenoliths include garnet and spinel peridotites, with eclogite in some locations. They commonly show gradients near 5 °C/km or slightly higher in the upper 100 km of the mantle. The estimation of ancient mantle thermal gradients poses more problems than the determination of modern ones or continental gradients of any age. One of the greatest difficulties is that the upper mantle under the continents has been thoroughly metasomatized and/or intruded since it was formed (Section 2.3). Geochronology of xenoliths rarely shows ages greater than middle to late Proterozoic, and most are younger. Thus, thermal gradients shown by xenoliths represent gradients at the time of modification rather than at the original age of formation.

- The position of the solidus of continental crust depends strongly on the composition of the crust and the types of volatiles present. Maximum melting temperatures (at the solidus) are in dry and mafic rocks. Fig. 2.6 shows a $P–T$

region above the solidus in which crustal melting should occur regardless of rock type or nature of the volatiles. At the solidus, some melting would produce granite magmas and leave a more refractory residue. At temperatures greatly above the solidus, however, melting would be so extensive that the crust would effectively disappear. This mechanism probably acted as a significant control on the thickness of continental crust that could develop in the earlier stages of earth history.

- The ratio of komatiites to basalts in the geologic record is high in many Archean terranes and decreases to zero in the Proterozoic. Thus, some investigators regard komatiites as the 'basalts of the Archean' and consider them to be the equivalent of modern MORB (mid-ocean-ridge basalts; see discussion on p. 83). Both the production and eruption of komatiites pose difficulties of interpretation. The most peridotitic komatiites have MgO contents >30%, which require melting temperatures to be >1600 °C, perhaps more than 1800 °C at depths of 100 to 150 km in the mantle. These temperatures require thermal gradients that are higher than modern ones (Fig. 2.6). Once produced, eruption of komatiites is hindered by the high density of high-Mg magmas and the necessity of maintaining a sufficiently high temperature that they do not crystallize as they rise to the surface. Some investigators have proposed a komatiite 'magma ocean' during the Archean at depths of ~100 km, from which the eruptive magmas rose rapidly toward the surface.

- Diamonds are brought to the surface by sudden rise of kimberlite (mica peridotite) from mantle depths in regions that have been stable for long periods of time. The restriction to stable areas presumably results from the necessity of mantle thermal gradients to pass below the diamond–graphite phase boundary (Fig. 2.6). Although most diamonds were erupted during the late Phanerozoic, earlier examples have been found, including Archean dates for mineral inclusions in some diamonds. The presence of Archean diamonds indicates that upper-mantle $T–P$ conditions in at least some continental areas were as 'cold' as they are under modern continental crust.

With our lack of knowledge of so much fundamental information, we can provide only a very general summary of thermal evolution of the earth. Average thermal gradients have decreased through time, but the heat loss that made this decrease possible occurred mostly through the oceans by the ascent of hot mantle to oceanic spreading centers. Times of more rapid heat release, primarily the Archean, were characterized by at least one (probably both) of the following processes: 1) the earth's ocean basins contained a greater length of ridges releasing heat from smaller convection cells; and/or 2) spreading rates were more rapid along individual ridges. Subcontinental and suboceanic upper

mantles may have had even larger temperature differences in the Archean than they do today, permitting komatiites to be produced in oceanic areas and diamonds in continental areas.

[**References** – Discussions of the general thermal history of the earth are in McKenzie and Weiss (1975) and Davies (1990). The problem of the Archean thermal regime is discussed by Bickle (1978), Richter (1985) and Ashwal et al. (1987). Inferences concerning the mantle are based on discussions in volumes edited by Hawkesworth and Norry (1983), Kornprobst (1984) and Nixon (1987) and the paper by Herzberg (1983). The early evolution of the earth is in papers in Newsom and Jones (1990); continents are discussed by Ridley and Kramers (1990). Estimates of heat production from radioactive isotopes are from Faul (1954).]

2.3 Evolution of the crust and mantle

THE SOLID earth is heterogeneous. Below some 200 km depth, the heterogeneity is primarily radial, partly caused by gravitative compositional differentiation (see introduction to this chapter) and partly caused by phase changes. Both vertical and lateral variations occur shallower than ~200 km, caused by: 1) extreme gravitative fractionation to form the continental crust; 2) phase changes; and 3) different geologic histories of different parts of the outer earth (e.g. the effects of subduction, plume activity, etc.). Our survey of the history of the earth requires an understanding of the evolution of its heterogeneity. First, however, we must discuss one of our most important tools – the evolution of Sr, Pb and Nd isotopes. Then we discuss the differences among various domains in the outer 200 km of the earth and, finally, the source and rate of evolution of continental crust.

Sr, Nd and Pb isotopic systems

The general principles for using Sr, Nd and Pb isotopic systems for age dating are discussed in Section 1.1 (Figs 1.3 to 1.6). Here, we concentrate on the use of isotopic information for deciphering the history of rocks and the environments that they represent. Much of this history is contained in 'initial' ratios such as $({}^{87}Sr/{}^{86}Sr)_i$, with the symbol IR_{Sr}, and $({}^{143}Nd/{}^{144}Nd)_i$, with the symbol IR_{Nd} (Figs 1.3 and 1.4). Rocks inherit these ratios from their sources at the time when they form. For example, magmatic rocks acquire initial ratios equal to

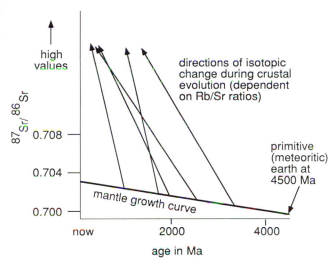

Fig. 2.7. Sr isotopic evolution. The mantle growth curve has a low slope because of preferential removal of Rb into the crust and consequent depletion of the mantle. Studies of ocean-ridge basalts and related mantle-derived rocks indicate that the depleted mantle has attained ^{87}Sr/^{86}Sr ratios of approximately 0.703 at the present time. Present old continental crust has very high ^{87}Sr/^{86}Sr ratios because it has had high Rb/Sr ratios for a long time. □

the ratios in their source regions at the time of partial melting, and these ratios provide estimates of the Rb/Sr, Sm/Nd and/or U/Pb ratios in those sources over some period of time before melting.

The Rb–Sr system is particularly useful for recognition of rocks that represent the upper continental crust. Because Rb is an LIL element (Section 2.0) and Sr is not, fractionation generally causes upper crustal rocks to have high Rb/Sr ratios (Fig. 2.7). Evolution of Sr isotopes in the upper crust, therefore, produces material with higher ^{87}Sr/^{86}Sr ratios than can be attained in other parts of the earth, such as the mantle or lower crust. This upper-crustal ratio is shown by high values of IR_{Sr} in metasediments and in igneous rocks crystallized from magmas formed by partial melting of crustal materials.

The Rb–Sr system also provides information on the evolution of the mantle (Fig. 2.7). Beginning from an initial ^{87}Sr/^{86}Sr (estimated from meteorites) at 4500 Ma, continuous decay of ^{87}Rb has produced a slow increase in ^{87}Sr/^{86}Sr in basalts from oceanic ridges and apparently related sources. We will see below, however, that different parts of the mantle can be distinguished by differences in the IR_{Sr} of magmatic rocks produced from them.

The Sm–Nd system is characterized largely by the ratio of ^{143}Nd/^{144}Nd (Fig. 1.4). Because Sm and Nd are

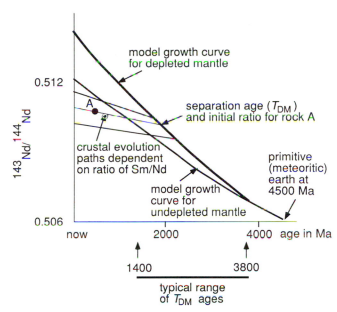

Fig. 2.8. Nd isotopic evolution. The earth's mantle is 'depleted' in Nd relative to an unfractionated earth. Thus, it has higher Sm/Nd ratios than the primitive earth and produces higher ^{143}Nd/^{144}Nd ratios. The beginning of formation of the growth curve for a depleted mantle is estimated as approximately 3800 Ma, close to the age of the oldest continental crust (see text). Progressive depletion has occurred as Nd-enriched continental crust continued to separate from the mantle.

The growth curves for crustal rocks have lower slopes than those of any mantle curves because the crust has lower Sm/Nd ratios. In order to clarify the Nd isotopic system in the crust, let point A represent the ^{143}Nd/^{144}Nd ratio in a crustal rock of known age. An isotopic growth curve, whose slope is determined by the Sm/Nd ratio of the sample, passes through point A and intersects the growth curve of the mantle from which the rock was derived. Extrapolation along this curve yields an age and initial ^{143}Nd/^{144}Nd ratio for separation of the rock from the depleted mantle if the sample has not undergone further fractionation in the crust (e.g. by partial melting). The time of separation from the depleted mantle is designated T_{DM}. If point A represents the partial melt of a crustal protolith derived from the mantle, then model separation ages from the mantle can be estimated if both the effects of separation and melting on the Sm/Nd ratio can be inferred.

Typical T_{DM} ages for separation of continental crust from the mantle are in the range of 3800 Ma to 1400 Ma, with a concentration at ~1800 Ma (further discussed in text). □

both rare-earth elements (REE), they react similarly during chemical processes and undergo less fractionation from each other than Rb and Sr. Separation of Sm and Nd almost certainly did not occur during accretion of the earth, and the initial ^{143}Nd/^{144}Nd ratio in the bulk-earth was presumably the one shown for meteorites in Fig. 2.8. Some fractionation, however, caused

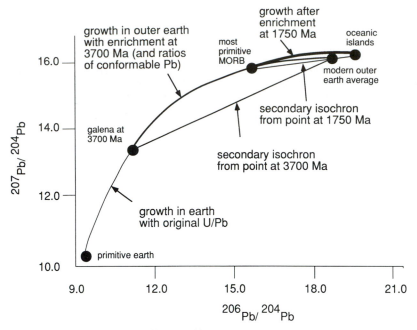

Fig. 2.9. Pb isotopic evolution. The radiogenic [207]Pb and [206]Pb are shown as ratios to the non-radiogenic [204]Pb. The point for the primitive earth is from U-free phases in meteorites. The point for the modern outer earth is largely based on modern oceanic sediments, which presumably are a representative sample of the 'accessible' part of the earth (the part that interacts with the surface and affects samples that we collect).

The original growth curve in an unfractionated earth is shown as a thin line. Enrichment of the entire accessible earth in U apparently occurred at ~3700 Ma, or possibly over a long period of time early in earth history, and led to the [207]Pb/[204]Pb and [206]Pb/[204]Pb ratios of different parts of the accessible earth. These ratios form a 'secondary isochron' (see Fig. 1.6) extending from a theoretical galena (U-free) sample formed at 3700 Ma to the modern outer-earth average.

Conformable Pb ores are galena samples apparently derived from the mantle. The growth curve of their [207]Pb/[204]Pb and [206]Pb/[204]Pb ratios is shown by a thicker line than that of the primitive earth and demonstrates that the mantle source region had a higher U/Pb ratio than that of the primitive earth (see text discussion of mantle evolution).

Further U/Pb fractionation of the accessible earth apparently occurred at ~1750 Ma, forming a new growth curve (thick line). This enrichment formed modern [207]Pb/[204]Pb and [206]Pb/[204]Pb ratios in rocks ranging from the most primitive MORB to more-enriched basalts of oceanic islands. □

preferential extraction of the relatively low-density (larger-radius) Nd relative to the heavier Sm from the mantle into the crust. The removal of Nd and other low-density (LIL) elements from the mantle leaves a 'depleted mantle', in which the increased Sm/Nd has produced progressively higher [143]Nd/[144]Nd ratios over time than could have been produced in an undepleted mantle (Fig. 2.8). Because separation of magmas from the mantle does not cause any fractionation of Nd isotopes during melting, the curve for this depleted mantle can be verified by measuring the IR_{Nd} of mantle-derived rocks that have not been contaminated by the crust.

Fig. 2.8 shows the method of estimating T_{DM}, the time of separation of a material from the depleted mantle. The determination of T_{DM} is complicated by

the fact that most continental igneous rocks are not derived directly by partial melting of the mantle but are formed from magmas generated by partial melting of a crustal 'protolith' (parent rock). If the protolith was formed by crystallization of a mantle-derived magma, however, values of T_{DM} for the protolith may be estimated from the initial [143]Nd/[144]Nd in the igneous rock and the inferred Sm/Nd ratio of the protolith.

Pb isotopic evolution in the earth presumably started from values similar to those found in U-free phases in meteorites (Section 2.1; Figs 1.6 and 2.9). From this beginning, decay of U in an unfractionated earth would have produced a series of [207]Pb/[204]Pb and [206]Pb/[204]Pb ratios that plot along some curve whose position is determined by the initial ratio of U/Pb in

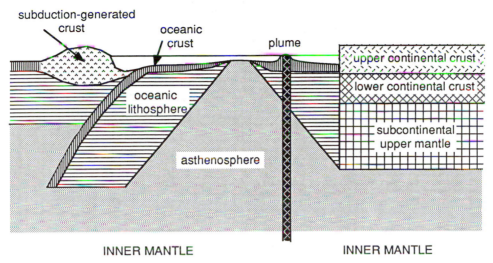

Fig. 2.10. Diagrammatic representation of various domains in the outer part of the earth. The *inner mantle* is a zone below the asthenosphere that is probably represented at the earth's surface primarily by magmas generated by *plumes*. The *asthenosphere* is a zone of apparent mobility that is separated from the overlying, more-rigid, *lithosphere*. In oceanic regions, the asthenosphere is the source of MORB at spreading ridges. The lithosphere–asthenosphere boundary in oceanic areas is probably the level at which partial melting occurs in the mantle. The lithosphere–asthenosphere boundary beneath continents may be partly compositional, with the lithosphere (*subcontinental upper mantle*) more enriched in LIL elements than underlying parts of the mantle. *Oceanic crust* consists largely of basalts (MORB) formed at spreading ridges. *Subduction-generated crust* represents magmatic rocks of island arcs and continental margins plus sediments and other materials (e.g. ophiolites) accreted into them. *Continental crust* is separated from the subcontinental upper mantle by the MOHO, a very sharp seismic discontinuity. The boundary at the MOHO is mostly compositional, with the continental crust considerably more siliceous and richer in LIL elements . A slight seismic discontinuity (Conrad) between *upper and lower continental crust* has been recognized in several areas. The lower crust is certainly more highly metamorphosed, and may be more mafic, than the upper crust.   □

the earth. Higher U/Pb values would generate higher values of $^{207}Pb/^{204}Pb$ and $^{206}Pb/^{204}Pb$ at any given time, and the more-rapid decay of ^{235}U to ^{207}Pb than of ^{238}U to ^{206}Pb produced a convex-upward curve in which $^{207}Pb/^{204}Pb$ ratios increased rapidly at first and at a slower rate as time progressed (Fig. 2.9). Regardless of the precise U/Pb of the bulk earth, all Pb isotope ratios in an unfractionated earth must lie on some curve that extrapolates back through the meteoritic starting point.

Fractionation in the earth changed the pattern of Pb isotopic evolution as follows. Separation into phases of different U/Pb at the time the earth formed, with no further chemical fractionation, would have formed a series of isotopic growth curves leading to present earth materials with $^{207}Pb/^{204}Pb$ and $^{206}Pb/^{204}Pb$ ratios that plot along the geochron (Fig. 1.6). If the U/Pb ratios of various parts of the earth were modified by increase or decrease in U and/or Pb at some time, the isotopic evolution would have proceeded along a series of new growth curves radiating from the initial point on the old growth curve and forming a series of present $^{207}Pb/^{204}Pb$ and $^{206}Pb/^{204}Pb$ ratios along a secondary isochron (Figs 1.6 and 2.9). In later parts of this section, we discuss the possibility that at least two major fractionations affected rocks that we sample in the outer ('accessible') part of the earth – one at ~3700 Ma and one at ~1750 Ma.

Because U is an LIL element, it has a strong tendency to concentrate in the upper crust. This concentration causes crustal rocks to develop high ratios of $^{207}Pb/^{204}Pb$ and $^{206}Pb/^{204}Pb$. Later processing, such as partial melting, of these rocks generates magmas or other starting materials that have the same high initial ratios. This information is very useful in identifying the source regions of igneous and other rocks.

Domains in the outer part of the earth

Fig. 2.10 shows the outer part of the earth divided into several 'domains' including: inner mantle, suboceanic asthenosphere, oceanic crust, subduction-generated crust, subcontinental upper mantle (perhaps of two types), and both lower and upper continental crust. The terminology is inconsistent among investigators; for example, some workers regard 'asthenosphere' as equivalent to 'depleted mantle' and refer to the 'inner mantle' as 'asthenosphere'. We discuss these domains separately below (oceanic crust is discussed in more completely in Section 4.2 and plumes in Section 4.4).

The lowest domain shown in Fig. 2.10 is termed *'inner mantle'*. It represents material below the depleted asthenosphere (see below), and no readily definable boundary with the asthenosphere has been detected. The difference between the inner mantle and the asthenosphere is shown by the difference in the compositions of basalts formed at spreading ridges (derived from the asthenosphere) and the igneous products of plumes and possibly other magmas that originate deep in the mantle (Section 4.4). Products of the deep mantle are considerably enriched in LIL elements relative to MORB (see below). Enrichment in LIL elements could be the result of melting of sources with high LIL contents and/or melting of very small fractions of the source. For the inner mantle, long-term high LIL contents are shown by the radiogenic isotope ratios of magmas derived from it. Compared to MORB, plume-generated rocks have high initial $^{87}Sr/^{86}Sr$, $^{207}Pb/^{204}Pb$, and $^{206}Pb/^{204}Pb$ ratios and low $^{143}Nd/^{144}Nd$ ratios, implying high Rb/Sr and U/Pb and low Sm/Nd in their source regions.

The *suboceanic asthenosphere* overlies the inner mantle and has a highly irregular upper boundary with the *lithosphere*. The boundary appears as a seismic low-velocity zone, probably associated with partial melting of the mantle beneath the more-rigid lithosphere. The depth to the asthenosphere is almost certainly controlled by the thermal gradient, with greater depth caused by deeper intersection of the gradient with the mantle solidus (Section 2.2). Thus, the lithosphere thickens from zero depth at spreading ridges to 100 km to 200 km under old oceanic crust. The conversion of asthenosphere to lithosphere by cooling is responsible for the virtually identical relationships between age and depth in all modern oceans (Section 4.2). Because their boundary in oceanic areas is purely thermal, we do not expect any compositional differences between oceanic lithosphere and asthenosphere.

Our best indication of the nature of the suboceanic asthenosphere comes from the study of volcanism at spreading centers, where the principal product is mid-ocean-ridge basalt (MORB). MORB is tholeiitic basalt that is compositionally homogeneous in most major elements but shows some variation in the abundances of minor elements and in isotope ratios. The most 'primitive' MORB contains very low concentrations of LIL elements (e.g. 0.1% to 0.2% K_2O), very low ratios of $^{87}Sr/^{86}Sr$, and high ratios of $^{143}Nd/^{144}Nd$. The $^{207}Pb/^{204}Pb$, and $^{206}Pb/^{204}Pb$ ratios in MORB are lower than in other parts of the outer earth. Thus, the asthenospheric reservoir of MORB must have been LIL-depleted for considerable periods of time.

Despite general compositional homogeneity, the source regions for MORB produce a range of radiogenic isotope values. A complete trend from primitive MORB erupted along mid-ocean ridges to ocean-island basalts (possibly from plumes) is shown by the $^{207}Pb/^{204}Pb$ vs. $^{206}Pb/^{204}Pb$ diagram (Fig. 2.9). Fig. 2.9 can be interpreted in at least two ways. One is that the line joining the various isotope ratios is an isochron resulting from the production of separate asthenospheric reservoirs with different U/Pb ratios during the time of ~2000 Ma to ~1500 Ma (see explanation of the Pb isotopic system above). A second interpretation is that the line simply represents a mixture of materials whose isotopic systems were established at different times. Either conclusion implies a general enrichment of the uppermost part of the suboceanic mantle in U at some time or times in the past. This enrichment is in addition to the general increase in U/Pb ratios for the entire outer part of the earth at ~3700 Ma (see below).

The *oceanic crust* consists of MORB, its intrusive equivalents, and oceanic sediments in a column with a total thickness averaging 6 km to 8 km down to the oceanic mantle (either asthenosphere or lithosphere). The base of the crust is defined as the Mohorovicic discontinuity (mercifully abbreviated to MOHO), where seismic velocities increase sharply downward to values of ~8 km/s for P waves. Magmas that rise along ridges result from the melting of peridotite, a process that preferentially removes clinopyroxene and leaves a residue in the mantle depleted in LIL elements and Fe. These residues are the same types of rocks that lie at the base of ophiolites (Section 4.2), generally referred to as harzburgite (a refractory mixture of olivine and orthopyroxene).

Subduction-generated crust is shown in Fig. 2.10 as an island arc, although similar lithologies are produced along continental margins (Section 4.1). This domain represents magmas derived from subduction zones and emplaced into, or at the base of, the overlying crust. Most magmas directly generated by subduction are basalts to basaltic andesites, with more silicic varieties evolved by some complex combination of fractional melting, crystal fractionation, magma or source mixing, crystal cumulation, etc.

Subduction-zone magmas have compositional properties that distinguish them from magmas of other environments. For reasons that have eluded consensus in the petrologic community, subduction-zone magmas contain lower concentrations of 'high-field-strength' (HFS) elements. These elements include Ti, Zr and Nb, all of which have high valences and comparatively low radii and, thus, have high ratios of charge/radius. Another characteristic of subduction-zone igneous rocks is isotopic ratios that suggest evolution of the magmas in source volumes that had higher concentrations of LIL elements than the asthenosphere.

The *subcontinental upper mantle*, with the marvelous acronym of SCUM, provides more difficulty in distinguishing lithosphere and asthenosphere than the suboceanic mantle, and only within the last decade has adequate evidence of a seismic low-velocity layer been found below continents. This layer has an upper 'lid' at ~150 km depth, which is at least partly a thermal boundary that forms the base of the moving, rigid, lithosphere. The subcontinental upper mantle, however, also exhibits vertical compositional variations. For example, extraction of continental crust from the underlying mantle may have left a gradation from depleted harzburgite at the top of the mantle to more LIL-rich, 'fertile', peridotite below. Thus, in contrast to suboceanic mantle where the lithosphere is simply cold asthenosphere, the subcontinental lithosphere/asthenosphere boundary may be partly compositional.

Compositional boundaries in the upper mantle may not occur at the same depth as thermal ones, permitting the identification of both a 'compositional lithosphere' and a 'thermal lithosphere'. The amount of lithosphere may: 1) increase with time because of continued separation of basalt from the asthenosphere and underplating on, or intrusion into, the lithosphere; and 2) decrease with time as it becomes more dense because of cooling and/or conversion to eclogite and founders ('delaminates') into the asthenosphere. Delamination of dense lithospheric mantle downward into less-dense asthenosphere may also remove some attached continental crust, which is recycled into the mantle and contributes to mantle heterogeneity.

In addition to vertical compositional variations, the subcontinental upper mantle may exhibit lateral differences. A major mineralogical distinction is the more magnesian composition (higher enstatite content) of orthopyroxenes in xenoliths from mantle underlying Archean crust than from mantle under either the oceans or under continental crust of Proterozoic and Phanerozoic age. This compositional difference may result from the extraction of komatiites from Archean mantle (Section 3.5) and basalts from younger mantle. The very Mg-rich komatiites form by melting of large volumes of peridotite and cause extraction of most of the Fe in the mantle, whereas extraction of basalts requires lesser amounts of melting. Because the extraction of Fe from peridotite produces a residue that has a lower density than that of the parent, the Archean mantle should be relatively buoyant and may have kept Archean shields exposed throughout much of earth history.

The mineralogically refractory nature of mantle beneath Archean shields does not signify 'primitive' characteristics of its present chemical components. For example, initial $^{87}Sr/^{86}Sr$, $^{207}Pb/^{204}Pb$, $^{206}Pb/^{204}Pb$, and $^{143}Nd/^{144}Nd$ ratios for basalts (particularly plateau basalts) that contain components from the mantle lithosphere show a broad range of values. Apparently, the subcontinental mantle has been variably affected (veined) by intrusion of magmas and by metasomatizing fluids of different sources and compositions.

The subcontinental upper mantle provides Pb isotopic evidence for the process that probably affected the entire outer, 'accessible', part of the earth (see Fig. 2.9 and discussion of Pb isotopes near the beginning of this section). Volcanogenic deposits of galena in sedimentary rocks are regarded as 'conformable' if they 'conform' (are parallel) to the structure of their host rocks, have the same age as the surrounding sediments, and apparently contain Pb derived from the mantle and uncontaminated by crustal Pb (which is more radiogenic). Because the galenas do not contain U, the Pb isotopic values are the same as the isotopic values of the mantle at the time the deposits formed.

Conformable Pb isotopic ratios commonly lie on a curve that passes through the point for the modern outer earth (Fig. 2.9). This continuity of isotopic values along one (secondary) growth curve lends confidence to the interpretation that this curve represents the development of the accessible part of the bulk earth following its fractionation from primordial U/Pb ratios. The secondary curve shown in Fig. 2.9 can be extrapolated back to intersect the Pb isotopic growth curve of an unfractionated primitive earth at an age of approximately 3700 Ma, which is presumably a time of increase in U/Pb ratio in the outer part of the earth. This age is also approximately the time at which significant amounts of continental crust began to develop (see below).

The boundary between upper mantle and *continental crust* is sharply defined by an increase of ~1 km/s in P-wave velocities over a distance of a few kilometers, the MOHO. Recent investigations show considerably more complexity to the MOHO than had been recognized previously, and intermingling of 'mantle' and 'crust' probably occurs over a depth range of several kilometers. The MOHO must be largely a compositional break, with the more-silicic rocks of the crust lying on some variety of ultramafic rocks. Parts of the upper mantle and lower crust, however, are likely to be eclogites (high-pressure basalts), which have seismic properties nearly identical to those of peridotites. Thus, basalt–eclogite phase transitions may cause some of the seismic changes near the MOHO in addition to the compositional variations.

Continental crust above the MOHO has thicknesses ranging from 70 km in two modern orogenic zones (Himalayas and Andes; Chapter 9) to perhaps less than 20 km in some rifted continental margins (Section 4.2). Average thicknesses are 40 km or more in areas stabilized in the Proterozoic and Phanerozoic and 35 km in Archean shields; (see preceding discussion of Archean and younger upper mantle). The crust can be subdivided into upper and lower parts, generally marked by a slight increase in seismic-wave velocities downward. This discontinuity is referred to as the 'Conrad discontinuity' by many investigators (Fig. 2.10).

The division between upper and lower crust may be either of the following transitions or a combination of two or more: 1) a change from brittle behavior in the upper part to ductile behavior in the lower part, caused primarily by increase in temperature downward; 2) a transition from amphibolite-facies rocks (characterized by the hydrous mineral amphibole) in the upper part to granulite-facies rocks (characterized by anhydrous pyroxenes) in the lower part; and/or 3) a change from

siliceous, granitic, rocks in the upper part to rocks of basaltic composition below. All transitions may be present, but not at exactly the same depth.

The *lower continental crust* is one of the most enigmatic parts of the earth. Because the transition from amphibolite to granulite facies is at ~6 kbar (0.6 GPa), roughly mid-crustal, we could assume that exposed granulite-facies rocks represent uplifted and/or exhumed lower crust. The importance of structural movements on the evolution of the lower crust is shown by the abundance of granulite-facies metasediments, which presumably reached high pressure as a result of covering by thrust sheets. Different granulite exposures have compositions ranging from gabbroic to granitic, probably with an average approximating that of tonalite (andesite). This evidence suggests a lower crust compositionally similar to the upper crust. Conversely, crustal xenoliths in igneous rocks are predominantly gabbroic and indicate equilibration pressures of 10 kbar to 15 kbar (35 km to 50 km depth). This mafic material apparently is not capable of uplift as coherent blocks but may constitute a large part of the true lower crust.

The composition of the lower crust is partly constrained by observations of surface heat flow. The average heat flow in continental areas is ~60 mW/m^2 (milliWatts per square meter; ~1.4×10^{-6} cal/cm^2/s); Archean shields are colder, with an average of ~40 mW/m^2. Based on abundances of heat-producing elements in near-surface rocks, nearly one half of continental heat flow can be accounted for by heat generated in the upper crust. The remainder of the heat flow (the 'reduced heat flow') is derived from the lower crust and mantle. Any reasonable assumptions about mantle heat production leave virtually no heat production attributed to the lower crust, and for this reason we assume that the lower crust is greatly depleted in heat-producing (and other LIL) elements. Exposed granulite-facies rocks confirm this depletion.

Depletion of the lower crust in heat-producing elements probably occurs through a series of processes that affect different areas to different extents. One process is melting of granitic magmas out of the lower crust, leaving a depleted residue underlying an upper crust enriched in granitic components. Residues consisting largely of plagioclase and pyroxene (including anorthosites) may constitute large parts of the lower crust. A second process of depletion is mobilization of elements by passage of fluids through the crust. Many granulite terranes show the effect of 'drying out' by displacement of H_2O by CO_2 rising through the crust; the displaced water carries LIL elements upward with it and contributes to the mobilization of granitic melts (e.g. southern India; Section 3.3).

The *upper continental crust* is the area that geologists can investigate. As such, it constitutes much of the subject of the remainder of this book and need not be discussed in detail here. Major components of the upper crust include Archean 'gray gneisses' (Section 3.5), younger metasediments and meta-igneous rocks, and granites. The average composition is probably granodioritic to granitic. The upper crust has been the recipient of LIL elements, volatiles and possibly other components removed from the lower crust and underlying mantle.

Growth of continental crust

We finish this section with a discussion of the segregation of continents from a primordial mantle and the production of the residual mantle that the earth has today. We do so in three ways. One is by preparing a mass balance of the distribution of K (the only major element in the LIL category) between the crust and mantle. This distribution permits us to make some inferences about the nature of the original earth and the present mantle and the rates at which continental crust separated from it. Our second approach is to study the effects of continental segregation on both the Pb and Nd isotopic systems and to infer both the time at which major fractionation began and the ensuing rates of continental growth. Third, we compare these conclusions with inferences that can be drawn from studies of sediments derived by continental erosion.

Continents clearly are the low-density, SiO_2-rich, LIL-rich fraction of the silicate component of the earth. Thus, to a first approximation, the mass of each element in the crust plus the mass in the mantle must equal the mass of that element in the original, unfractionated, silicate part of the earth. (Exceptions to this simple condition involve only volatile elements that have been lost to outer space, such as hydrogen, plus elements that have been almost completely concentrated in the atmosphere or oceans, such as nitrogen.) In theory, with many assumptions, we can measure the concentrations of elements in the various continental and mantle domains, multiply those concentrations

by the masses of the domains, and obtain a satisfactory mass balance against the primitive earth.

The table below provides very generalized estimates of the abundances of K and the masses of present crust and undifferentiated (primitive) mantle. These estimates can be used to calculate the unknown value y, the % K_2O in the present 'average' mantle. (We could also estimate this value and calculate some other variable as the unknown.) The composition for the primitive mantle assumes that K was lost from meteorites during condensation of the original earth; (CI carbonaceous chondrites have K_2O contents of ~0.15%; see Sections 2.1 and 2.2). All estimates are 'averaged' from various sources, and we could easily choose ones that are quite different.

mass of continental crust	0.02×10^{27}g
mass of mantle	4×10^{27}g
K_2O in primitive mantle	0.02%
K_2O in continental crust	2.0%
K_2O in present mantle	y%

The value of y is calculated by the equation $(4 \times 10^{27}\text{g}) \times 0.02\% = (0.02 \times 10^{27}\text{g}) \times 2\% + (4 \times 10^{27}\text{g}) \times y\%$, which yields a value of $y = 0.01\%$. This value is consistent with estimates made from weighted averages of various parts of the mantle (considering the range of estimates, that statement is not particularly meaningful). The calculation does, however, illustrate the important fact that approximately one half of the original amount of K in the earth has been extracted from the mantle into the crust. Other LIL elements yield comparable estimates of their extraction from the earth's interior, demonstrating the high efficiency of fractionation in the earth's gravity field (see Section 2.0).

Continental evolution can also be investigated by a mass balance of K_2O based on the rate of generation of new crust at oceanic spreading centers. At present, oceanic ridges have a total length of ~55 000 km, an average spreading rate (total of both sides) of ~6 cm/yr, and produce crust with a thickness of ~6 km. These figures permit calculation of the amount of continental crust that could have formed during the past

4000 m.y. based on three assumptions: 1) constant rates of generation of new oceanic crust; 2) constant composition of new oceanic crust; and 3) no destruction and recycling of continental crust by foundering into the mantle and erosion into ocean basins, where the sediment is ultimately subducted. If typical MORB has a density of ~3 g/cm^3 and contains 0.2% K_2O, then crust formed each year at oceanic ridges contains ~1.2×10^{14} g of K_2O. Using figures for continental crust in the table above, we calculate that continents contain ~4×10^{24} g of K_2O. Thus, if all of the K_2O in new crust formed at oceanic spreading centers were ultimately extracted into continental crust, formation of the present continents by this method alone would have required $(4 \times 10^{24} g)/(1.2 \times 10^{14} g/yr)$, = ~$3.3 \times 10^{10}$ years.

Although the preceding calculations are very simplified, a period of 3.3×10^{10} years is nearly ten times the age of the earth and indicates that one or more (probably all) of our assumptions are incorrect. One conclusion could be that much of the existing continental crust grew at a very early age, when rapid segregation from the mantle was possible because of high temperature gradients and resultant high rates of convective overturn (Section 2.2). Possibly continents attained virtually their present total volume by the end of the Archean (Section 3.1). An alternative conclusion is that much of continental crustal growth does not result from accretion in orogenic belts but occurs by upward movement of LIL and related components beneath stable continental interiors. This material would be delivered to the crust without passing through oceanic spreading centers and later subducted.

A more sophisticated method for investigating the extraction of continental crust from the mantle is by the use of Nd and Pb isotopes (see the first part of this section). We limit our discussions to the first-order fractionation of primitive earth into mantle and continental crust. This fractionation probably follows a multi-stage process of initial removal of basalt or basaltic andesite and further melting of that rock to form silicic magmas. The exact process, however, is not significant in the following discussion.

We can draw two general conclusions from Nd isotopic information interpreted (modeled) according to Fig. 2.8. The first is based on the time of intersection of the depleted-mantle curve with the curve for unfractionated mantle. According to one model, the separation of the two curves occurred at the time of the earliest significant production of continental crust and complementary depletion of the mantle. This age is controversial but is no younger than 3700 Ma to 3800 Ma and possibly older. It is consistent with interpretations from Pb isotopes (see below) and with the observation that the oldest crustal rocks were formed in the range of 3800 Ma to 4000 Ma (Chapter 3). Zircons older than 4000 Ma (Chapter 3) demonstrate the existence of granites of that age, but apparently these crustal materials had a small volume and were short-lived.

The second interpretation that can be made from Fig. 2.8 is the rate of separation of continental crust. In theory, T_{DM} can be calculated for all igneous rocks formed by partial melting of protoliths (parents) that were derived directly from the mantle. The distribution of these T_{DM} values is essentially a determination of crustal growth rates. Where adequate data are available (not a very large part of the earth), distribution of T_{DM} indicates that most of the earth's continental crust separated from the mantle between ~3800 Ma and ~1400 Ma, with virtually complete separation by ~1800 Ma. This 1800-Ma age is a time of formation of large volumes of continental crust and also is a significant time of transition for both internal and external processes in the earth, a topic discussed more completely in Section 5.8.

Interpretation of Pb isotopic data also indicates general early fractionation of the earth. As discussed above, the entire earth appears to have undergone fractionation into high-U/Pb regions, which are accessible to isotopic investigation, and low-U/Pb regions, which do not interact with the earth's surface (Fig. 2.9). Although this early fractionation may have occurred over a long period of time, the best estimate of a single age is ~3700 Ma, a time that coincides approximately with the age of the oldest preserved continental crust (Chapter 3). Another major worldwide event recorded by the Pb isotopic system was enrichment of the suboceanic mantle in U, probably in the approximate age range of 2000 Ma to 1500 Ma (see discussion of suboceanic asthenosphere above). The present mantle appears to contain several reservoirs of different U and Pb concentrations and Pb isotopic characteristics.

Any discussion of continental evolution is constrained by variations in the composition of sedimentary rocks through time. Many clastic sediments whose compositions have not been highly modified after deposition represent comprehensive samples of the earth's continents. Methods of averaging compositions are controversial, but even the largest discrepancies between estimates are smaller than a major compositional change that seems to have occurred across the Archean–Proterozoic boundary. Generally, Archean sediments are more mafic (have more Mg, Cr, etc., and less K and other LIL elements) than sediments deposited in Proterozoic and younger times. We infer that some major period of segregation of continental crust occurred approximately 2.5 billion years ago and, for the first time in earth history, created a crust sufficiently rich in granite and poor in mafic rocks (gabbros, basalts) that it could produce clastic debris rich in K feldspar and related minerals.

To summarize, various types of evidence yield moderately consistent conclusions about the evolution of continental crust. The oldest preserved (preservable?) crust probably formed about 3800 Ma or slightly older. Much of the present crustal volume was attained by some time in the range of 2500 Ma to 1800 Ma. Since that time, the major processing of continental crust may have consisted of recycling through the mantle, perhaps more than once. This recycling has caused continual replenishment of the mantle in crustal components.

[**References** – Isotopic information is provided by: general, Moorbath and Taylor (1981), Faure (1986); Sr, Faure and Powell (1972); Nd, Jacobsen and Wasserburg (1980), DePaolo (1988), DePaolo, Linn and Schubert (1991); Pb, Doe (1970), Stacey and Kramers (1975), Tatsumoto (1978). Compositional variation among major earth domains is discussed by Zindler and Hart (1986). Discussions of the Archean earth are provided by Nisbet and Walker (1982) and Nisbet (1987). Interpretations of the evolution of the mantle are in the volume edited by Menzies (1990) and papers by Boyd (1989) and McDonough (1990). Data concerning continental evolution are summarized by Taylor and McLennan (1985) and also in volumes edited by Ashwal (1989), Leven *et al.* (1990), Durrheim and Mooney (1991) and Bohor *et al.* (1992). Particular problems of the lower crust are discussed by Kay and Kay (1981), Bohlen and Mezger (1989), papers in the volume edited by Mereu, Mueller and Fountain (1989), Blundell (1990), and Kay and Mahlburg-Kay (1991). Calculations of rates of crustal evolution are based on information in Crisp (1984), Ellam and Hawkesworth (1988), Schubert and Sandwell (1989) and von Huene and Scholl (1991). Bird (1979) discusses delamination. Newton (1989) summarizes fluid movements in the crust.]

2.4 Sealevel and continental freeboard

THE RELATIONSHIPS of sealevel to continental elevations, and the variations in continental elevations themselves, pose a tangled set of problems. In this section we discuss primarily the issue of continental 'freeboard' and its variation through the entire geologic history of the earth. In a maritime sense, freeboard is the distance between the waterline and the deck of the ship. Geologically, freeboard refers to the extent to which ocean water has covered the surface of continental crust.

Changes in sealevel occur over various time intervals (Section 8.3). The most rapid is glacially controlled ('glacio-eustatic'), with periodicities of 10^4 to 10^5 years. Variations caused by increase and decrease in mean elevation of oceanic ridges occur over periods of 10^6 to 10^8 years. These two causes can account for the ~500 m of sealevel change that can be demonstrated for the Phanerozoic (this range is the difference between a highstand of 300 m to 400 m above present sealevel in the Cretaceous, and possibly slightly higher in the Ordovician, and the lowstand of ~150 m below present level in the late Pleistocene – see Sections 8.3 and 8.4).

Variations in mean sealevel over time spans of hundreds of millions to billions of years are very problematic. We can recognize four controlling factors.

- Radius of the earth. Decrease in the earth's radius diminishes the surface over which water can be distributed, thus forcing it farther up over continents. Although the radius may have decreased because of continued cooling of the earth, the change is presumably minor (Section 2.2).
- Volume of ocean water. Escape of water from the earth's mantle by volcanic emissions has presumably increased the volume of sea water through time. One problem, however, is our inability to discern the percentage of volcanic water that is derived from the mantle, and may never have been on the earth's surface before eruption, from the percentage that is simply recycled from the surface. Surface water entrained in volcanic eruptions on land can be recognized by its depletion in the heavy ^{18}O isotope. No such convenient isotopic distinction is available, however, for ocean water carried back down a subduction zone in wet sediments and recycled through the mantle. Some investigators have concluded that the volume of ocean water has remained approximately constant throughout much of geologic time, but others propose a steady increase in oceanic volume from very small amounts in the early Archean to the present.
- Volume of continents. At present, continents contain ~7.5×10^9 km^3 of material distributed over 25% to 30% of the earth's surface with an average thickness of 35 km to 40 km (the ranges of data result from questions such as the extent to which continental shelves extend out from land, whether submerged plateaus are continental or oceanic, etc.). Clearly a large amount of continental crust existed by the end of the Archean, ~2500 million years

ago, and some investigators propose that almost the entire volume had developed by that time. Alternatively, the volume of continental material derived ultimately from the mantle may have increased steadily toward the present, although destruction of crust by erosion of the surface and by subcrustal reincorporation into the mantle may equal that rate of production. Changes in the volume of continental crust affect sealevel into two contrasting ways – vertical growth (thickening) raises average continental elevation relative to the oceans, whereas horizontal growth ('continental outgrowth') fills in the ocean basins and forces water up over the continents.

• Relative thicknesses and densities of continental and oceanic lithosphere and the asthenosphere. Asthenospheric densities probably have increased toward the present as average thermal gradients have declined. An increase in the density of the asthenosphere without changes in the densities or thicknesses of continental or oceanic lithosphere increases the difference between the elevations of continental and oceanic crust. Thus, continued mantle cooling would require an increase in volume of ocean water if freeboard remained the same and no other variables changed. Lithospheric thickening, however, would certainly occur as thermal gradients declined, with an unclear effect on relative elevations of continental and oceanic crust.

The fossil-based stratigraphy of the Phanerozoic provides an opportunity to investigate variations in sealevel during the past 500 m.y. to 600 m.y. Probably the most useful information is estimates of the percentage of continental area that was flooded at various times. These estimates, however, are affected by such factors as the degree of erosion that sediments of different ages have undergone and the distribution of elevations on the continents (their hypsometries). For example, continental areas recently affected by rifting have lower average elevations, and hence are more broadly covered by transgressive seas, than continental areas recently affected by orogeny.

Because the entire ~500-m variation in Phanerozoic sealevel can be explained by glacial and ridge-crest activity (see above), no long-term trend in continental freeboard has been demonstrated. Many investigators presume that this long-term sealevel constancy would be a remarkable coincidence if any of the controlling factors varied because variation in one factor (e.g. water volume) would have to be precisely compensated by variation in other factors (e.g. continental volume). One explanation, therefore, is that no significant changes occurred in volume or thickness of continents, volume of sea water, lithospheric densities, etc.

Conversely, variation in more than one variable could be linked. For example, increase in the area of continents might be accompanied by a compensating increase in continental thickness, resulting in no change in freeboard.

Precambrian sealevels are more difficult to investigate than younger ones. The existence of stromatolites from the early Archean to the present demonstrates shallow-water covering of continents during almost all of geologic time. We do not know, however, whether Archean stromatolites represent local accumulations on small 'islands' of continental crust or erosional remnants of large banks. Certainly by the Proterozoic, however, broad shallow-water basins had developed over many continents.

[References – A basic discussion of continental freeboard is in Wise (1974). More-recent studies of sealevel and freeboard are in Schubert and Reymer (1985), Wyatt (1987), Algeo and Wilkinson (1991) and the volume edited by Cloetingh (1991). Discussions of hypsometry are in Harrison *et al*. (1983) and Cogley (1985).]

2.5 Summary

THE PHENOMENA discussed in this chapter are primarily the result of the two major factors that control processes in the interior of the earth: 1) gravity, with the thermal effects caused by loss of potential energy as materials move to positions of greater mechanical equilibrium; and 2) radioactivity, which produces heat that enhances the heat generated by loss of potential energy.

The earth accreted as a result of gravitational attraction, with the gravity field growing stronger as the planet grew larger. Virtually all of the accretion was in the earliest stages of earth history, and now the earth accumulates only minor amounts of cosmic dust and sporadic objects large enough to be referred to as meteorites. The frequency with which meteorites of different size strike the earth is controversial, and collisions with large ones have been proposed to be responsible for such major events as biotic extinctions.

In addition to accretion, gravitational forces also control the relative motions of the earth, moon and sun. Frequencies of orbital irregularities (Croll–Milankovitch cycles) may be partly responsible for climatic changes recorded in sediments, and transfer of angular momentum from the earth to the moon has caused gradual reduction in rotational velocity of the earth and consequent increase in the length of the day.

All gravitative segregation releases potential energy, but the only significant contributions to the earth's heat flow were its initial accretion and the separation of the core. These sources constitute a large, but unknown, part of the earth's total heat flow. The other contributor to heat flow is the decay of radioactive elements, of which the only significant ones are Th, U and K. Both the mechanical heat of segregation and the heat produced by radioactivity have decreased toward the present, but uncertainty in the ratio of the two types of heat production makes an accurate calculation of the rate of decline impossible.

Regardless of the exact rate of decrease in heat flow, thermal gradients and heat release clearly were higher during the early history of the earth than they are now. As at present, these high gradients were most characteristic of oceanic regions. The high heat flow could only have resulted from high rates of convection in the mantle, and convection cells may have been smaller and more irregularly organized in the early history of the earth than they are now.

The ability of the earth to fractionate light material upward and dense material downward in its gravity field is the primary cause of the establishment of 'domains' of different rock types. This separation is shown primarily by downward segregation of the dense Fe–Ni core (inner core and outer core; ~12% of the volume of the earth) and upward segregation of the light crust (~1% of the volume of the earth). The first-order subdivision can be refined in the upper part of the earth into recognizably different parts of the mantle and crust. We have chosen a classification into eight categories: 1) an inner mantle, possibly rich in LIL elements and postulated as the source of plumes; 2) an asthenospheric mantle, relatively mobile and depleted in LIL elements; 3) oceanic lithosphere, of considerable rigidity, overlying the asthenosphere but having the same composition as the asthenosphere; 4) island-arc and other primitive crust generated by subduction and derived from some combination of sources in the mantle and descending lithospheric slab; 5) oceanic crust, mostly MORB, forming the uppermost part of the oceanic lithosphere; 6) subcontinental upper mantle, probably consisting of mafic lithosphere enriched in LIL relative to oceanic mantle or crust; 7) lower continental crust, composed of LIL-depleted sialic rocks that form a relatively mobile zone beneath a rigid upper crust; and 8) upper continental crust, consisting of LIL-enriched silicic rocks that form a brittle zone above the Conrad discontinuity.

Gravitational segregation to form the domains listed above operates through a variety of mechanisms. One is magmatic fractionation, in which the molten magma (of low density) moves upward through surrounding wall rocks and solidifies at shallower levels. Separation of all forms of crust from the mantle is clearly dependent on this mechanism, although the magmatic process may involve repeated sequences of fractional melting, crystallization, and remelting. The magmatic separation process leaves refractive residues in the mantle (and possibly lower crust). A second mechanism of gravitative separation is metasomatic transfer of light elements (or ionic radicals) upward. This transfer is probably accomplished largely by upward movement of fluids that percolate through the rocks and exchange ions with them.

The combination of magmatic and metasomatic processes causes the various domains to have greatly different average compositions. Primary variation is almost certainly the result of magmatism, which generates 'granitic' continental and basaltic oceanic crust and large volumes of subduction-related ('calcalkaline') igneous rocks. The depletion of the asthenosphere in LIL elements probably was mostly the result of magmatic removal of mobile elements into the crust, leaving an impoverished residue. Depletion of the lower continental crust may occur through some combination of magmatism and metasomatism, with the relative importance of each process unclear. Enrichment of the subcontinental upper mantle in LIL elements has been proposed to be largely a result of metasomatism, although magmatic transfer may also be important.

The lightest materials in the earth are air and water, and the segregation of the atmosphere and oceans was as much a consequence of gravitative fractionation as separation of the crust. Compositional evolution of atmosphere and oceans is discussed in several later chapters. In this chapter, we considered the relationship between the elevation of the surface of the oceans and the elevation of continents – the 'freeboard' of the continents through time.

Relative elevation of oceans and continents is controlled by numerous factors, including volumes of ocean water and of continents, thickness and density of oceanic and continental lithosphere, and total volume of continental crust. Except for the present, none of these variables are accurately known, and their variation through time is conjectural. Shallow-water marine deposits are known from the early Archean through all of geologic time, but the extent of broad continental shelves is uncertain prior to approximately the middle Proterozoic. During the Phanerozoic, sealevel fluctuated through a range of approximately 1000 m because of changes in volumes of ridge crests, variations in amount of ice accumulated on continents, and possibly epeirogenic movements of continents (Chapter 7), and this range may also have characterized the Precambrian.

References

Algeo, T. J. & Wilkinson, B. H. (1988). Periodicity of mesoscale Phanerozoic sedimentary cycles and the role of Milankovitch orbital modulation. *Journal of Geology*, **96**, 313–22.

Algeo, T. J. & Wilkinson, B. H. (1991). Modern and ancient continental hypsometries. *Geological Society of London Journal*, **148**, 643–53.

Anders, E. & Ebihara, M. (1982). Solar-system abundances of the elements. *Geochimica et Cosmochimica Acta*, **46**, 2363–80.

Ashwal, L. D., ed. (1989). Growth of the continental crust. Special Issue of *Tectonophysics*, **161**, 143–352.

Ashwal, L. D., Morgan, P., Kelley, S. A. & Percival, J. A. (1987). Heat production in an Archean crustal profile and implications for the flow and mobilization of heat-producing elements. *Earth and Planetary Science Letters*, **85**, 439–50.

Berger, A., Imbrie, J., Hays, J., Kukla, G. & Saltzman, B. (1984). *Milankovitch and Climate – Understanding the Response to Astronomical Forcing*. Dordrecht: D. Reidel, 895 pp.

Bickle, M. J. (1978). Heat loss from the Earth: a constraint on Archaean tectonics from the relation between geothermal gradients and the rate of plate production. *Earth and Planetary Science Letters*, **40**, 301–15.

Bird, P. (1979). Continental delamination and the Colorado plateau. *Journal of Geophysical Research*, **84**, 7561–71.

Blundell, D. J. (1990). Seismic images of continental lithosphere. *Geological Society of London Journal*, **147**, 895–913.

Boher, M., Abouchami, W., Michard, A., Albarede, F. & Arndt, N. T. (1992). Crustal growth in West Africa at 2.1 Ga. *Journal of Geophysical Research*, **97**, 345–69.

Bohlen, S. R. & Mezger, K. (1989). Origin of granulite terranes and the formation of the lowermost continental crust. *Science*, **244**, 326–9.

Boyd, F. R. (1989). Compositional distinction between oceanic and cratonic lithosphere. *Earth and Planetary Science Letters*, **96**, 15–26.

Cloetingh, S., ed. (1991). Long-term sea-level changes: special section of *Journal of Geophysical Research*, **96**, 6583–949.

Cogley, J. G. (1985). *Hypsometry of the Continents*. Zeitschrift für Geomorphologie Supplementband 53, 48 pp.

Crisp, J. M. (1984). Rates of magma replacement and volcanic output. *Journal of Volcanology and Geothermal Research*, **20**, 177–211.

Croll, J. (1875). *Climate and Time in their Geological Relations – A Theory of Secular Changes of the Earth's Climate*. New York: D. Appleton, 577 pp.

Davies, G. F. (1990). Mantle plumes, mantle stirring and hotspot chemistry. *Earth and Planetary Science Letters*, **99**, 94–109.

DePaolo, D. J. (1988). *Neodymium Isotope Geochemistry*. Berlin: Springer-Verlag, 187 pp.

DePaolo, D. J., Linn, A. M. & Schubert, G. (1991). The continental crustal age distribution: Methods of determining mantle separation ages from Sm–Nd isotopic data and application to the southwestern United States. *Journal of Geophysical Research*, **96**, 2071–88.

Dodd, R. T. (1981). *Meteorites – A Petrologic-Chemical Synthesis*. Cambridge: Cambridge University Press, 368 pp.

Doe, B. R. (1970). *Lead Isotopes*. Berlin: Springer-Verlag, 137 pp.

Durrheim, R. J. & Mooney, W. D. (1991). Archean and Proterozoic crustal evolution: Evidence from crustal seismology. *Geology*, **19**, 606–9.

Ellam, R. M. & Hawkesworth, C. J. (1988). Is average continental crust generated at subduction zones? *Geology*, **16**, 314–7.

Faul, H. (1954). *Nuclear Geology*. New York: John Wiley, 414 pp.

Faure, G. (1986). *Principles of Isotope Geology*, 2nd edn. New York: John Wiley, 589 pp.

Faure, G. & Powell, J. L. (1972). *Strontium Isotope Geology*. Berlin: Springer-Verlag, 188 pp.

Fischer, A. G. & Bottjer, D. J., eds. (1991). Orbital Forcing and Sedimentary Sequences. Special issue of *Journal of Sedimentary Petrology*, **61**, 1063–252.

Goudie, A. S. (1977). *Environmental Change*. Oxford: Oxford University Press, 244 pp.

Grieve, R .A. F. (1987). Terrestrial impact structures. *Annual Reviews of Earth and Planetary Sciences*, **15**, 245–70.

Harrison, C. G. A., Miskell, K. J., Brass, G. W., Saltzman, E. S. & Sloan, J. L., II (1983). Continental hypsography. *Tectonics*, **2**, 357–77.

Hawkesworth, C. J. & Norry, M. J., eds. (1983). *Continental Basalts and Mantle Xenoliths*. Cheshire: Shiva Publishing, 272 pp.

Herzberg, C. T. (1983). Solidus and liquidus temperatures and mineralogies for anhydrous garnet lherzolite to 15 GPa. *Physics of the Earth and Planetary Interiors*, **32**, 193–202.

Jacobsen, S. B. & Wasserburg, G. J. (1980). Sm–Nd isotopic evolution of chondrites. *Earth and Planetary Science Letters*, **50**, 139–55.

Kay, R. W. & Kay, S. M. (1981). The nature of the lower continental crust: Inferences from geophysics, surface geology, and crustal xenoliths. *Reviews of Geophysics and Space Physics*, **19**, 271–97.

Kay, R. W. & Mahlburg-Kay, S. (1991). Creation and destruction of lower continental crust. *Geologische Rundschau*, **80**, 259–78.

Kornprobst, J., ed. (1984). *Kimberlites – II: the Mantle and Crust-Mantle Relationships*. Amsterdam: Elsevier, 392 pp.

Lambeck, K. (1980). *The Earth's Variable Rotation: Geophysical Causes and Consequences*. Cambridge: Cambridge University Press, 449 pp.

Leven, J. H., Finlayson, D. M., Wright, C., Dooley, J. C. & Kennett, B. L. N., eds. (1990). Seismic probing of continents and their margins. Special Issue of *Tectonophysics*, **173**, 641 pp.

McDonough, W. F. (1990). Constraints on the composition of the continental lithospheric mantle. *Earth and Planetary Science Letters*, **101**, 1–18.

McKenzie, D. & Weiss, N. (1975). Speculations on the thermal and tectonic history of the Earth. *Geophysical Journal of the Royal Astronomical Society*, **42**, 131–74.

Menzies, M. A., ed. (1990). *Continental Mantle*. Oxford: Oxford University Press, 184 pp.

Mereu, R. F., Mueller, S. & Fountain, D. M., eds. (1989). *Properties and Process of Earth's Lower Crust*. American Geophysical Union Geophysical Monograph 51, 338 pp.

Milankovitch, M. (1941). *Kanon der Erdbestrahlung und seine*

Anwendung auf das Eiszeitproblem. Royal Serbian Academy Special Publication 133, 633 pp.

Moorbath, S. & Taylor, P. N. (1981). Isotopic evidence for continental growth in the Precambrian. In *Precambrian Plate Tectonics*, Developments in Precambrian Geology 4, ed. A. Kroner, pp. 491–525. Amsterdam: Elsevier.

Newsom, H. E. & Jones, J. H., eds. (1990). *Origin of the Earth*. New York: Oxford University Press, 375 pp.

Newton, R. C. (1989). Metamorphic fluids in the deep crust. *Annual Reviews of Earth and Planetary Sciences*, **17**, 385–412.

Nisbet, E. G. (1987). *The Young Earth – An Introduction to Archaean Geology*. Boston: Allen and Unwin, 402 pp.

Nisbet, E. G. & Walker, D. (1982). Komatiites and the structure of the Archean mantle. *Earth and Planetary Science Letters*, **60**, 105–13.

Nixon, P. H., ed. (1987). *Mantle Xenoliths*. Chichester: John Wiley, 844 pp.

Richter, F. M. (1985). Models for the Archean thermal regime. *Earth and Planetary Science Letters*, **73**, 350–60.

Ridley, J. R. & Kramers, J. D. (1990). The evolution and tectonic consequences of a tonalitic magma layer within Archean continents. *Canadian Journal of Earth Sciences*, **27**, 219–28.

Rosenberg, G .D. & Runcorn, S. K. (1975). *Growth Rhythms and the History of the Earth's Rotation*. London: John Wiley and Sons, 559 pp.

Schubert, G. & Reymer, A. P. S. (1985). Continental volume and freeboard through geological time. *Nature*, **316**, 336–9.

Schubert, G. & Sandwell, D. (1989). Crustal volumes of the continents and of oceanic and continental submarine plateaus. *Earth and Planetary Science Letters*, **92**, 234–46.

Sharpton, V. L. & Ward, P. D., eds. (1990). *Global Catastrophes in Earth History*. Geological Society of America Special Paper 247, 631 pp.

Silver, L. T. & Schultz, P. H., eds. (1982). *Geological Implications of Impacts of Large Asteroids and Comets on the Earth*. Geological Society of America Special Paper 190, 528 pp.

Stacey, J. S. & Kramers, J. D. (1975). Approximation of terrestrial lead isotope evolution by a two-stage model. *Earth and Planetary Science Letters*, **26**, 207–21.

Tatsumoto, M. (1978). Isotope composition of lead in oceanic basalt and its implication to mantle evolution. *Earth and Planetary Science Letters*, **38**, 63–87.

Taylor, S. R. & McLennan, S. M. (1985). *The Continental Crust: Its Composition and Evolution*. Oxford: Blackwell Scientific, 312 pp.

Turcotte, D. L. & Schubert, G. (1982). *Geodynamics – Applications of Continuum Physics to Geological Problems*. New York: John Wiley and Sons, 450 pp.

von Huene, R. & Scholl, D. W. (1991). Observations at convergent margins concerning sediment subduction, subduction erosion, and the growth of continental crust. *Reviews of Geophysics*, **29**, 279–316.

Wise, D. U. (1974). Continental margins, freeboard and the volumes of continents and oceans through time. In *The Geology of Continental Margins*, ed. C. A. Burk & C. L. Drake, pp. 45–58. New York: Springer-Verlag.

Wyatt, A. R. (1987). Shallow water areas in space and time. *Geological Society of London Journal*, **144**, 115–20.

Zindler, A. & Hart, S. (1986). Chemical Geodynamics. *Annual Reviews of Earth and Planetary Sciences*, **14**, 493–571.

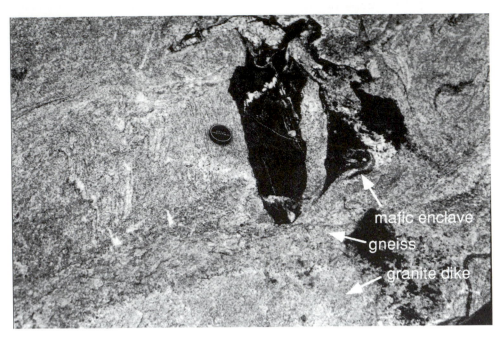

A capsule of Archean history in one outcrop (from the Western Dharwar Craton of southern India). Gneisses with ages of >3000 Ma incorporate enclaves of old, mafic, supracrustal rocks. Some of the veining in the gneisses is probably related to a widespread metamorphic/metasomatic event at ~3000 Ma. Post-orogenic granitic dikes were intruded at ~2500 Ma. □

3

THE ARCHEAN

3.0 Introduction

THE HISTORY of the earth begins with its accretion from some type of dispersed material in the solar system, probably about 4500 to 4600 million years ago. The first several hundred million years of the earth's existence are referred to as the 'Hadean'. It is an apt term – from a human point of view, the earth was a hellish place. The atmosphere presumably contained ammonia, methane, nitrogen, water vapor, and probably other compounds. This combination of gases retained much of the solar radiation, yielding a hot surface. Water may, or may not, have been able to condense. In the earliest times, the entire earth may have been molten, the result of heat generated by accretion, by separation of the core, and by the rapid rate of radioactive decay (including the decay of short-lived isotopes no longer present on the earth). The moon may initially have been part of the accreted earth and later separated.

No material that can now be sampled on the earth's surface provides a record of those earliest times. That history must be inferred from theoretical and laboratory studies, by comparison with other planets, and by studies of meteorites. Because of these uncertainties, and because we have four billion years of history to compress into a book of finite length, we shall leave the Hadean earth roiling in its enigmatic vapors and begin our story with the oldest material that we can actually find.

At the time of writing (early 1992), the oldest known minerals on the earth are zircons from water-deposited quartzites in western Australia that have nearly concordant ages in a range from ~4000 Ma to as old as ~4300 Ma. Because zircons crystallize only sparsely in magmas containing less than about 65% SiO_2, we can infer that silicic igneous rocks had formed on the earth at some time about 4300 Ma or earlier. None of the source rocks of the zircons have been preserved, testifying to the tendency of the early crust of the earth to be re-incorporated into younger rocks by some combination of remelting and/or partial assimilation. The zircons occur in metasediments older than about 3800 Ma, and some of the grains appear to have been rounded by water transport. Thus, running water must have occurred on the earth's surface before ~3800 Ma, and perhaps considerably earlier.

The oldest known rocks at the time of writing are tonalitic to granitic gneisses that have an age of ~3960 Ma based on the crystallization ages of zircons. These gneisses are apparently originally magmatic rocks now deformed into the eastern part of the Wopmay orogen (Section 5.2) along the western edge of the

ARCHEAN TERRANES IN THE NORTH ATLANTIC REGION

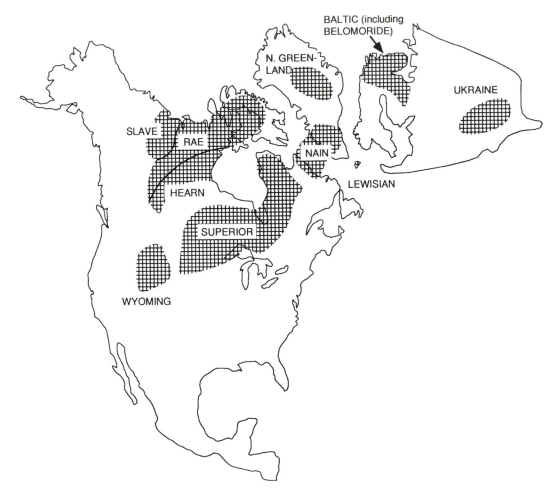

Fig. 3.1. Archean areas in North Atlantic region (North America, Greenland, Baltica, northern Scotland and parts of Russian platform).

Archean Slave province. Ages greater than ~3100 Ma have not been found within the Slave province, and whether the ~3960-Ma gneisses were originally part of that province or not is unclear.

Rocks and minerals with isochron ages in the range of ~3800 Ma are now known to be common in many of the world's Archean shields. For this reason, the Archean is defined by many geologists as a period of time beginning at ~4000 Ma, the time of first separation and consolidation of siliceous materials on the earth's surface. We do not know how widespread this early crust was, but presumably it was considerably smaller than the continental area existing at the end of the Archean. Most investigators believe that the very

old sialic rocks are the survivors of numerous small nuclei that fractionated from the primitive mantle but were largely destroyed by later events.

The areas of major outcrop of Archean rocks are shown in Figs 3.1 to 3.4. If this book had been written several years ago, the areas shown would have been considerably larger. The reason is that many 'crystalline' terranes once regarded as Archean simply because they 'looked old' are now known to contain rocks of younger ages. In particular, large areas of apparent Archean tonalite-trondhjemite gneisses (the 'typical Archean gray gneiss') have recently been found to have undergone reactivation and addition of new material in the period from 2500 Ma to 2000 Ma (Section 5.2).

ARCHEAN TERRANES OF SOUTH AMERICA, AFRICA, AND ARABIA

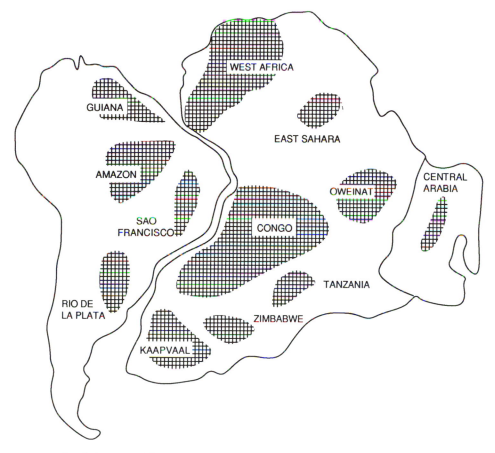

Fig. 3.2. Archean areas in South America, Africa and Arabia.

Typical relationships between Archean rock suites are shown in Fig. 3.5. Gneiss terranes and old, volcanic-dominated, supracrustal suites have an uncertain age relationship to each other, with the supracrustal rocks commonly separated from the surrounding gneisses by intrusive suites and/or faults. Younger supracrustal assemblages are dominated by graywackes and local volcanic piles, and some suites appear to lie unconformably on gneissic basement. Silicic intrusive rocks occur diapirically in the 'granite–greenstone' terrane and in lesser amount in the gneisses.

The 'end' of the Archean is a difficult concept. Most investigators regard the Archean as a period of time and designate its end at ~2500 Ma. This age is chosen because it represents the age of the last major thermal event in much of the Canadian shield (Superior and Slave provinces), where it is indicated by intrusion of diapiric (massive) granites and by establishment of whole-rock K/Ar isochrons. For this reason, we will use 2500 Ma as the chronologic younger boundary of the Archean.

The Archean may also be defined in terms of tectonic processes. The distinction is between a highly mobile ('permobile') crust and lithosphere in the Archean and a more rigid crust and lithosphere in the Proterozoic. In this concept, the end of the Archean is a time at which significant areas of continental crust became 'stabilized' ('cratonized'). Following stabilization, continental crust can undergo rifting or tectonic overprinting in local orogenic zones, but broad-scale remobilization does not occur.

The age of stabilization of a crust can be inferred from several types of observations. A major one is the

ARCHEAN TERRANES IN INDIA, ANTARCTICA, AND AUSTRALIA

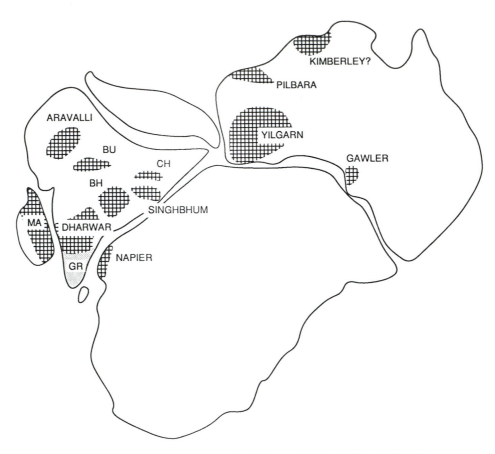

Fig. 3.3. Archean areas in India, Australia and East Antarctica. Symbols are: BU, Bundelkhand; CH, Chhotanagpur; BH, Bhandara; MA, Madagascar; GR, Granulite Terrane. ☐

age of the oldest sediments on continental platforms or in intra-cratonic basins. The area of stable continental crust capable of supporting cratonic basins has generally increased through time. Other criteria for the age of stabilization are: 1) the age of emplacement of post-orogenic, diapiric intrusive rocks, which generally occur in basement only slightly older than the cover sequences; and 2) absence of whole-rock isochrons in basement rocks that show any ages younger than the age of the oldest materials in the overlying sedimentary sections.

In the sense of the preceding paragraphs, the end of the Archean is not a time but a change in tectonic condition and may occur at different times in different places. It is at approximately 3000 Ma in the Kaapvaal

and Western Dharwar cratons and possibly older in the Pilbara craton (Figs 3.2 and 3.3). In most of the world's other shield areas, a tectonic stabilization at ~2500 Ma can be reasonably inferred, and a 2500-Ma event is important in the Kaapvaal craton (shown by an unconformity) and the Western Dharwar craton (shown by metamorphism). The major exception to stabilization at ~2500 Ma is the central Nubian–Arabian shield, which passed through its entire development from oceanic crust to stable continental platform during the approximate period of 1000 Ma to 500 Ma.

The remainder of this chapter discusses the history of this difficult-to-define Archean. First, we ask the question 'How different is the Archean?' The high

ARCHEAN TERRANES IN ASIA

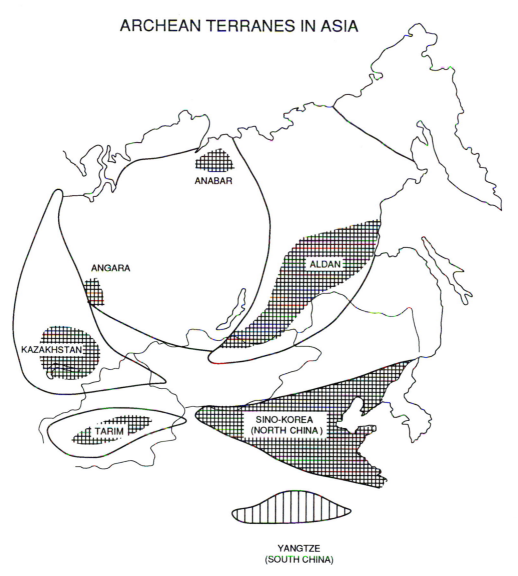

Fig. 3.4.   Archean areas of eastern Asia. The area shown includes terranes accreted as recently as Mesozoic. The Yangtze craton is shown with a different pattern because the oldest known rocks in it are Proterozoic. Boundaries are the same ones that delineate tectonic regions in Fig. 7.5.

internal heat flow and high surface temperatures, the scarcity of continental crust, the atmospheric gases inherited from planetary condensation, the possibility that the oceans held only a small volume of water (perhaps in isolated basins), the lack of life – all of these factors indicate an earth in the Archean, or at least an early Archean, that was very different from the one that we know today. But how different? By the middle of the Archean, and possibly earlier, stromatolites were emitting oxygen and accumulating calcite in their intricate layers. Streams were washing sediment into areas of shallow water, from which it was locally redistributed by turbidity currents. Basalts approximately similar in composition to modern ones were being erupted into shallow basins. Possibly the differences are not as large as we initially imagine. We discuss the apparent differences in internal processes in Section 3.1 and surface processes in Section 3.2.

The following sections present more detailed information on the Archean record. Section 3.3 provides a general description of three typical shields and identifies the lithologic suites that constitute them. The

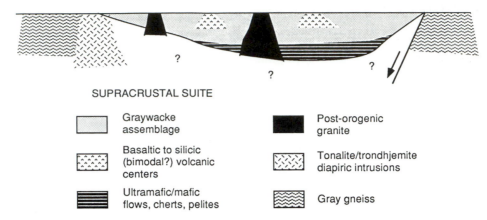

SUPRACRUSTAL SUITE

	Graywacke assemblage		Post-orogenic granite
	Basaltic to silicic (bimodal?) volcanic centers		Tonalite/trondhjemite diapiric intrusions
	Ultramafic/mafic flows, cherts, pelites		Gray gneiss

Fig. 3.5.   Generalized relationships among major suites in Archean terranes. The supracrustal suite (greenstone belt) is shown with a lower unit of mafic/ultramafic volcanic rocks overlain by a clastic suite consisting mostly of graywacke. The greenstone belt also contains localized volcanic centers. The belt is bordered by terranes of 'gray gneiss' on both sides, with the margins of the belt either occupied by batholithic intrusions or faulted against the gneisses. The greenstone belt, and locally the surrounding gneiss terrane, is intruded by late-tectonic (post-orogenic) granites.   □

immediately following sections discuss these components in more detail (supracrustal rocks in Section 3.4; gray gneisses and other intrusive igneous/metamorphic rocks in Section 3.5). The last section discusses the origin of organic life (Section 3.6).

[**References** - Zircon ages from Australia are reported by Froude *et al.* (1983), Compston and Pidgeon (1986), and Kinny *et al.* (1988). The currently oldest rock was analyzed by Bowring, Williams and Compston (1989). Discussion of the significance of the Archean and Proterozoic, particularly in the Kaapvaal craton, is provided by Cheney, Hoering and Winter (1990) and Master (1990). Nutman and Collerson (1991) discuss old Archean crust in the North Atlantic area.]

3.1 How different was the Archean? (Internal processes)

GIVEN the high heat flux in the early earth (Section 2.2) and the evidence of instability of continental crust, is it possible to describe the Archean in terms of the same processes of seafloor spreading, rifting, and subduction that we use for more recent events? The answer is not clear, and we discuss each process below.

First, what was the nature of Archean seafloor spreading? One very pertinent observation has been made in continental rocks. Heat dissipation through oceanic crust must have been much higher than at present (Section 2.2). This high heat loss, and the presum-

ably small scale of fast-moving mantle convection cells, means that the Archean oceanic lithosphere was recycled very rapidly through a series of closely spaced spreading 'centers'. The pattern of spreading was probably more similar to the diffuse, broad-area spreading characteristic of modern back-arc basins (Section 9.3) than to the system of linear ridges of modern large oceans.

The rapid overturn of small convection cells may have been responsible for the development of the world's earliest clastic sediments. The oldest rocks in greenstone belts (Section 3.4) include mafic/ultramafic volcanic rocks and clayey sediments rich in mafic components. They did not evolve by erosion of continental crust, but they did require a mechanism of crustal downwarp to form the depositional basin and a mechanism of flank uplift to form the source terrane. Some investigators have proposed that these earliest sedimentary belts developed in meteorite impact structures, but it is possible that they formed in the downgoing areas of convection cells adjacent to nearby rising margins.

The nature of the Archean oceanic lithosphere is controversial. A proposed 3500-Ma ophiolite sequence in South Africa contains a crust only 3 km thick (thinner than modern oceanic crust) consisting of intermingled komatiites and tholeiites overlying peridotite and capped by chert and shale. This proposed ophiolite, however, contains the type area of komatiites, and many geologists do not regard the sequence as a cross section of Archean oceanic lithosphere.

Our second question about Archean tectonics concerns rifting. The major problem with this question is the difficulty of locating any structures or lithologies equivalent to those of modern rifts. Individual faults are difficult or impossible to locate. Arkosic suites are rare, but their absence may be more the result of the lack of high-standing, erodable, granitic crust than the absence of rift valleys. Alkaline magmatic assemblages of the type found in some modern rifts also are rare in the Archean. Their absence, however, may simply reflect the high Archean thermal gradient that prevented formation of magmas at depth in the mantle and their penetration to the surface without equilibration and mixing at shallow depths; (evidence for such shallow-level equilibration is the complete absence of alkali-olivine basalts in Archean suites).

The lack of lithologic criteria means that the question of Archean rifting must be resolved on more inferential bases. One observation is that continental rifting and fracturing clearly has been an important process from the early Proterozoic to the present (Section 5.3). This evidence suggests that older crust was simply too hot and/or too thin to fracture rigidly. Thus, extensional stresses in the Archean presumably caused broad-scale thinning and stretching of the lithosphere but not the development of organized rifts (the exception is those areas in the Kaapvaal and Dharwar cratons that became 'tectonically Proterozoic' at ~3000 Ma; see above and Section 3.3).

Our final question is whether subduction occurred in the Archean. Evidence for it is considerably firmer for the latter part of the era (<~3000 Ma), which contains accretionary prisms recognized by their sedimentary/volcanic assemblages and thrust-faults. By analogy with modern subduction zones, these volcano–sedimentary wedges presumably developed in response to subduction of oceanic lithosphere.

Theoretical arguments can be constructed both for and against Archean subduction. For example, subduction generally requires, and may partly be driven by, a higher density in the subducting slab than in the surrounding mantle (Section 4.1). If that high density is caused partly by conversion of oceanic basalt to eclogite, and if Archean thermal gradients led to such high T/P conditions that they prohibited the conversion, then subduction was inhibited. Conversely, because

the asthenosphere into which an Archean slab descended was also hotter than present asthenosphere, and thus less dense, the density increase in the slab could have been less in the Archean than at present and still have permitted subduction.

One answer to the problem of Archean subduction is that the entire problem may be partly semantic. Clearly, the Archean surface foundered. All of the small pieces of continental crust formed before about 3900 Ma (and most of the ones afterward) plus all Archean oceanic crust has disappeared from the earth's surface down into the mantle. The unresolved issue is whether this foundering occurred along definable descending slabs, as at present, or in a small-scale pattern exemplified (on an even smaller scale) by reincorporation of the cooled crust of a lava flow into its molten interior. Do we call the latter process 'subduction' or reserve some other term for it? Does the terminology matter?

[**References** – Archean thermal regimes are discussed by Watson (1978), and Richter (1985). A possible Archean ophiolite is described by De Wit, Hart and Hart (1987). Considerable information on the Archean is in the special issue edited by Jahn and Moorbath (1990).]

3.2 How different was the Archean? (Surface processes)

THE EARTH'S ATMOSPHERE, ocean water and surface rocks form a complex chemical system. Despite the inter-relationships of that system, we discuss the Archean atmosphere and oceans somewhat separately below.

Atmosphere

The quantity of atmospheric gases is controlled by the amount of volatiles left at the surface of the earth after planetary condensation, by the rate of increment from the interior of the earth, by the rate of temporary withdrawal into sea water, and by the rate of semi-permanent withdrawal into sedimentary rocks, possibly including recycling into the mantle. Except possibly in the earliest stages of earth evolution, when surface temperatures were very high, the only gaseous species that have been able to escape from the earth's gravity field are H_2 and He; Ne, with an atomic weight of 18, does not escape, and the lightest of the major atmospheric gases have similar molecular weights (NH_3 – 17; H_2O –18: CH_4 – 16). Thus, the quantity of volatile or potentially volatile material in the whole earth has not changed since the earliest times except for loss of hydrogen.

Continued increment of gases to the atmosphere from the mantle and crust is demonstrated by the occurrence of ^3He in the atmosphere. ^3He escapes rapidly into outer space, and although some ^3He may be added to the atmosphere by non-terrestrial processes (e.g. cosmic-ray production), that increment could not account for the amount of ^3He lost. The conclusion is that ^3He could not occur in the atmosphere in its present abundance unless it was continually replenished from a source within the earth.

Extraction of gases by deposition or reaction with surface rocks is most important for CO_2. An approximate inventory of the world's surface carbon indicates that, at present, more than 90% is stored in carbonate rocks, and most of the remainder is dissolved in the oceans. This figure raises the question of the distribution of carbon in an Archean earth that contained only minuscule amounts of precipitated carbonate. Was the atmospheric CO_2 content very high, perhaps more than one order of magnitude greater than present atmospheric level (PAL)? Conversely, has the abundance of mobile carbon on the surface remained reasonably constant by an equality of the rate of increment from the mantle with the rate of precipitation as carbonates? A balance between mantle release of CO_2 and precipitation of carbonates could be maintained because higher atmospheric CO_2 causes more rapid weathering, thus greater availability of Ca and Mg in ocean water and more rapid precipitation.

Two principal arguments show that the earliest atmosphere was almost certainly reducing. One is the necessity of synthesizing organic molecules that led to the formation of life (Section 3.6). Mixtures of reduced gases (NH_3, CH_4, H_2O, and H_2) have been shown to form complex organic compounds when they are activated either photolytically (by the sun's radiation) or electrically (by lightning). Oxidized atmospheres containing H_2O, N_2, O_2, and CO_2, however, do not produce these organic molecules, and life may not have been able to evolve if the early earth had contained an oxidized atmosphere.

A second argument for the presence of a reducing earliest Archean atmosphere is the likelihood of equili-

bration with the earliest mantle. If gases were added to the atmosphere by the mantle, or at least were cycled through the upper mantle, then their redox condition would have been buffered by the mantle. Virtually all of the redox buffering in the solid earth is by the three oxidation states of iron (Fe^0, Fe^{II} and Fe^{III}). These states provide two levels of oxidation: 1) buffering by Fe^0 (metallic Fe) and Fe^{II} minerals, which yields a low oxygen fugacity in equilibrium with an atmospheric mixture of H_2, H_2O, CO, N_2 and minor NH_3 and H_2S; 2) buffering by minerals containing Fe^{II} and Fe^{III}, which yields a higher oxygen fugacity and an atmosphere dominated by H_2O, CO_2, N_2, O_2 and minor SO_2. Because of the buffering by iron, the components of the Archean atmosphere were partly dependent on the rate of removal of metallic Fe from the upper part of the mantle. This removal could have been caused by oxidation of Fe^0 and formation of ferrous silicates or, much more likely, by gravitative sinking of the metallic Fe toward the core (Section 2.0).

The two lines of reasoning followed above indicate that the earth's earliest atmosphere probably was reducing but that gases such as CH_4 and NH_3 were converted into CO_2, H_2O and N_2, with a net loss of H_2, at some time not long after the start of the Archean. Thus, we can infer, perhaps not too convincingly, that the atmosphere through much of the Archean differed from the present by its very low content of O_2 but not with regard to other components.

Oceans

At some time after the accretion of the earth, the surface cooled sufficiently to permit accumulation of liquid water. The finding of abraded (transported) zircons in metasediments older than ~3800 Ma (Section 3.0) indicates that this condensation probably occurred only a few hundred million years after the earth's formation. Condensation of a liquid phase did not require temperatures below 100 °C, partly because of the ability of dissolved salts to raise boiling temperatures and partly because total atmospheric pressures may have been high. Nevertheless, temperatures significantly above 100 °C would have been impossible, and limited oxygen-isotope data indicate sea water temperatures of ~70 °C at about 3500 Ma.

The initial volume of sea water was probably less than that of present oceans, although some investigators have proposed a world-encircling ocean without exposed land during much of the Archean. Possibly the initial 'ocean basins' were small, shallow areas isolated from other basins and with water compositions differing from each other. Increment of water to the earth's surface could have been caused by continued condensation from the atmosphere or by volcanic and other degassing of the mantle.

Compositions of the Archean sea water must be discussed in terms of its contents both of dissolved solids and dissolved volatiles. With respect to solids, rapid creation of new oceanic lithosphere at hot spreading centers and its interaction with sea water would have caused considerable buffering of oceanic composition by the lithosphere. The effects are not certain, but one of the major results of this process in modern oceans is the extraction of Mg from sea water into altered, hot, oceanic crust. This extraction balances the amount of highly soluble Mg delivered to the oceans by river runoff. We might infer that a similar process of Mg extraction occurred even more completely in the Archean, but many Archean metapelites are very rich in Mg (Section 3.4), indicating a Mg-rich ocean. Possibly Archean sea water reacted with basalts (komatiites) with such high Mg concentrations that the equilibration was reversed and Mg was added to sea water by reaction with oceanic crust.

Extensive modification of sea water by mafic/ultramafic lithosphere may have been enhanced by volcanothermal 'exhalation'. This process presumably was similar to ones that have produced sediment-hosted massive sulfides throughout geologic history. Elements that could have been added to the surface environment by such processes include Cr and Ni, both of which are enriched in a variety of Archean sediments, including cherts and pelites (Section 3.4).

The abundances of P and N in Archean sea water raise issues directly related to the development of life. Life as we know it cannot exist without nitrogen, a component of all amino acids. The problem with amino acids is that they are difficult to synthesize and maintain in water unless the water contains significant amounts of ammonium (presumably as NH_4^+). Gaseous NH_3 and its dissolved form, however, are not very stable under conditions of the earth's surface, largely because atmospheric NH_3 is easily decomposed by radiation unless the H_2 content is extremely high. If the dissolved NH_4^+ cannot be formed by solu-

tion of atmospheric NH_3, then some unknown mechanism must be responsible for its synthesis.

Phosphorus is particularly important for the evolution of life, partly because DNA and RNA contain it and partly because the major methods of storing energy in cells require the generation and decomposition of unstable phosphates (Section 3.6). In modern oceans, the PO_4^{3-} concentration is maintained at a low level by precipitation in apatite. In Archean oceans, the Ca^{2+} concentration was probably as high as in present oceans (possibly higher), indicating a similar absence of phosphate mobility. This low mobility of phosphate would have made any original biotic synthesis very difficult (Section 3.6).

The difficulties that we have outlined in estimating the composition of Archean sea water are compounded by the possibility that early oceans were small, unconnected, bodies of water. Under these conditions, the compositions of the water could have varied from basin to basin because of variation in ridge-crest and volcanothermal activity and possibly in the nature of surrounding subaerial terranes. For example, life may have evolved only in a few places, possibly followed by extinction in most of them, before ultimately being integrated into a global ocean.

[**References** – Discussions of the atmosphere are mostly from Walker, Hays and Kasting (1981), Walker (1982) and Holland (1984). The history of sea water is discussed by Hargraves (1976), Knauth and Lowe (1978), Holland (1984) and Veizer *et al.* (1989). Studies of alteration of modern sea water are in Humphris and Thompson (1978), Edmond *et al.* (1979) and Staudigel and Hart (1983). Discussions of the compositions of Archean sediments are in Schreyer, Werding and Abraham (1981), Martyn and Johnson (1986), chapter 2 of Naqvi and Rogers (1987), and Janardhan, Shadhakshara Swamy and Capdevila (1990).]

3.3 Three Archean shields

MUCH of the Archean record has been blurred by metamorphism or completely destroyed. As an introduction to it, we describe three of the better-known shield areas. Our purpose is to discover whether some general sequence of events leads to the development of all (or most?) shields despite differences in their detailed histories. For this purpose, the shields discussed here have been chosen for their differences: the Pilbara craton of western Australia (Fig. 3.3) is domi-

PILBARA CRATON

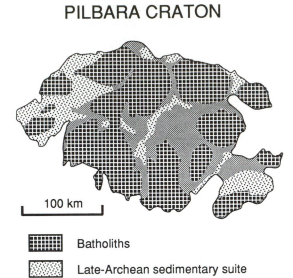

100 km

▦ Batholiths

▨ Late-Archean sedimentary suite

▨ Early-Archean volcanic suite

Fig. 3.6. Map of Pilbara craton, Western Australia (omitting westernmost part). Silicic intrusive rocks of ~3500- to ~3400-Ma age intrude a sequence of mafic to andesitic volcanic rocks. Older basement has not been recognized with certainty. The intrusive and volcanic activity may be nearly synchronous. Younger supracrustal assemblages contain both volcanic and sedimentary rocks apparently formed in troughs between the exposed plutons. The banded iron formations and other supracrustal rocks of the Hamersley basin overlie the southern margin of the craton (Sections 5.5 and 5.6). Map adapted from Bickle *et al.* (1983).   □

nated by intrusive rocks formed in the early Archean; southern India (Fig. 3.3) contains a diverse set of terranes, perhaps stabilized as early as 3000 Ma, and broadly affected by granulite metamorphism at ~2500 Ma; and the Superior province (Fig. 3.1) appears to have undergone a mostly late-Archean development leading to stabilization at ~2500 Ma. If a common history can be ascertained for shields that are so different, then perhaps we may conclude that the growth of continental crust follows a common path regardless of where or when it occurred. This topic is discussed after the individual reviews.

Pilbara craton

The Pilbara craton is bordered on the south by the very thick accumulation of iron formations and other rocks of the Proterozoic Hamersley basin (Section 5.5), on the east and north by other Proterozoic suites, and on the

west by the rifted continental margin of the Indian Ocean (Fig. 3.6). The craton differs from most of the world's continental blocks in its 'moth-eaten' or 'Swiss-cheese' appearance, which results mostly from the invasion of all of the craton's older supracrustal rocks by magmatic intrusions. Older gneisses are present only as enclaves within the plutonic areas, and some investigators regard much of the gneiss as deformed equivalents of the intrusive suite. The only contacts between supracrustal and crystalline rocks are intrusive or tectonic, and no gneissic basement has been located.

The greenstone belts of the Pilbara craton represent several unconformity-bounded assemblages deposited over a time range from older than 3500 Ma to late Archean (approximately 2800 Ma). The oldest suites, engulfed by the plutonic rocks, consist mostly of mafic volcanic rocks with some intermediate varieties. The intermediate ('andesitic') rocks are compositionally similar to the invading batholiths, leading to the proposal that the greenstone successions and the plutonic suites were coeval, possibly with the magmas intruding their own volcanic piles. This possibility is reinforced by nearly identical ages of 3500 to 3400 Ma in both the oldest greenstones and the adjacent plutonic rocks.

Younger supracrustal rocks tend to occupy the same inter-batholith screens as the older greenstones. Lithologies in these younger rocks are the typical assemblage of mafic and silicic volcanic rocks, metapelite and graywacke, plus minor iron formation, chert, and carbonates. The supracrustal suites have been proposed to represent a craton-wide stratigraphy, with present outcrops occurring merely as erosional relicts between the intrusive complexes. This stratigraphic continuity, however, has been challenged and the present outcrop pattern attributed to deposition in individual basins with only the appearance of stratigraphic correlation between outcrop areas. Regardless of their origin, localized preservation of supracrustal rocks older than 3500 Ma indicates that the areas of older greenstones remained 'downwardly mobile' for several hundred million years after they were engulfed by the batholithic suite.

The batholithic part of the Pilbara craton is a variety of rocks ranging from tonalite to granodiorite, with ages from ~3500 Ma to ~3200 Ma. Some investigators have regarded the plutonic suite as compositionally

SOUTHERN INDIAN (DRAVIDIAN) SHIELD

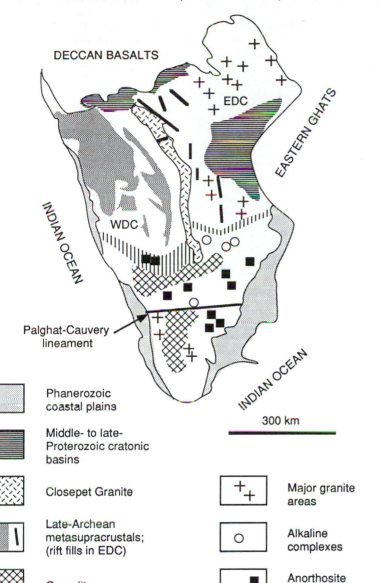

Fig. 3.7. Map of southern India (Dravidian shield). The gneiss–granulite transition zone separates a Granulite terrane to the south from the amphibolite-facies Dharwar craton to the north (which includes the EDC, Eastern Dharwar craton, and the WDC, Western Dharwar craton). Progressive increase in metamorphic grade southward through much of southern India culminates in the Granulite terrane, which consists largely of high-grade granulites (mostly charnockites) and lower-grade gneisses, partly formed by retrogression from the granulites. The Granulite terrane also contains the only anorthositic and alkaline igneous complexes in southern India. Much of the metamorphic activity along the transition zone was accompanied by a flux of CO_2 upward in the crust. The Dharwar craton is separated into two parts by the Closepet Granite, of ~2500-Ma age. The gneisses of the EDC were extensively invaded by granites at 2500 Ma, but the WDC apparently was stabilized at ~3000 Ma and became so 'chemically inert' that it did not undergo this 2500-Ma plutonism. The Eastern Ghats is a middle-Proterozoic granulite terrane that bounds the Dharwar craton on the east. Adapted from Naqvi and Rogers (1987). □

equivalent to modern 'calcalkaline' assemblages (Section 3.5). The remarkable nature of this intrusive suite cannot be overemphasized – over an area of some 500 km by 200 km, silicic magmas apparently synchronously intruded an unknown basement, its cover supracrustal rocks, and/or its own volcanic pile! What thermal process in the earth could mobilize the production of silicic magmas over an area of roughly 100 000 km²? What was the source of these magmas? Was

there some pre-existing sialic crust in the Pilbara area, or did all of the supracrustal suite rest on oceanic basement?

Another unknown aspect of the Pilbara craton is whether it is unique among the world's Archean terranes or whether comparable cratons exist elsewhere. Clearly it differs from areas dominated by old gneisses or in which the late-stage intrusive suite is small. Possibly its unusual features are merely the result of its

Fig. 3.8. Amphibolite-facies gneiss converted to charnockite (granulite) along zones of weakness that permitted CO_2 to 'dry out' the gneiss as it moved upward through the crust (Kabbaldurga, southern India).   ☐

evolution prior to ~3000 Ma, a time when the earth's thermal gradients were too high to achieve any structural stability. The exposed part of the craton, however, is relatively small, approximately the size of the large Abitibi granite–greenstone terrane of the Superior province (see below). Possibly further exposure of the Pilbara outcrop area would reveal greater similarities to other shields.

Southern India (Dharwar craton; Dravidian shield)

The southern peninsula of India consists of an old continental shield uplifted by the collision of India with the southern margin of Asia (Section 8.1). The area appears to have been a coherent block since about 2500 Ma, and the term 'Dravidian shield' has been applied to it (Fig. 3.7). The northern margin of the shield is mostly delimited by the Cretaceous/Tertiary Deccan basalts (Section 8.7). Other borders are mostly the rifted margins of India, and the original extents of features shown in Fig. 3.7 are unknown. The area is partly covered by middle-Proterozoic and younger platformal and basinal sediments (Section 5.5).

The southern Indian (Dravidian) shield contains extensive suites of granulite-facies rocks (including the type locality of charnockite) separated from a dominantly amphibolite-facies terrane to the north by a transition zone. The transition zone represents equilibration

pressures of about 6 kbar (0.6 GPa) and is simply one part of a gradational increase in metamorphic grade southward across the entire shield. The area to the north of the transition zone is referred to as the 'Dharwar craton', which can be subdivided into an eastern and western block by the ~2500-Ma Closepet Granite.

The Western Dharwar craton (WDC) contains the oldest rocks in the shield, gneisses with ages >3400-Ma. Stabilization of the WDC occurred at ~3000 Ma, accompanied by the emplacement of undeformed (diapiric) trondhjemite plutons and a wash of metasomatizing fluids through the crust. These fluids removed heat-producing and other LIL elements from the lower crust (and upper mantle?) and enriched middle and upper parts of the crust in uranium (and possibly other elements). The depletion of the lower crust in radioactive elements was so complete that modern heat flow in the WDC is very low compared with heat production in surface rocks. In a sense, the WDC appears to have been 'chemically inert' for the past 3000 Ma. The stabilization of the WDC at ~3000 Ma formed a platform on which graywacke assemblages, possibly rift-related, were deposited between 3000 Ma and 2500 Ma. These supracrustal rocks were deformed at 2500 Ma, but no whole-rock isotopic modification occurred in the crystalline basement since 3000 Ma.

The history of the Eastern Dharwar craton (EDC) differs greatly from that of the WDC. The EDC appears generally to be younger, with linear greenstone belts

SUPERIOR PROVINCE

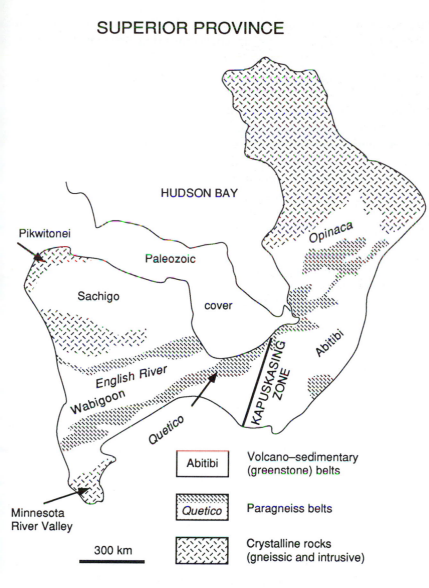

HUDSON BAY

Opinaca

Pikwitonei

Paleozoic

Sachigo

cover

English River

Wabigoon

Quetico

KAPUSKASING ZONE

Abitibi

Minnesota
River Valley

300 km

Abitibi	Volcano–sedimentary (greenstone) belts
Quetico	Paragneiss belts
	Crystalline rocks (gneissic and intrusive)

Fig. 3.9. Map of Superior province, Canadian shield. The map shows the distribution of the three major lithologic suites that constitute the craton – greenstone belts, paragneiss belts and gneiss/intrusive-rock ('crystalline') terranes. The greenstone belts are volcano–sedimentary assemblages whose relationships to surrounding gneisses are controversial (e.g. see Fig. 3.5). The paragneiss belts are metamorphosed and deformed clastic suites, possibly formed as accretionary wedges along rifted margins. The crystalline terrane consists mostly of deformed gneiss ('gray gneiss') but also includes minor late- and post-orogenic plutons. The Kapuskasing zone is an area of uplift of lower crust (see text). Map adapted from Card (1990). ☐

(mostly rift-related tholeiitic suites) formed between gneissic blocks and the whole terrane accreted at ~2500 Ma. At the time of accretion, the EDC was penetrated by abundant granites that have been attributed to mobilization by hydrous fluids pushed out of the lower crust by CO_2 during granulite metamorphism at depth.

The transition zone represents granulite-facies metamorphism at ~2500 Ma (Fig. 3.8). Passage of large quantities of CO_2 through the rocks is recorded by fluid inclusions. Rocks south of the transition zone are a mixture of granulite- and amphibolite-facies suites formed at various times throughout the Proterozoic, with both prograde and retrograde reactions widespread. Exposed granulites (including charnockites) are mostly silicic and greatly impoverished in LIL elements.

The granulite terrane is the only part of southern India that contains either anorthosite complexes or alkaline igneous suites. The alkaline complexes are generally undated but probably middle Proterozoic and younger; the reason for their restriction to the granulite terrane is unknown. The anorthosite complexes, which do not have accompanying rapakivi granites (Section 5.4), appear to have formed at ~2500 Ma, the same time as the granulite metamorphism along the transition zone.

Superior craton (Superior province)

The map pattern of the Superior province (Fig. 3.9) is completely different from that shown for the Pilbara area and is much more characteristic of other Archean cratons. The three major lithologies are: 1) volcano–sedimentary belts intruded by diapiric granites (granite–greenstone terranes); 2) silicic terranes consisting of gneisses and tonalite–trondhjemite plutonic suites in either amphibolite or granulite facies; and 3) paragneiss belts developed by metamorphism of rocks possibly deposited as accretionary prisms. The greenstone and paragneiss terranes alternate along approximately east–west strips (Fig. 3.9). The craton is completely bordered by middle- to late-Proterozoic orogenic belts or by Phanerozoic onlap sequences (Fig. 3.1).

The major greenstone suites of the Superior province appear to have formed in the general age range of 3000 Ma to 2700 Ma. The belts contain supracrustal assemblages of predominantly bimodal (silicic-basaltic) volcanic rocks and sediments consisting of turbidites plus ironstones, cherts and minor carbonates. Individual volcanic centers show fractionation from basalt to basaltic andesite and commonly are topped by silicic suites, but they do not contain a complete sequence along a modern 'calcalkaline' trend from basalt to rhyolite. The volcanic centers are surrounded by aprons of pyroclastic and epiclastic debris, and much of the silicic component of the greenstone belts is volcanically derived. The margins of the greenstone belts are commonly tectonic or are occupied by tonalite/trondhjemite intrusive suites, and they do not show unconformable relationships at the (generally unknown) base of the sedimentary assemblage.

The terranes of crystalline (silicic igneous and metamorphic) rocks represent a variety of rock types. The majority of the area mapped as 'crystalline rocks' is some mixture of old tonalite/trondhjemite gneisses ('typical Archean gray gneisses') and somewhat younger intrusive suites ranging in composition from tonalite to end-Archean granite. The oldest suites are gneisses of ~3800-Ma age in the Minnesota River Valley, but they are separated from the rest of the craton by a major structural zone (suture?). In most of the craton, rock types and ages range from gneisses at ~3100 Ma to more massive tonalites at about 2800 Ma and finally to isolated bodies of granite at about 2600 Ma. The granites are approximately the same age as granites penetrating the greenstone belts. Some areas of crystalline rocks near the border of the craton are overlain by rocks thrust onto the craton from adjoining Proterozoic mobile belts, and granulite-facies assemblages are common in these locations.

A major feature of the Superior craton is the Kapuskasing structural zone (Fig. 3.9), which crosscuts all other trends in the craton. The eastern margin is primarily a westward-dipping thrust fault, raising rocks with equilibration pressures up to 9 kbar (~30-km depth) against the lower-grade Abitibi terrane. The high-grade rocks are silicic to mafic granulites plus a layered anorthosite complex, and over a distance of approximately 250 km they grade westward through amphibolite facies into greenschist assemblages. This gradation apparently exposes the former Conrad discontinuity (Section 2.3) and provides a transect from lower to upper Archean crust.

A common history for the shields?

Despite obvious specific differences, all three shields described above exhibit some features of a common development. We can list them briefly.

- All of the shields contain very old supracrustal sequences, many of which are dominated by mafic and/or ultramafic volcanic suites (including komatiite). Mantle-derived volcanic rocks clearly have been major contributors to the growth of continental crust.
- Sedimentary and metasedimentary rocks were deposited in a variety of environments, generally accompanied by volcanism and other indications of tectonic instability.
- The oldest exposed crust contains highly deformed gneiss as a major component (although recognition of these rocks is difficult in the Pilbara craton). These gneisses probably formed by separation of large quantities of silicic magma out of the mantle by some set of processes.
- The deformation of the gneisses and accompanying supracrustal suites demonstrates that the growth of shields is accompanied by lateral compression before, and perhaps during, the stabilization of the crust.
- All of the areas ultimately became tectonically stable, although the onset of stability apparently occurred at different times in each shield (Section 3.0).
- Shields may undergo compositional modification even after they appear to attain stability. This process is best displayed in southern India.
- Stabilization of shields forms a basement on, and around, which sediments may accumulate. Evidence for the existence of a sialic basement for supracrustal rocks older than the apparent age of stabilization is more apparent in the relatively young Superior craton than in the very old Pilbara craton.

In brief summary, we can conclude that the development of continental crust from oceanic areas is a fundamental aspect of earth history. Although differing in detail from place to place and time to time, all shields have properties sufficiently similar that a process of 'shield evolution' is as important in the history of the earth as the process of biologic evolution.

[**References** – References for the shields are: Pilbara, Bickle *et al.* (1983), Horwitz (1990), and Williams and Collins (1990); southern India, Naqvi and Rogers (1983, 1987); Superior, Card (1990).]

3.4 Archean supracrustal sequences

WHEN geologists noticed that many Archean supracrustal rocks had a greenish tinge, they named them 'greenstones'. The green color results from low-grade metamorphism (mostly greenschist facies) that developed chlorite and epidote (plus serpentine in some of the mafic/ultramafic suites). In rocks where higher grades of metamorphism replaced these minerals with biotite, muscovite, and amphibole (among others), the green color disappeared. This metamorphic process commonly destroyed structures that indicated a sedimentary parentage of the rock, and hence the green color remains primarily associated with obvious Archean supracrustal suites.

Archean sedimentary rocks show elements of similarity and dissimilarity to younger suites. Both clastic and precipitated rocks are present, although some varieties are difficult to place in a particular category. Basinal, shallow-water, and subaerial deposition can be demonstrated. Sedimentary structures are broadly similar to modern ones. Despite these similarities, however, many differences exist. Source terranes contained high proportions of mafic and ultramafic volcanic rocks, commonly in a bimodal association with silicic varieties. Some suites have high concentrations of Mg, Cr and Ni, whose abundances are difficult to explain.

This section discusses the nature of Archean supracrustal suites and, at the end, speculates on the nature of Archean provenances. We can recognize three categories. The first is *coherent volcano–sedimentary belts (greenstone belts)*, which consist of large, commonly highly deformed, areas of ultramafic to silicic volcanic rocks interbedded with a predominantly clastic sedimentary suite. The second category is *linear metasedimentary belts*, probably related to rifting and/or evolution of continental margins. The third category is *supracrustal suites dispersed in gneisses*, either as highly deformed enclaves or, in the extreme, as small zones of apparent metasedimentary origin within the typical gray gneiss terrane.

Coherent volcano–sedimentary belts ('greenstone belts')

Coherent Archean greenstone belts are areas ranging in size up to more than 10 000 km². They are characterized by a mixture of volcanic centers, volcanic rocks possibly erupted along fissures and/or spreading centers, clastic sediments derived by erosion of adjacent gneiss terranes, volcaniclastic sediments that grade into tuffs and agglomerates, and varieties of chemical and/or biochemical precipitates (chert, carbonates, etc.). Metamorphism is commonly at greenschist facies (hence the name 'greenstone belt').

A highly idealized representation of a coherent belt is shown in Fig. 3.5. The diagram is overly simplified, both because no single belt contains all of the relationships shown and also because most belts do not display the stratigraphic continuity shown. Typically, large belts that contain a wide range of rock types have a more volcanic-rich (and komatiitic) older sequence overlain by a more clastic, sedimentary suite. Older sequences, both volcanic and clastic, appear to be largely subaqueous, implying that the belts were initiated on sea floor rather than in some shallow-water basin. Some old belts consist almost exclusively of volcanic/komatiitic rocks, whereas some young belts are simply a thick sequence of graywackes, siltstones, and shales. Most of the clastic sedimentary assemblages are turbidite deposits ('resedimented' in the sense that turbidites are formed by rapid transport of clastic material that had already been deposited in some shallower-water environment). Directly deposited sediments (e.g. fluvial and deltaic complexes) are also present in many belts.

The greenstone belts were depositional centers for millions or tens of millions of years and commonly evolved from deep- to shallow-water areas. This progressive emergence is shown by several factors. One is the occurrence of subaerial volcanic rocks, including airfall tuffs, in upper parts of belts or in younger belts. Another is the greater abundance of shallow-water features (tidal ripple marks, mud cracks, stromatolites, etc.) in the upper parts of sedimentary sequences.

Fig. 3.10. Photo of komatiite, showing 'spinifex texture' of the olivine; field of view ~1 cm wide. (The term signifies that the olivine blades have the appearance of the Australian spinifex plant, and STPK denotes 'spinifex-textured peridotitic komatiite'.) Most investigators attribute spinifex texture to rapid quenching of a hot peridotitic magma formed by high-percentage partial melting of the mantle. An alternative explanation is that some spinifex-textured rocks are cumulates in Archean ophiolites. (Thin section from J. R. Butler.) □

Estimates of growth rates are difficult to make, with proposals ranging from accretion rates similar to those of modern arcs to rates nearly one order of magnitude greater.

Efforts to determine the size and shape of the original depositional basins of the coherent belts yield conflicting results. Some studies show directional structures requiring transport of debris into the basin from nearly all sides, implying that the present belt has approximately the shape of the original depositional basin. Conversely, some geologists have proposed that the stratigraphic sequences in different belts can be correlated across intervening gneiss terranes, with the present outcrops merely erosional remnants of once-continuous, large basinal or platformal deposits.

Progressive growth of the volcano–sedimentary belts is consistent with the difficulty of finding a 'basement' for the greenstone belts. In some of the younger, clastic-dominated, belts a gneissic basement is shown both by an underlying unconformity and by gneissic debris in the sediments. The more volcanic-rich belts, however, rarely expose the base of any suite, and the early eruption of vast amounts of mafic and komatiitic rocks implies an oceanic (ensimatic) origin. One reason for inferring a basement of oceanic crust is that the high density of the komatiitic magmas would prevent their rise through continental crust.

The volcanic centers of the coherent belts commonly show one or more cycles of variation from basaltic to andesitic or more silicic eruptives. Most investigators have regarded the evolution as a result of crystal fractionation, but more recent proposals that the compositional variation of subduction-zone suites can be caused by contamination of later eruptive rocks by earlier magma batches from the same source may also apply to these suites (Section 4.1). The volcanic centers commonly emerged above sealevel at some time during their growth, at which time silicic pyroclastic rocks became important members of the surrounding sedimentary suites.

Many greenstone belts contain the ultramafic volcanic rock komatiite (Fig. 3.10), which was first recognized in the Barberton Mt. belt of southern Africa in the 1960s. The identifying feature is the dendritic 'spinifex texture', generally regarded as the result of rapid quenching of olivine because of eruption of very high-temperature lava; (the abbreviation STPK stands for 'spinifex-textured peridotitic komatiite'). Some investigators attribute komatiite magmas to partial melting of the mantle at depths of several hundred kilometers. A minority viewpoint is that some komatiites owe their texture to formation of olivine cumulates as part of ophiolite complexes.

The margins of the coherent belts are locations of

great controversy. Some of the relatively young clastic suites encroach unconformably over sialic basement, but most belts are isolated from their surrounding gneiss terrane by some combination of fault or ductile shear zone and/or by a rim of intrusive rocks slightly younger than the supracrustal suite (Fig. 3.5). The common occurrence of shear zones may be explained by the greater density of the 'greenstones' relative to the surrounding gneisses, which permitted the supracrustal assemblage to collapse through the hot, weak, Archean basement. Such collapse would explain the enigmatic juxtaposition of greenschist-facies supracrustal rocks against amphibolite-facies gneisses, a relationship that has puzzled geologists since the first recognition of greenstone belts. Tectonic juxtaposition of old gneisses and greenstone belts could explain the absence of gneissic debris in sedimentary suites, although an alternative explanation is that the gneissic crust was so thin that it did not protrude above water and could not have been eroded as a source of sediment.

The common presence of intrusive rocks along the margin between supracrustal belts and gneiss terrane has been explained in a variety of ways. One is that the supracrustal and calcalkaline (batholithic) suites were formed above a single subduction-zone, with the marginal batholiths merely the intrusive products related to the volcanism in the belt. This explanation could be valid for all margins of the belt only if the belt had evolved by subduction in more than one direction. Another possibility is that the marginal intrusive rocks were formed by remelting of deeply buried (metamorphosed) supracrustal assemblages. The problem with this explanation is the difficulty of distinguishing an amphibolitic source of the magmas within the belt from a lithologically similar source not related to the supracrustal suite. Both sources should be capable of melting to form at least a trondhjemitic trend and possibly a calcalkaline trend toward granodiorite and granite (Section 3.5).

One of the continuing issues about the coherent supracrustal belts is the tectonics of the environment in which they formed. Many investigators have regarded it as comparable to a modern subduction zone. In this situation, the supracrustal suite was either formed in an arc-trench environment, a back-arc setting, or a combination of the two. Subduction directions have been proposed for individual belts on the basis of variation in composition of volcanic rocks and strati-

graphic patterns of the entire supracrustal complex. Another possible environment is in intra-continental rifts not related to subduction. Geologists who doubt that modern plate tectonic processes, with definable subduction and rift zones, can be applied to the Archean (Section 3.1) propose that the 'greenstone' belts were merely the result of 'sag' in the weak Archean crust. The proposed sags may have been located between mantle diapirs.

Metasedimentary (paragneiss) belts

Linear belts of paragneisses more than 1000 km long have been described in the Superior province of the Canadian shield but not elsewhere. As in so many other situations, we do not know whether this localization results from the fact that such belts occur only in the Superior province or whether they occur elsewhere and simply have been described in other terms by geologists working in those areas.

The paragneiss belts in the Superior province are metamorphosed sandstones, siltstones, and shales aligned in an east–west orientation, parallel to the orthogneiss terranes and the major volcano–sedimentary belts (Fig. 3.9). They appear to have formed in the period of 3000 Ma to 2700 Ma, overlapping the evolution of the coherent greenstone belts. Although generally metamorphosed to amphibolite facies, the dominantly siliceous lithology of the paragneiss belts preserves some directional and other sedimentary structures that indicate that the belts developed as accretionary prisms, perhaps on the side of emergent greenstone terranes. The siliciclastic components appear to have been derived largely by erosion of the silicic rocks of bimodal volcanic suites rather than by erosion of orthogneisses. Magmatism in the paragneiss belts is slightly younger than in adjacent greenstone belts, supporting the concept that the entire deposition, deformation and metamorphism of the paragneisses was younger than the activity in the greenstones.

Magmatic rocks in the paragneiss belts are largely anatectic migmatites and larger peraluminous granitic intrusions. These granites imply high thermal gradients, consistent with the high-T/P metamorphism that characterizes the metasedimentary rocks. Whether these gradients can be related to Archean subduction or to post-subduction processes is unclear. If the paragneiss belts represent accretionary wedges, then

ISUA SUPRACRUSTAL SEQUENCE

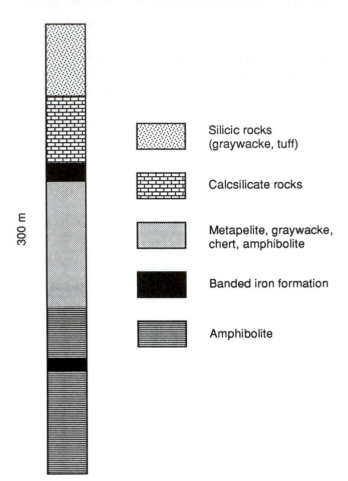

300 m

Silicic rocks
(graywacke, tuff)

Calcsilicate rocks

Metapelite, graywacke,
chert, amphibolite

Banded iron formation

Amphibolite

Fig. 3.11. Generalized stratigraphic section at Isua, Greenland. The total thickness of ~300 m was presumably much greater before thinning by deformation. The section at Isua is the best-known early-Archean supracrustal suite and has an age of at least 3800 Ma. Although most of the lithologies are typical of Archean greenstone belts, the Isua section contains metamorphosed carbonate/clastic rocks and rounded zircons in some clastic suites, indicating shallow-water deposition and erosion of a gneissic/granitic terrane. □

the Superior province can be regarded as a set of accreted terranes (Fig. 3.9).

Supracrustal suites in gneisses

The large areas of gray gneisses that constitute so much of Archean outcrop contain enclaves of highly metamorphosed supracrustal assemblages with sizes ranging from tens of kilometers down to individual bands on a hand-specimen scale. In fact, the smallest scale of intermingling may be individual grains and chemical components thoroughly mixed into the gneisses as 'cryptic' sedimentary components.

The most famous of the large supracrustal terranes is the Isua sequence exposed for an along-strike distance of more than 30 km in gneisses on the central west coast of Greenland. It has been dated by a variety

of techniques as at least 3800 Ma old. A composite, generalized, stratigraphic section at Isua is shown in Fig. 3.11. The thicknesses shown in the section have been greatly reduced by deformation from those of the original sediments.

Lithologies at Isua closely approximate those of typical greenstone belts. The lower part of the section is dominated by amphibolites with a basaltic composition, and other metabasaltic rocks and meta-ultramafic rocks intrude the section. Ironstones and cherts are present, mostly in the lower to middle parts of the section. The upper part of the Isua sequence contains metamorphosed pelites and graywackes plus calcilicates (metamorphosed sandy/shaly carbonates). The lithologies clearly indicate shallow-water deposition. The nature of the basement (mafic or continental) is unclear, but the presence of rounded zircons in some

Isua rocks may signify erosion of continental gneissic/granitic precursors.

Supracrustal suites smaller than Isua commonly do not exhibit a coherent stratigraphy but consist of similar lithologies. Major rock types include: quartzite–pelite suites; quartzite–ironstone associations; calcsilicates; mica schists; and amphibolites/serpentinites. The quartzite–pelite suites are almost certainly clastic, with both quartz and clay derived by erosion of pre-existing rocks. The quartzites associated with ironstones, however, are more problematic. As preserved, the quartz-rich layers of the ironstones consist of fine-grained quartz that could have been detrital or could have been deposited as chert and metamorphosed to coarser grain size (the probable precipitated origin of these quartzites is discussed below). Calcsilicates and mica schists are metamorphosed siliceous carbonates and metapelites, respectively, and the meta-mafic/ultramafic rocks have a volcanic parentage. Zircons occur in some siliceous rocks, but whether they are rounded (and thus transported) or oddly shaped because of metamorphic growth is controversial. If the quartz and zircon are detrital, then presumably some of the 'enveloping' gneiss is older and formed a basement and/or source area for detritus.

Sources of Archean sediments

Early Archean sediments are similar to modern ones in being both detrital and precipitated. Inferences derived from the precipitated rocks concerning the nature of Archean and Proterozoic sea water are discussed in Section 5.8, and here we concentrate on the clastic rocks. Compositions of these rocks were controlled by the compositions of their sources and possibly by volcanogenic ('exhalative') processes.

Most of the provenance for Archean supracrustal rocks seems to be similar to modern sources, but some differences can be ascertained. Siliceous volcanic rocks provided SiO_2, feldspar, and clays and were the primary source of the limited amount of quartz in older sediments. Detrital quartz is scarce or absent in sedimentary suites until the late Archean, when it was derived partly from silicic volcanics and partly from crystalline terranes of gray gneiss and intrusive rocks. The high Mg content of metapelites (mica schists) in older parts of coherent greenstone belts and metasedimentary enclaves in gneisses probably was the result of mechanical erosion of mafic/ultramafic sources, with the Mg retained largely in chloritic clays. With that exception, mafic and ultramafic debris is sparse in many clastic suites, implying weathering and general atmospheric modification approximately similar to modern processes.

High contents of Cr have been found in numerous Archean metasediments, particularly quartzites and mica schists. One unusual variety is a quartzite containing high quantities of fuchsite (a chrome mica), which for some unknown reason has been reported only from Archean metasediments in the Gondwana shields (India, South Africa, and Australia). An origin of high-Cr metasediments by accumulation of detrital chromite is unlikely because of the high Cr concentrations in minerals such as kyanite and the constancy of the Cr/Ni ratio in various rocks and mineral phases. Apparently, this constant Cr/Ni ratio was established by some igneous, hydrothermal, and/or metamorphic process rather than by clastic sedimentary accumulation.

One model for the origin of pelites rich in Mg, Al, Cr, and Ni is clastic deposition of Al-rich clays with a high Mg-chlorite component plus some type of volcanogenic (exhalative) addition of Cr, Ni and other metals (Section 3.2). Neither Cr nor Ni are significant components of massive sulfide (volcanogenic) ore deposits that characterized the later Archean to the present. Thus, their abundance in older sediments apparently indicates a change in mantle-derived (and crustally modified) fluids, possibly because of further fractionation of the mantle and consequent loss of Cr and Ni during geologic time.

[**References** – A major summary of greenstone belts is by Condie (1981). General aspects of Archean sedimentation are from Naqvi and Rogers (1983, 1987), volumes edited by Ayres *et al.* (1985) and Gaal and Groves (1990), a paper by Horwitz (1990), and a summary by Ojakangas (1990). Accretionary wedges are discussed by Percival (1989) and Percival and Williams (1989). The discussion of the Isua sequence is from Nutman *et al.* (1984, 1989). Specific discussions of fuchsite quartzites are in Schreyer, Werding and Abraham (1981) and Martyn and Johnson (1986).]

3.5 Archean plutonic suites

ARCHEAN SUPRACRUSTAL assemblages are underlain by, surrounded by, and/or intruded by a variety of plutonic rock suites, including: broad ter-

Fig. 3.12. Archean gray gneiss of Baltic shield (Finland). Most Archean gray gneisses are largely meta-igneous but contain some metasedimentary component and have been deformed more than once.   □

ranes of gray gneisses, less-deformed intrusive rocks, anorthositic layered complexes, and diapiric (post-tectonic) granites. We discuss them in that sequence.

Gray gneisses

Many geologists have felt that, once they have seen one Archean gray gneiss, they have seen them all (Fig. 3.12). The attitude is understandable after days and weeks of walking across seemingly endless expanses of highly deformed quartz-plagioclase–biotite gneisses. Variations among gneisses consist of subtle compositional differences and are commonly impossible to detect in outcrop. More than any other single rock type, the 'typical Archean gray gneiss' appears to be the dominant silicic component in continental crust and the major constituent of old shields. Their ubiquitous occurrence in Archean terranes implies that they must have formed by a 'simple' process that either occurred worldwide at one time or was repeated on numerous occasions at various places. The problem is at the heart of the controversy over the rate of growth of continental crust, and the two endmember models are early separation of most of the world's sial or continued separation through time.

Models of early formation of a worldwide sialic 'scum' a few kilometers thick are no longer as favored as they were a decade ago. Under this concept, frac-tionation of the earth during the early Archean produced a tonalite/trondhjemite liquid that crystallized over the entire surface and was deformed while still hot and plastic. Later deformations complicated the original gneissic pattern and/or obliterated part of it. The observation that some gray gneiss terranes underwent lateral shortening during the early to middle Archean strengthens (but does not prove) a model in which the continents were formed by compressional stacking of the thin early crust into continental blocks of smaller area but greater thickness.

A more popular current model is that the world's continents evolved as separate nuclei of tonalite/trondhjemite composition and grew to their present size mostly by lateral accretion. This explanation for the origin of continents is strengthened by recent work showing the suturing of early-Archean terranes during the later Archean. The discovery of gneisses with ages of 3800 Ma to 3900 Ma in so many Archean terranes suggests that many nuclei formed at about this time during some worldwide event in the mantle (Section 2.3). If rocks similar to gray gneisses were the provenance for the oldest transported zircons (see Section 3.0), then the separation of gray gneiss could have begun as early as 4200 to 4300 Ma.

Unquestionably, the gray gneisses consist primarily of metamorphosed igneous rocks (and some directly emplaced ones), although other components may be present. In many areas the silicic gneisses are intimately admixed with metabasaltic rocks, formed either as an initial bimodal magmatic association or later injected as dikes into the gneissic terrane. The mafic rocks are invariably highly deformed, indicating their emplacement into hot, mobile, rocks. Sedimentary components may also be present in the silicic rocks, and many gneisses have undergone compositional modification late in their history (see below).

Gneiss terranes apparently include rocks derived by partial melting at a variety of depths. A remarkably consistent distinction in all of the world's Archean terranes is between 'high-Al' and 'low-Al' gneisses, with the separation at approximately 14% to 15% Al_2O_3. One explanation for the compositional differences is that the low-Al gneisses formed by partial melting at shallow depths, in which plagioclase could have been retained in the solid phase, whereas high-Al gneisses represent melting at depths below the stability field of plagioclase (greater than about 50 km).

The source material for the igneous component of gneisses is not particularly controversial (an astonishing situation that will presumably change as more data become available). Trondhjemitic/tonalitic magmas are not in equilibrium with mantle peridotites. Consequently, the magmas must have been derived by melting of some material removed from the mantle. Isotopic studies show that the typical gneiss source regions did not contain high concentrations of LIL elements (e.g. K, U, Rb) for long periods of time between separation from the mantle and remelting to form the trondhjemitic/tonalitic magmas. The most likely source with these various properties is a basaltic layer several tens of kilometers thick 'floating' near the top of the Archean mantle and separated from the mantle only a short time before partial melting to yield the gneisses. This conclusion is enhanced by laboratory studies that show that hydrous basalts (amphibolites) melt to trondhjemitic/tonalitic magmas at almost any pressure.

Metasedimentary components in the gray gneisses are difficult to detect because commonly they are completely mixed into the igneous host. One index is the presence of rounded zircons, which presumably resulted from transportational abrasion before deposition in a sediment. A second indication of sedimentary material is a high content of Al_2O_3, which may result from the incorporation of clays into the trondhjemitic/tonalitic magmas.

A model of initial crustal genesis based on the preceding information is: 1) partial melting of basalt from the mantle and crystallization at the base of the crust; 2) hydration of the basalt; 3) 'subduction' or other means of placing the hydrated basalt to depths of several tens of kilometers; 4) partial melting of the hydrous basalt to form trondhjemite/tonalite magmas; 5) intrusion or extrusion, possibly with assimilation of some surface materials; and 6) deformation during and after emplacement, probably on several occasions, to form the intricate structures now shown by the gneisses.

Many of the gneiss terranes have been eroded to levels that expose rocks of granulite facies, commonly at equilibration pressures of 6 kbar to 8 kbar (mid-crustal). Conversion to granulite facies has been enhanced in many areas by 'drying' of the rocks by introduction of CO_2, which aids the conversion of hydrous minerals, such as amphibole, to orthopyroxene, the characteristic mineral of granulites. The areas affected by the drying can be several hundred kilometers in diameter (e.g. southern India), implying large-scale escape of CO_2 from the mantle.

Conversion of the lower crust to granulite is commonly associated with metasomatic mobility of LIL elements, which dissolve in water and move upward in the crust. This type of transfer has left granulite-facies rocks generally impoverished in elements (such as K, Rb, Ba, U and Th) and upper levels of the crust enriched in those elements. Because all of the heat-producing elements are LIL elements, this type of transfer reduces heat generation in the lower crust and consequently lowers heat flow in exposed Archean cratons.

Intrusive complexes

In addition to highly deformed gneisses, many shields contain younger trondhjemite/tonalite suites that are either massive (structureless) or show only broad-scale structures. These suites range from batholith-scale to small bodies only a few kilometers in diameter and commonly occupy the border zones between Archean greenstone belts and surrounding gray gneisses (Fig. 3.5). Their magmas probably were derived from sources similar to those of the gray gneisses.

The trondhjemite/tonalite suites show significant

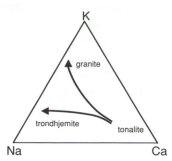

Fig. 3.13.   KCN diagram. The diagram shows the difference between the tonalite/trondhjemite trend (Na enrichment) and the tonalite/granite trend (calcalkaline, K enrichment). The Na-enrichment trend is dominant in Archean suites, with calcalkaline magmatism dominant at least from the middle Proterozoic to the present (Section 5.8).   □

compositional differences from modern subduction-zone (calcalkaline) batholiths. The principal difference is that most Archean assemblages are relatively deficient in K, whereas the Proterozoic and Phanerozoic suites from continental margins follow a normal path of K enrichment from gabbro to granite (Fig. 3.13). Modern magmas follow a trondhjemitic trend only in intra-oceanic subduction zones (island arcs).

The reason for the differences between Archean and younger continental batholiths is not clear but probably is related to the difference in composition of material added to the crust from the mantle. One explanation for this difference is the temperature of the subducted mantle slab. In modern plate configurations, this mantle is likely to have travelled considerable distances from a mid-ocean ridge. Consequently it is subducted cold, melting occurs at depths below the stability range of K-bearing minerals such as amphibole, and K is released into the melts. In the Archean, however, small-scale, rapid convection presumably forced the descent of hotter mantle that would have melted at shallower depths and retained K-bearing minerals in the solid residue.

Layered anorthositic complexes

Anorthosites and related rocks are widely distributed throughout Archean terranes. The oldest occurrence (in the Yilgarn shield of Western Australia; Fig. 3.3) has an age of nearly 3800 Ma, and the Fiskenaesset complex of western Greenland is only slightly

younger. Older suites tend to be fragmented, but some are more coherent. Archean anorthosites are areally associated with layered gabbros and/or layered ultra-mafic rocks (including chromite-rich varieties) and presumably formed by a process similar to the one that leads to the development of Proterozoic layered mafic/ultramafic complexes (Section 5.4). One difference from Proterozoic complexes, however, is that plagioclase grains in Archean anorthosites typically have average compositions from 70% to 80% anorthite, considerably more calcic than those of Proterozoic rocks.

The importance of anorthosites has been enhanced by the discovery that lunar highlands consist largely of rocks with anorthositic compositions. This occurrence implies that separation of anorthosite is a major process during planetary fractionation, and some investigators have proposed that the primitive (Hadean) earth contained large areas of anorthositic crust formed by floating of plagioclase on dense komatiitic melts and Fe-rich basaltic melts.

One indication of the importance of anorthosites in development of Archean crust is that mafic dikes containing large (anorthosite-type) plagioclase crystals intrude Archean gneisses in all cratons. The term 'leopard-rock' has been applied to them because of their spotted appearance (although the rock is white on black). These dikes presumably crystallized from basaltic magmas produced in a crustal chamber rich in previously crystallized plagioclase, and their withdrawal from the chamber may have left large areas of Archean crust underlain by anorthositic residue.

Diapiric (post-tectonic) granites

In most shields, intrusion of diapiric potassic granites is the final magmatic episode before the deposition of undeformed platform sediments over the stabilized craton. Mostly they intrude greenstone belts, in some places causing the belts to occupy 'embayed' patterns around the granites. In very old shields, such as the Western Dharwar craton of souther India (Section 3.3), the position of these granites may be occupied by diapiric trondhjemites.

Beginning at ~2500 Ma and extending through the Phanerozoic, post-tectonic potassic granites are one of the harbingers of crustal stability. Their abundance in the Superior province (Kenoran granites) was one of the reasons for the designation of the end of the

Archean at ~2500 Ma, although more recent dating has indicated that these granites are more correctly placed in the age range of 2600 Ma to 2700 Ma.

Because of their high content of K and related LIL elements, the granites cannot be derived from the same mafic crustal source that provided older magmatic components of the shield. Thus, they probably represent partial melts of the older gneissic terrane or, in some cases, silicic members of the greenstone assemblages. The relatively non-radiogenic character of the granites is consistent with derivation from lower crusts consisting of gneiss/granulite depleted in LIL elements during earlier granulite metamorphism.

[**References** – Gneiss terranes are discussed by Bridgwater, McGregor and Myers (1974), Glikson (1979), Drummond and Defant (1990), Ridley and Kramers (1990) and Nutman and Collerson (1991). Specific information on trondhjemites and tonalites is provided by the volume edited by Barker (1979), and the paper of Winther and Newton (1991). Anorthosites are discussed by Myers (1988), Phinney, Morrison and Maczuga (1988), Ashwal *et al.* (1989) and Lindsley and Simmons (1990). The discussion of post-orogenic granites is from Gower, Crocket and Kabir (1983), Gariépy and Allègre (1985) and Rogers and Greenberg (1990). Evolution of the crust of southern Africa is discussed in the volume edited by Tankard *et al.* (1982).]

3.6 The beginning of life

WHEN ASKED the question 'What is the oldest life on earth?', we commonly think of the oldest form of recognizable organism. At present, that organism is blue-green algae (cyanobacteria), which developed stromatolites at least as old as 3500 Ma. Blue-green algae, however, and other 'primitive' varieties of life are actually complex biological and chemical systems, possessing many of the properties that characterize all organic life, including the most advanced (presumably us?). The evolution of bacteria and the precursors to higher organisms are discussed in Section 5.7, and here we investigate the highly contentious issue of the biological and chemical developments that preceded even these earliest forms of life.

The chemical requirements for life

All organisms can be viewed as a collection of complex molecules embedded in water. These complex molecules are all constructed from four kinds of simple organic molecules, which thus form the basis of life. These simple compounds are as follows.

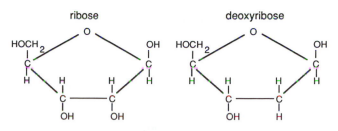

Fig. 3.14. Structure of ribose and deoxyribose. □

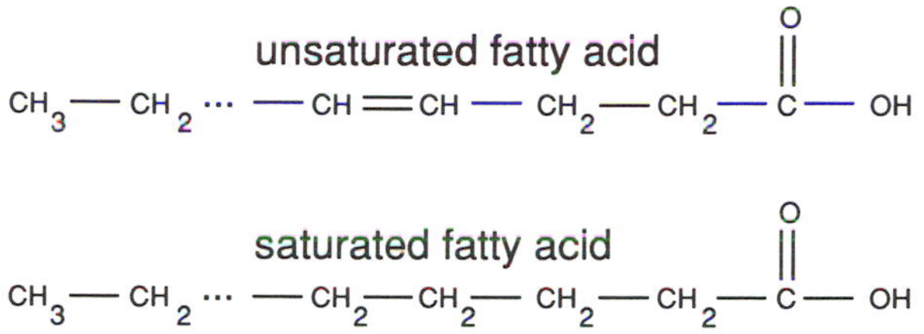

Fig. 3.15. Structure of fatty acids. □

- *Carbohydrates (sugars).* Carbohydrates have the general chemical formula $(CH_2O)_n$, where the carbon atoms are arranged in rings and n is commonly from 4 to 7 in biologically important compounds. The sugars ribose and deoxyribose contain a ring of four carbon atoms joined by an oxygen and attached to another C group (Fig. 3.14); they are necessary constituents of DNA and RNA (see below). Sugars can polymerize into molecules of thousands of individual rings to form polysaccharides, such as starch, which can then combine with other members of the four basic compounds to form even more complex organic molecules. Polysaccharides and their complex combined forms provide reserves of carbon and energy and also are major components of cell walls.

- *Fatty acids.* Organic acids contain a COOH group in which the H is weakly ionizable. In fatty acids, the C of the COOH group is joined to a long chain of carbon atoms and attached hydrogen (Fig. 3.15). Several of these chains can be linked in the same molecule with alcohols such as glycerol and with other elements such as N and P to form complex compounds. The entire group is known as fats, or lipids, because of their solubility in organic solvents. Lipids are major components of cell membranes.

- *Amino acids.* Amino acids are organic acids (see above) in which the C is joined to another C that, in turn, is attached to: a H atom, another H or organic group, and to a NH_2 group (Fig. 3.16). The diversity of amino acids is provided by differences in the side groups, and approximately twenty different acids occur in modern organisms. Polymerized amino acids are referred to as polypeptides, and linkages of polypeptides form proteins. Proteins serve as catalysts (enzymes) that promote reactions in cells and also as structural components of various parts of the cell.

AMINO ACID

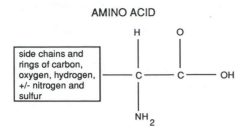

Fig. 3.16.   Structure of amino acids.   □

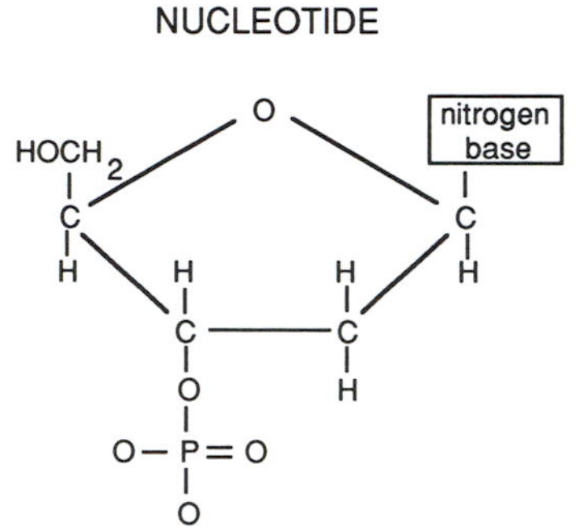

NUCLEOTIDE

Fig. 3.17.   Structure of nucleotide.   □

ISOMERIC PAIRS

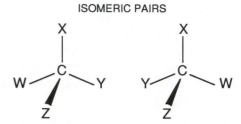

Fig. 3.18.   Isomeric pairs shown in tetrahedral arrangement around carbon (C) atom. The two arrangements produce different rotations of polarized light. The W, X, Y and Z represent elements or groups attached to the central carbon atom.   □

• *Nucleotides.* Nucleotides are combination of three units. One unit is phosphate. The second unit is a sugar, either ribose or deoxyribose. The third unit is a base consisting of a C–N ring or fused rings; five of these bases occur in nature. Nucleic acids form by polymerization of nucleotides, using the phosphate and sugar parts of the molecules to form a 'backbone' for a sequence of the nucleotide bases (Fig. 3.17). Polymers based on ribose form a single chain known as ribonucleic acid (RNA), and polymers based on deoxyribose form the famous double helix of deoxyribonucleic acid (DNA). DNA contains four of the nucleotide bases, and RNA contains three of the bases found in DNA plus one other. The sequence of bases forms a complex pattern that is replicatable and preserves the genetic integrity of the organism.

The chemical commonality of living organisms is shown by several consistencies between all forms of life. One is that virtually all organisms contain necessary elements in the ratio C:N:P of 105:15:1 (the 'Redfield ratio'). Apparently, all life on earth has evolved from some initial life form that used compounds in this ratio. A second indication of commonality is the near identity of types of molecules in all forms of life. All organisms contain polysaccharides, lipids, proteins and nucleic acids. Naturally, these compounds vary in detail from one type of organism to another; for example, the lipids are joined as ethers in the cell walls

of archaebacteria and as esters in the cell walls of eubacteria (Section 5.7). A third similarity is that virtually all organisms use the energy-rich molecule adenosine triphosphate for energy transfer (see below).

The fourth, and perhaps most remarkable, line of evidence for inter-relationship of all organisms consists of the identical direction in which crystalline forms of some of their compounds rotate polarized light. Carbon atoms bond covalently with four other elements (other C atoms, H, etc.) in a tetrahedral coordination. Those C atoms that are bonded to four different groups can arrange the bonds in two different configurations known as stereoisomers (Fig. 3.18). Crystals of these different isomers rotate light in different directions and, hence, are termed dextro-rotary (D forms) and laevo-rotary (L forms). The different isomers have the same thermodynamic properties and should be present in equal amounts. In fact, individual isomers naturally rearrange themselves into mixtures of the two forms over periods of time, a process termed 'racemization' and one that has been used for dating of organisms. Despite this identity of stability and ease of synthesis of the two isomers, all living organisms use only one isomer of many compounds. For example, only the L forms of amino acids and the D forms of sugars are metabolizable, and organisms contain enzymes that convert the rare opposite isomers of these compounds into ones that can be used.

The biological requirements for life

All organisms must meet certain criteria in order to be considered alive. Because the most primitive life form is a cell, we list some very generalized requirements for living cells.

- The cell must be separated from its environment by some type of membrane, possibly including a protective cell wall. (Without a membrane, how does one distinguish a cell from the rest of the universe?) Membranes must be semi-permeable, accepting necessary nutrients to the cell and passing waste products out. For example, photosynthesizing cells that produce oxygen must accept CO_2 and H_2O and eliminate O_2. Thus, membranes and cell walls can be constructed only by some of the specialized forms of organic compounds described above.
- The cell must have a method of storing and using energy obtained from its nutrients. Metabolic processes are controlled by a large number of proteins serving as enzymes. Energy can be stored in carbohydrates, but the cell needs some compound that absorbs energy during its formation and releases it quickly when it is triggered to do so. In all modern organisms, this energy transfer agent is some form of adenosine phosphate, commonly adenosine triphosphate (ATP). ATP is a nucleotide of the base adenine. Thus complex interaction of proteins and ATP, plus other molecules, is necessary for maintenance of the cell.
- The cell must have a method of replicating (reproducing) itself and of generating the proteins typical of that particular organism. The ability to perform these operations is stored in the DNA. In eukaryotes (Section 5.7) the replicating functions are in the nucleus, whereas in prokaryotes they are stored in DNA floating somewhere in the cell. DNA also occurs in mitochondria (bodies that control part of the metabolic process) and in the chloroplasts of plants (which contain green chlorophyll).

The replication process requires numerous steps, and here we discuss only those essential aspects that pertain to the evolution of life. The information in DNA is stored by sequences of the four nucleotide bases along the double strand. This sequence is the genetic code that is unique to each variety of organism. A series of three nucleotide bases is a code for one of the amino acids utilized in the production of proteins. The DNA replicates corresponding nucleotides onto the single-strand RNA, which, through a series of intermediate steps, produces proteins in bodies termed ribosomes. (Ribosomes range from simple combinations of RNA and protein to more complex bodies.) The ability of the DNA and RNA to replicate with virtually no error enables organisms to reproduce themselves. Errors introduced either spontaneously or by external activity (e.g. radiation) are called mutations. Most mutations are fatal, but some lead to new viable strains of organisms, presumably a major process in evolution.

Pre-biotic synthesis of molecules required for life?

The preceding discussion demonstrates the complexity of the molecules that are necessary for life as we know it and the complexity of the interactions between them. Without these basic chemicals, life cannot exist. In modern organisms, elaborate mechanisms control the synthesis of the necessary molecular components within the cell itself. What happened before cells existed? What degree of molecular complexity could have been synthesized by non-biotic processes?

The issue that we discuss here is somewhat akin to the old question of the chicken or the egg. Is a chicken merely an egg's way of making another egg? (We might also ask if the chicken is satisfied with its role in preserving the genetic code.) More specifically, if life cannot exist without complex molecules, and complex molecules can be synthesized only by living organisms, then how did the first complex molecules form? If RNA preserves the very nature of an organism and controls synthesis of its most important components, could life have developed without some 'pre-biotic' synthesis of RNA? The problem is extremely controversial, not only scientifically but also philosophically and theologically.

The problem of pre-biotic synthesis involves three difficulties. The first is sufficient energy to activate the various reactions necessary to produce complex organic molecules. Although internal heat sources in the earth (plus meteorite impact) were far more intense in the early Archean (and Hadean) than they are now, the only noticeable surface effects would have been excess volcanism or volcanothermal exhalation. This energy source could have been important for biotic evolution at the sites of submarine vents, perhaps similar to present vent communities. These communities generate their energy by chemosynthesis and are characterized by varieties of archaebacteria that tolerate very high temperatures (see Section 5.7 for further discussion of these concepts). Except in these special locations, the earth's internal heat probably could not have contributed significantly to general surface energy.

The other major source of necessary energy was the sun. Some of its sunlight could have been used directly in reactions promoted by radiant energy (photolysis). Some might have been used indirectly through the generation of atmospheric electricity (lightning). The intensity of high-energy radiation would have been enhanced by the scarcity of oxygen in the early atmosphere, which should also signify a scarcity of ozone and an absence of the ozone shield that screens out ultraviolet radiation in the modern earth. If the high-energy radiation were too intense, however, DNA and other complex molecules might have been destroyed or greatly modified. The supply of energy probably was not a significant problem in pre-biotic synthesis and may have been so high that the evolution had to occur in sheltered environments, such as deep water.

The second difficulty with synthesis of complex molecules outside of a living cell is the series of reactions needed to form them and the conditions under which those reactions occur. At present, several of the basic building blocks of the molecules necessary for life have been made in the laboratory, but virtually none of the complex molecules themselves. Amino acids, the nucleotide bases and a large variety of sugars have been produced artificially. The production of the base adenine also permits the formation of ATP (adenosine triphosphate).

Each of the series of reactions for producing the necessary compounds raises questions concerning its applicability to the problem of pre-biotic synthesis. Reactions to form the amino acids and nucleotide bases commonly involve HCN and may require solutions of relatively high pH (>7). The availability of HCN and water of this pH in the early earth is questionable. Sugars commonly require formaldehyde as a precursor; formaldehyde has been detected in interstellar gases but may not have been a common molecule on the earth's surface. Furthermore, the most necessary sugar for the development of life is ribose, a component of DNA and RNA, and ribose is more unstable than most other sugars.

Experimental production of amino acids has been accomplished by excitation of 'atmospheres' of NH_3, CH_4, H_2 and H_2O. On the earth, an atmosphere of this reduced state could exist only by equilibration with a mantle containing metallic Fe (Fe^0; Section 3.2), largely because of the tendency of H_2 to escape from the earth and decrease the reductive capacity of the surface environment. We cannot be sure that such an atmosphere ever existed on the earth, and if one did, it probably disappeared within a few hundred million years (or less) after the earth's condensation. Amino acids

similar to those produced experimentally have been found in meteorites, supporting the possibility of their pre-biotic synthesis on earth.

Even if the early earth developed conditions that permitted the synthesis of amino acids, ribose and nucleotide bases, these compounds could not develop a living organism. They must be combined to form proteins, polysaccharides, lipids (with fatty acids) and DNA and RNA. No experiment outside of a living cell has been able to produce these molecules, and until such a synthesis is demonstrated, the conditions for the inception of life must remain mysterious.

The third difficulty in the origination of a living organism from an abiotic world is probably the most serious – specifically, the development of some mechanism by which the organism can replicate itself. If DNA and RNA did not exist before organisms learned to make them, how did the first form of 'life' preserve any consistency in the synthesis of its own molecules and in the matching of those molecules with other living individuals? Many investigators have concluded that RNA must have been available to the first organism, although they have not determined how it could have been.

In the absence of RNA and DNA, other forms of reproducible 'templates' have been suggested. These templates include the surfaces of such minerals as clays or pyrite (which are common in submarine vent communities). The theory is that the atomic arrangements on these surfaces, particularly where lattice imperfections disturb a complete symmetry, provide the same type of replicating structure that is now supplied by DNA and RNA. Possibly an organism could 'learn' to reproduce sufficient biologically necessary molecules from a mineral surface that it would eventually produce a genetic code. The concept is fraught with controversy.

Our discussion of the origin of life has yielded few conclusions in which we can have confidence. The problem is more complicated than envisioned by Charles Darwin when he proposed that life developed in a 'warm little pond' provisioned with the necessary chemicals. We cannot even be sure that life evolved at one time or in one place. Possibly it evolved on many occasions followed by extinction after all but the last one. Most fundamentally, we are still left with the problem of the chicken and the egg. We will not solve it here.

[**References** – A general discussion of the requirements of life is provided by Brock and Madigan (1988) and De Duve (1991). General summaries of the origin of life are in Clark (1988), Miller and Bada (1988), Oro, Miller and Lazcano (1990) and Margulis and Olendzenski (1992). Specific topics include: importance of clays in evolution, Cairns-Smith and Hartman (1986); importance of sulfur in evolution, Wachtershauser (1988, 1990), Kimioto and Fujinaga (1990); and the effects of impacts, Maher and Stevenson (1988).]

References

Ashwal, L. D., Jacobsen, S. B., Myers, J. S., Kalsbeek, F. & Goldstein, S. J. (1989). Sm–Nd age of the Fiskenaesset Anorthosite Complex, West Greenland. *Earth and Planetary Science Letters*, **91**, 261–70.

Ayres, L. D., Thurston, P. C., Card, K. D. & Weber, W., eds. (1985). *Evolution of Archean Supracrustal Sequences*: Geological Association of Canada Special Paper 28, 380 pp.

Barker, F., ed. (1979). *Trondhjemites, Dacites, and Related Rocks*. Amsterdam: Elsevier, 659 pp.

Bickle, M. J., Bettenay, L. F., Barley, M. E., Chapman, H. J., Groves, D. I., Campbell, I. H. & de Laeter, J. R. (1983). A 3500 Ma plutonic and volcanic calc-alkaline province in the Archaean East Pilbara block. *Contributions to Mineralogy and Petrology*, **84**, 25–35.

Bowring, S. A., Williams, I. S. & Compston, W. (1989). 3.96 Ga gneisses from the Slave province, Northwest Territories, Canada. *Geology*, **17**, 971–5.

Bridgwater, D., McGregor, V. R. & Myers, J. S. (1974). A horizontal tectonic regime in the Archaean of Greenland and its implications for early crustal thickening. *Precambrian Research*, **1**, 179–97.

Brock, T. D. & Madigan, M. T. (1988). *Biology of Microorganisms*, 5th edn. Englewood Cliffs, New Jersey: Prentice Hall, 835 pp.

Cairns-Smith, A. G. & Hartman, H., eds. (1986). *Clay Minerals and the Origin of Life*. Cambridge: Cambridge University Press, 193 pp.

Card, K. D. (1990). A review of the Superior Province of the Canadian Shield, a product of Archean accretion. *Precambrian Research*, **48**, 99–156.

Cheney, E. S., Hoering, C. & Winter, H. de la R. (1990). The Archean–Proterozoic boundary in the Kaapvaal province of southern Africa. *Precambrian Research*, **46**, 329–40.

Clark, B. C. (1988). Primeval procreative comet pond. *Origins of Life and Evolution of the Biosphere*, **18**, 209–38.

Compston, W. & Pidgeon, R. C. (1986). Jack Hills, evidence for more very old zircons in Western Australia. *Nature*, **321**, 766–9.

Condie, K. C. (1981). *Archean Greenstone Belts*. Amsterdam: Elsevier, 434 pp.

De Duve, C. (1991). *Blueprint for a Cell*. Burlington, North Carolina: Neil Patterson Publishers.

De Wit, M. J., Hart, R. A. & Hart, R. J. (1987). The Jamestown Ophiolite Complex, Barberton mountain belt: a section through 3.5 Ga oceanic crust. *Journal of African Earth Sciences*, **6**, 681–730.

Drummond, M. S. & Defant, M. J. (1990). A model for trondhjemite–tonalite–dacite genesis and crustal growth via slab melting: Archean to modern comparisons. *Journal of Geophysical Research*, **95**, 21 503–21.

Edmond, J. M., Measures, C., McDuff, R. E., Chan, L. H., Collier, R., Grant, B., Gordon, L. I. & Corliss, J. B. (1979). Ridge crest hydrothermal activity and the balances of the major and minor elements in the ocean: the Galapagos data. *Earth and Planetary Science Letters*, **46**, 1–18.

Froude, D. O., Ireland, T .R., Kinny, P. D., Williams, I. S., Compston, W, Williams, I. R. & Myers, J. S. (1983). Ion microprobe iden-

tification of 4,100–4,200 Myr-old terrestrial zircons. *Nature*, **304**, 616–8.

Gaal, G. & Groves, D. I., eds. (1990). Precambrian Ore Deposits Related to Tectonics. Special Issue of *Precambrian Research*, **46**, 176 pp.

Gariépy, C. & Allègre, C. J. (1985). The lead isotope geochemistry and geochronology of late-kinematic intrusives from the Abitibi greenstone belt, and its implications for late Archaean crustal evolution. *Geochimica et Cosmochimica Acta*, **49**, 2371–83.

Glikson, A. Y. (1979). Early Precambrian tonalite-trondhjemite sialic nuclei. *Earth Science Reviews*, **15**, 1–73.

Gower, C. F., Crocket, J. H. & Kabir, A. (1983). Petrogenesis of Archean granitoid plutons from the Kenora area, English River subprovince, northwest Ontario, Canada. *Precambrian Research*, **22**, 245–70.

Hargraves, R. B. (1976). Precambrian geologic history. *Science*, **193**, 363–71

Holland, H. D. (1984). *The Chemical Evolution of the Atmosphere and Oceans*. Princeton, New Jersey: Princeton University Press, 582 pp.

Horwitz, R. C. H. (1990). Palaeogeographic and tectonic evolution of the Pilbara Craton, Northwestern Australia. *Precambrian Research*, **48**, 327–40.

Humphris, S. E. & Thompson, G. (1978). Hydrothermal alteration of oceanic basalts by seawater. *Geochimica et Cosmochimica Acta*, **42**, 107–25.

Jahn, B. M. & Moorbath, S., eds. (1990). Early developments of the earth and Archaean geochemistry. Special Issue of *Precambrian Research*, **48**, 193–418.

Janardhan, A. S., Shadhakshara Swamy, N. & Capdevila, R. (1990). Trace and REE geochemistry of pelites from the Sargur high-grade terrain, southern Karnataka. *Geological Society of India Journal*, **36**, 27–35.

Kimioto, T. & Fujinaga, T. (1990). Non-biotic synthesis of organic polymers on H_2S rich seafloor: a possible reaction in the origin of life. *Marine Chemistry*, **30**, 179–92.

Kinny, P. D., Williams, I. S., Froude, D. O., Ireland, T. R. & Compston, W. (1988). Early Archaean zircon ages from orthogneisses and anorthosites at Mount Narryer, western Australia. *Precambrian Research*, **38**, 325–41.

Knauth, L. P. & Lowe, D. R. (1978). Oxygen isotope geochemistry of cherts from the Onverwacht Group (3.4 billion years), Transvaal, South Africa, with implications for secular variations in the isotopic composition of cherts. *Earth and Planetary Science Letters*, **41**, 209–22.

Lindsley, D. H. & Simmons, E. C., eds. (1990). Anorthosites and associated rocks. Special Section of *American Mineralogist*, **75**, 1–58.

Maher, K. A. & Stevenson, D. J. (1988). Impact frustration of the origin of life. *Nature*, **331**, 612–4.

Margulis, L. & Olendzenski, L. (1992). *Environmental Evolution – Effects of the Origin and Evolution of Life on Planet Earth*. Cambridge, Massachusetts: The MIT Press, 400 pp.

Martyn, J. E. & Johnson, G. I. (1986). Geological setting and origin of fuchsite-bearing rocks near Menzies, Western Australia. *Australian Journal of Earth Sciences*, **33**, 1–18.

Master, S. (1990). 'Archaean', 'Proterozoic', and the Archaean-Proterozoic boundary: Semantic minefield in a Precambrian no-man's land. *South African Journal of Geology*, **93**, 417–9.

Miller, S. J. & Bada, J. L. (1988). Submarine hot springs and the origin of life. *Nature*, **334**, 609–11.

Myers, J. S. (1988). Oldest known terrestrial anorthosite at Mount Narryer, Western Australia. *Precambrian Research*, **38**, 309–23.

Naqvi, S. M. & Rogers, J. J. W., eds. (1983). *Precambrian of South India*. Geological Society of India Memoir 4, 575 pp.

Naqvi, S. M. & Rogers, J. J. W. (1987). *Precambrian Geology of India.* New York: Oxford University Press, 223 pp.

Nutman, A. P., Allaart, J. H., Bridgwater, D., Dimroth, E. & Rosing, M. (1984). Stratigraphic and geochemical evidence for the depositional environment of the early Archaean Isua supracrustal belt, southern West Greenland. *Precambrian Research*, **25.**, 365–96.

Nutman, A. P. & Collerson, K. D. (1991). Very early Archean crustal-accretion complexes preserved in the North Atlantic craton. *Geology*, **19**, 791–4.

Nutman, A. P., Fryer, B. J. & Bridgwater, D. (1989). The early Archaean Nulliak (supracrustal) assemblage, northern Labrador. *Canadian Journal of Earth Sciences*, **26**, 2159–68.

Ojakangas, R. W. (1990). Archaean sedimentation, Canadian shield. In *Precambrian Continental Crust and its Economic Resources: Developments in Precambrian Geology, vol. 8*, ed. S. M. Naqvi, pp. 179–202. Amsterdam: Elsevier.

Oro, J., Miller, S. L. & Lazcano, A. (1990). The origin and early evolution of life on earth. *Annual Reviews of Earth and Planetary Sciences*, **18**, 317–36.

Percival, J. A. (1989). A regional perspective of the Quetico metasedimentary belt, Superior Province, Canada. *Canadian Journal of Earth Sciences*, **26**, 677–93.

Percival, J. A. & Williams, H. R. (1989). Late Archean Quetico accretionary complex, Superior province, Canada. *Geology*, **17**, 23–5.

Phinney, W. C., Morrison, D. A. & Maczuga, D. E. (1988). Anorthosites and related megacrystic units in the evolution of the Archean crust. *Journal of Petrology*, **29**, 1282–323.

Richter, F. M. (1985). Models for the Archaean thermal regime. *Earth and Planetary Science Letters*, **73**, 350–60.

Ridley, J. R. & Kramers, J. D. (1990). The evolution and tectonic consequences of a tonalitic magma layer within Archean continents. *Canadian Journal of Earth Sciences*, **27**, 219–28.

Rogers, J. J. W. & Greenberg, J. K. (1990). Late-orogenic, post-orogenic, and anorogenic granites: Distinction by major-element and trace-element chemistry and possible origin. *Journal of Geology*, **98**, 291–309.

Schreyer, W., Werding, G. & Abraham, K. (1981). Corundum-fuchsite rocks in greenstone belts of southern Africa: petrology, geochemistry, and possible origin. *Journal of Petrology*, **22**, 191–231.

Staudigel, H. & Hart, S. R. (1983). Alteration of basaltic glass: Mechanisms and significance for the oceanic crust-seawater budget. *Geochimica et Cosmochimica Acta*, **47**, 337–50.

Tankard, A. J., Eriksson, K. A., Hunter, D. R., Jackson, M. P. A., Hobday, D. K. & Minter, W. E. L., eds. (1982). *Crustal Evolution of Southern Africa.* New York: Springer Verlag, 523 pp.

Veizer, J., Hoefs, J., Lowe, D. R. & Thurston, P. C. (1989). Geochemistry of Precambrian carbonates: II. Archean greenstone belts and Archean sea water. *Geochimica et Cosmochimica Acta*, **53**, 859–71.

Wachtershauser, G. (1988). Before enzymes and templates: theory of surface metabolism. *Microbiological Reviews*, **52**, 452–84.

Wachtershauser, G. (1990). The case for the chemautotrophic origin of life in an iron-sulfur world. *Origins of Life and Evolution of the Biosphere*, **20**, 173–6.

Walker, J. C. G. (1982). The earliest atmosphere of the earth. *Precambrian Research*, **17**, 147–71.

Walker, J. C. G., Hays, P. B. & Kasting, J. F. (1981). A negative feedback mechanism for the long-term stabilization of earth's surface temperature. *Journal of Geophysical Research*, **86**, 9776–82.

Watson, J. V. (1978). Precambrian thermal regimes. *Philosophical Transactions of the Royal Society of London*, **288**, 431–40.

Williams, I. S. & Collins, W. J. (1990). Granite–greenstone terranes in the Pilbara block, Australia, as coeval volcano–plutonic complexes: evidence from U–Pb zircon dating of the Mount Edgar batholith. *Earth and Planetary Science Letters*, **97**, 41–53.

Winther, K. T. & Newton, R. C. (1991). Experimental melting of hydrous low-K tholeiite: evidence on the origin of Archaean cratons. *Geological Society of Denmark Bulletin*, **39**, 213–28.

Margin of the mid-Atlantic rift as it crosses Iceland. The photograph is taken in the low spreading center and shows an uplifted flank consisting of rift-generated basalts. □

4

PROCESSES IN A RIGID LITHOSPHERE

4.0 Introduction

IN THE ARCHEAN, the earth may or may not have had a rigid lithosphere (Section 3.1). By the time the earth reached the age that we call the Proterozoic, it certainly did. What does this rigidity signify? In this chapter, we discuss the consequences of a rigid lithosphere for the history of the earth and identify some of the more pertinent aspects of 'plate tectonics'. First, however, we must discuss the history of the development of the concept of rigid lithospheric plates. It encompasses three main stages.

The first episode is the proposal that the continents moved around the earth rather than occupying fixed positions, which was first described effectively by A. Wegener in 1912. A key part of his evidence was the observation that the outlines of continents such as South America and Africa permitted them to fit together if the Atlantic Ocean were closed. Wegener proposed that all of the earth's continents could move horizontally, either splitting apart from each other or moving toward each other. The term 'Pangea' (Pangaea) has been used for a late-Paleozoic fusion of all of the earth's continental masses, from which the various modern continents separated and moved to their present positions.

Wegener's proposal was met by most of the scientific community with a mixture of reactions ranging from doubt to derision. As a very broad generalization, we can say that geologists in the northern hemisphere lived on continents whose evolution could be explained as a continued outgrowth from an old core. For example, the central (stable) part of North America was regarded as an old cratonic interior surrounded by the Appalachian, Ouachita, Cordillera, and the Innuitian and correlative orogens. Thus, geologists who worked in North America found no compelling evidence of continental fragmentation and tended to dismiss Wegener's hypothesis. In the southern hemisphere, however, geologists did not find continents so neatly circumscribed by their own borders. In South Africa, for example, structural and stratigraphic trends are commonly truncated at the edges of the continent, leaving the strong impression that some previously attached landmass had drifted away.

Regardless of the observational evidence, the basic issue facing proponents of continental drift remained the inability of any of them to propose a force large enough to move continents around the earth. Without such a force, all evidence of drift became essentially circumstantial, and the oceans and continents could be considered to have remained in the positions that they had always occupied. In this regard, one concept of

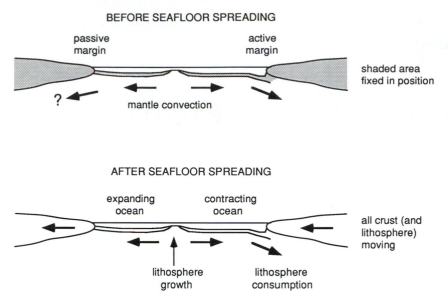

BEFORE SEAFLOOR SPREADING

passive
margin

active
margin

shaded area
fixed in position

?   mantle convection

AFTER SEAFLOOR SPREADING

expanding
ocean

contracting
ocean

all crust (and
lithosphere)
moving

lithosphere
growth

lithosphere
consumption

Fig. 4.1. Diagrammatic representation of tectonic concepts before and after the proposal of seafloor spreading. Before H.H. Hess introduced the idea of seafloor spreading in 1960, many geologists assumed that convection currents circulated beneath a rigid surface that underwent local deformation but no major movement. The concept of seafloor spreading includes creation of lithosphere at oceanic ridges, movement of the new lithosphere away from the ridges, and destruction of lithosphere in subduction zones The process of seafloor spreading provided a mechanism to explain continental drift and led geologists to a widespread belief in the mobility of surface 'plates'.   ☐

earth tectonics prior to 1960 is shown in the upper part of Fig. 4.1, in which mantle convection held oceanic ridges high above rising cells and then caused the descending parts of convection cells to return to the mantle somewhere else. The steeply dipping seismic zones under island arcs and continental margins (Wadati–Benioff zones; Section 4.1) were regarded as zones of faulting but not of large-scale consumption of oceanic lithosphere.

The older concept of tectonics was eternally changed in 1960 by H. H. Hess, with a refereed publication first appearing in 1962. Hess proposed 'seafloor spreading' around mid-oceanic ridges (lower part of Fig. 4.1). In this viewpoint, mantle convection cells did not simply circulate beneath a stationary crust but caused a continually moving crust; (modern terminology would refer to the interaction of convection cells with a rigid lithosphere rather than crust). This movement was accompanied by creation of new crust at spreading ridges and its destruction in descending slabs (the Wadati–Benioff zones). The concept of seafloor spreading answered many questions, such as the observation that the sedimentary sections in ocean basins are so thin that they could not contain any sediments older than Mesozoic; this observation was completely inexplicable if the oceans had always existed in fixed positions. Further evidence of spreading was provided soon after 1960 by the discovery that spreading centers preserved a symmetrical record of magnetic reversals in oceanic crust on either side of the ridge.

The concept of seafloor spreading also filled the conceptual void left by the absence of any force that could cause continental drift. Oceans that did not have descending slabs around their 'passive' margins could be regarded as simply expanding by creation of new crust at mid-ocean ridges. This expansion caused continental separation. Oceans surrounded by descending slabs ('active' margins) were contracting, and soon the Wadati–Benioff zones became referred to as 'subduction zones', where lithosphere could be destroyed at the same rate at which it was produced at the ridges.

The fusion of continental drift and seafloor spreading led to an examination of earth processes on a global scale. This effort was greatly aided in the 1960s by the development of international transportation networks that whisked geologists by jet speed to communicate closely with their counterparts in almost all countries and to conduct investigations worldwide. Gradually, the information gathered was collected into a concept known as 'plate tectonics'.

We must clarify the point that plate tectonics is a concept rather than a process. Whereas seafloor spreading, subduction and continental drift are actual processes, whose characteristics can be debated, plate tectonics is simply a method of codifying our thoughts in ways that make modern geology easier to understand. The concept is based on the evidence that the

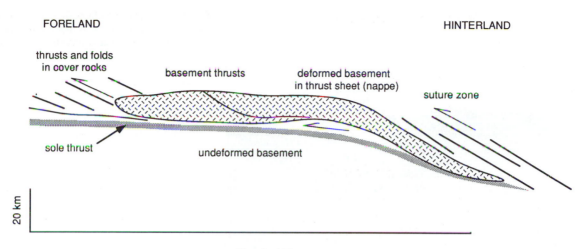

Fig. 4.2.   Generalized structural relationships in collision zones. Major compressional orogens commonly exhibit thrusts, tight folds, and overturned folds at various levels of exposure. Higher levels, and more brittle deformation, are generally toward the foreland of the orogen. Deeper levels of exposure and more ductile structures are common between the foreland and the suture zone. The hinterland is a region of comparatively undeformed rocks. The diagram is based largely on structures in the Moine thrust and comparable structures in the Appalachians. The basal shear zone is at a depth of approximately 20 km in many areas. ☐

earth contains a reasonably rigid outer part (the lithosphere) that can be subdivided into mobile zones, which are seismically active, and relatively stable plates between these zones. In this regard, plate tectonics built upon earlier concepts of localized mobility and stability; for example, a well-known book published by E.S. Hills in 1940 divided the earth into elongate 'mobile belts' (including geosynclines) and 'resistant masses' (blocks, shields) between them.

One of the principal codifications permitted by plate tectonics was a recognition of three types of plate margins. One type is a subduction zone, where lithosphere is consumed back into the mantle (Section 4.1). The second type of margin is accretionary, mostly spreading centers at mid-ocean ridges plus a few intracontinental rifts; we discuss this process in Section 4.2. The third margin is a zone of lateral (strike-slip) movement, where blocks move past each other along faults commonly referred to as 'transforms' (Section 4.3). At critical places in the earth, plate margins intersect to form 'triple junctions', which separate three lithospheric plates. A variety of triple junction termed 'R-R-R' is conceptually the simplest, consisting of three rifts that radiate from the junction (possibly a plume; Section 4.4). Section 4.5 summarizes the interaction of these various aspects of plate tectonics.

[**References** – The principal original paper on continental drift is Wegener (1912), and the first publication on seafloor spreading is Hess (1962). The reference to Hills is Hills (1940). One of the many important original papers establishing the concepts of plate tectonics is McKenzie and Morgan (1969). Stewart (1990) provides a review of the history of plate tectonics.]

4.1 Compression and subduction

COMPRESSION is responsible for some of the world's most magnificent scenery – from the Alps to the Antilles to the Andes; (possibly we should say from the Alps to the Zagros). The results of compression can be preserved only because the earth has a thick, rigid, lithosphere and a convecting inner mantle. If the earth was hotter and the lithosphere more mobile, or the earth so cold that interior convection did not occur, then compressional orogenic belts could not develop. We discuss compression in six parts: 1) the types of structures formed in compressional areas; 2) generation of magmas in subduction zones; 3) the nature of island arcs; 4) subduction along continental margins; 5) the problem of forming continents by accretion of arcs; and 6) a brief inquiry into the varieties of orogenic belts.

Fig. 4.3. Knockan Cliffs of Scotland, showing outcrop of the Moine thrust. Work in this area in the 1880s (e.g. by C. Callaway and C. Lapworth) first demonstrated the concept of overthrusting.   □

Compressional structures

The characteristic structure formed by lateral compression is a thrust fault and features related to it. The basic elements in a thrust complex are individual thrust sheets, bounded below, and commonly above, by thrust faults. Some investigators refer to these thrust sheets as 'nappes', although the term originally was applied to large recumbent folds thought to have formed by ductile flow in high-grade rocks.

Fig. 4.2 shows some of the elements of thrust complexes; it is based largely on work done along the Moine thrust of northwestern Scotland (Fig. 4.3) and studies of possibly correlative structures in the U.S. Appalachians. The principal structural element is a basal zone that separates deformed terranes above the zone from apparently undeformed rocks below it. This basal zone is described as a sole thrust, decollement, or a detachment, although some geologists limit 'detachment' to basal faults in extensional environments (Section 4.2). The concept of an undeformed region below a sole thrust is controversial, and many investigators propose a gradual dying out of brittle upper shear zones into a lower crust that has been pervasively ductilely deformed. Where observed in major continental orogenic belts, the basal thrust zone commonly is flat over large distances at depths of ~20 km or higher in the crust.

The basal thrust underlies a terrane of imbricated thrust sheets that rise from the base and steepen toward the surface. Most of these thrusts are 'forward' in the sense of having the same vergence (direction of movement of the upper block) as the basal thrust. Some 'backward' thrusting can develop locally, however. Depending on the relationship between the depth of each thrust and the topography of the underlying 'basement', these imbricated sheets may include supracrustal suites and/or basement rocks. In sets of imbricate thrusts, the earliest movement (oldest fault) is commonly on the highest fault, with progressively younger ages on progressively lower faults. This age sequence causes the older thrust sheets to be carried 'piggyback' on underlying thrusts and may isolate thick pods of crust above the youngest thrust surface (Fig. 4.4A). These structures are referred to as 'duplexes' and are responsible for many tectonically high areas that expose rocks of high metamorphic grades.

Many shallow thrusts in brittle rocks are characterized by ramps, or zones of higher dip than adjacent parts of the fault. A 'ramp-flat' geometry is shown in Fig. 4.4A. This configuration is particularly well displayed in sedimentary suites containing alternating resistant and non-resistant rocks (e.g. limestones and shales). In such suites, the thrust cannot propagate through resistant rocks except along frontal ramps with orientations that maximize the shear stress (i.e. generally at moderate dip angles). In addition to the frontal ramps, many thrusts exhibit lateral ramps along which sections of thrust sheets move past each other. These lateral ramps appear as 'tear faults' in map view (Fig. 4.4B).

A

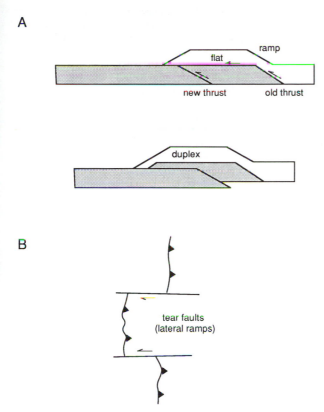

Fig. 4.4. (A) Representation of ramp-flat geometry and duplexes. Major thrusts commonly alternate between areas of low dip (flats) that follow bedding planes or other zones of weakness and zones of high dip (ramps) where the thrust cuts across resistant rocks. Duplexes are formed where one or more thrust sheets are stacked in sequence. (B) Strike-slip (tear) faults that offset major thrust zones commonly have steep dips and can be referred to as lateral ramps. □

Thrust complexes commonly develop in 'accretionary wedges' of volcano–sedimentary suites accumulated above subducting slabs (further discussion below). The wedges undergo compression or extension, or both synchronously, depending largely on the amount of resistance that the wedge encounters as it moves over the subducting slab (Fig. 4.5). Where resistance is high, accretion to the front of the wedge lengthens the wedge and reduces the 'angle of taper' (rate of thickening of the wedge away from the toe). This lower angle is unstable, and the wedge thickens by compression. The accreted material commonly forms structural fabrics subparallel to (synthetic with) the downgoing slab (Fig. 4.5), but reverse (antithetic) directions and areas of great complexity are also possible. Where resistance to subduction is low, the downgoing slab can carry sediment and (generally oceanic) crust deep below the wedge and add it to the wedge by 'underplating'. The resultant thickening increases the angle of taper of the wedge, causing instability and gravity-induced extensional faulting (Fig. 4.5).

Material carried to considerable depths in the subduction zone reappears at exposure levels in many orogenic belts that do not appear to have been deeply eroded. These blueschists (metasediments) and eclogites (metabasalts) mostly show equilibration temperatures of ~400 °C and pressures of 10 kbar to 15 kbar. They represent the low-T/P facies series of metamorphism that commonly forms paired metamorphic belts with higher-T/P series in Phanerozoic orogenic belts (e.g. see discussion of western-Pacific orogens in Section 9.3). Some blueschist/eclogite occurrences

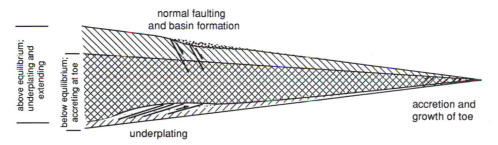

Fig. 4.5. Accretionary wedge at two elevations. Accumulation of sediment during compression forms a wedge that tapers toward the subduction zone. Accretion at the toe maintains a small taper angle and causes lateral growth of the wedge. A small taper angle permits underthrusting of sediment and volcanic rocks and resultant accumulation of these materials deep beneath the wedge ('underplating'). Underplating causes an increase in elevation and increase in the taper angle, which results in collapse of the wedge by normal faulting. The combination of collapse and growth maintains an approximate equilibrium in the angle of the wedge. □

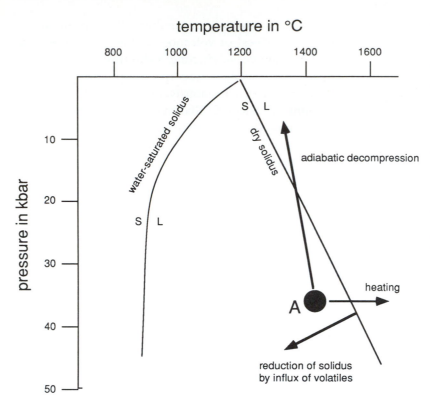

Fig. 4.6. Diagrammatic illustration of production of magmas in the mantle. The wet and dry solidus curves are generalized for different types of peridotite and variable results from different investigators. S = solid; L = liquid. Material starting at point A is below the mantle solidus and should be completely solid. Partial melting may occur under three conditions: 1) increase in temperature without change in depth – this process is unlikely because an increase in temperature commonly causes decrease in mantle density and uplift of the earth's surface, with erosion of the uplifted surface causing pressure decrease; 2) decompression melting as the material at point A moves adiabatically upward because of mantle instability (see Fig. 2.1); and 3) lowering of the solidus by influx of volatile material. The water-saturated solidus has a 'reversed' slope in which T declines as P increases because of the increasing effect of water vapor at higher pressures.   □

form linear belts subparallel to the orogen, but many are part of the 'melange' of chaotically jumbled material. Mechanisms that bring this deep material back to the surface are highly controversial, including flow of a fluidized shaly material back up the subduction zone (to form melange) or very deep normal faulting caused by large-scale underplating.

Generation of magmas

Magmas form by partial melting (rarely complete melting) of a source material when the T and P of the material are above its solidus (Fig. 4.6). Thermal gradients may be placed in a field of partial melting by three changes in condition. The first two are increase in temperature, decrease in pressure, or both. The third process is reduction in the temperature at which melting begins (lowering of the solidus), which is generally accomplished by increasing the volatile content (particularly water) of the material. Partial melting of 10% or more of the source material is probably required for separation of the magma from the source region. The resultant 'magma' contains silicate liquid, fragments of rocks and minerals from the source that were

entrained in the magma as it separates from the source region, early-formed crystals that have not separated out of the magma, and possibly an immiscible vapor phase. Proportions of these components change as the magma rises.

Once formed, a magma rises because of its low density. Magmas cannot rise higher than the equilibrium level permitted by their densities, but that level is controlled by numerous factors, including density of the magma, its rate of crystallization, and the densities and strengths of surrounding rocks. Dense magmas such as basalts tend to 'pond' at the base of the crust, although they erupt when the weight of the basalt column to some depth is less than the total weight of a column of surrounding rocks to the same depth.

Island arcs

Fig. 4.7 is a cross section of a typical island arc. The most seaward feature is a foredeep trench, formed where the oceanic lithosphere bends and begins its descent below the arc. The trench is very deep (>10 km) where the adjoining arc supplies only minor sedi-

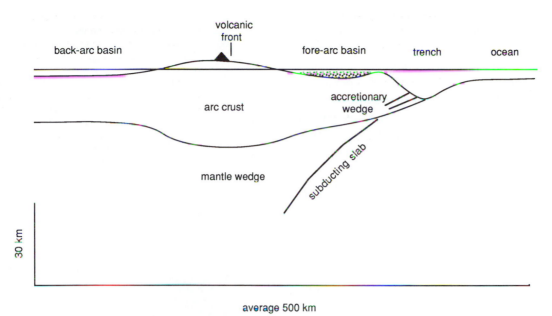

Fig. 4.7.   Diagram of major features of typical island arc. The accretionary wedge (accretionary prism) separates the foredeep trench (subduction zone) from the elevated island arc. The fore-arc basin is shown with an accumulation of sediments in the collapse structure on the top of the wedge. The mantle wedge is mantle between the subducted slab and the crust of the arc (including accreted material). The back-arc basin commonly overlies extensional parts of the subducted slab.   □

ment, but in some areas the trench is filled (choked) by influx of debris (e.g. northwestern Indonesia, Section 9.3; southern Lesser Antilles, Section 9.5). The filling of the trench by water or low-density sediment is associated with the characteristic negative gravity anomaly that indicates isostatic disequilibrium in the orogenic environment. The debris in the trench is a combination of oceanic sediment and basalt plus material eroded from the adjacent arc.

Accretion of sediment and volcanic rocks along the base of the sediment wedge in the trench characteristically occurs by sweeping of material against, and under, the tip of the wedge (Fig. 4.7). The resultant package is cut by imbricated thrusts subparallel to the descending slab. Some of the thrusts and volcano–sedimentary suites are large enough to be mapped as definable blocks, but in many locations the materials are so pervasively sheared that stratigraphic definition is nearly impossible. Where structures are so complex, and stratigraphy so impossible to define, the suites are referred to as 'melanges'.

Overthickening of an accretionary prism may result in collapse of part of the prism's surface between the trench and the emergent arc (Fig. 4.5; see discussion of

underplating above). This collapse forms a 'fore-arc' basin mostly filled with volcanogenic debris from the arc. Fore-arc basins exhibit coherent stratigraphy, and many are comparatively undeformed.

Most island arcs develop behind a 'magmatic front' that represents the point at which the descending slab encounters the hot asthenosphere, commonly at depths of ~125 km. Within the arcs, depths to some recognizable MOHO (not easy to distinguish in many areas) are generally in the range of 20 km to 25 km, although they are thicker where the arc extends over continental crust (e.g. eastern Aleutians or northern Okhotsk; Section 9.3). Volcanogenic debris is incorporated within the arc sequence, but unroofing and erosion of intrusive rocks generally occurs only in the later stages of an arc's history. Almost all of the rocks in the volcano–sedimentary sequence are clastic (breccias, tuffs, etc.), although flows are present locally.

Arcs undergo a sequential development from a more 'primitive' character to a more 'mature' one. The early stages of arc development are characterized by subaqueous accumulation of volcanic and sedimentary rocks. The volcanic materials are basalts to basaltic andesites (50% to 55% SiO_2) plus variable amounts of highly siliceous flow and pyroclastic rocks (rhyodacites and/or keratophyres). In some arcs, these mafic and silicic suites form a bimodal sequence. Variability in composition of the mafic rocks generally results from crystal fractionation of mantle-derived basalts. The silicic suites probably formed by partial melting of mantle peridotite under conditions of high water-vapor pressure.

Mature arcs are partly emergent and, thus, provide a mixture of subaerial and subaqueous environments. Virtually all of the rocks are clastic or volcaniclastic, but minor limestones develop in the shallow water of island rims and banks, yielding a stratigraphic sequence of volcanic rocks, volcanogenic sediment, and sporadic carbonate lenses. The volcanic suites are predominantly andesitic, forming a unimodal sequence with a mode of ~60% SiO_2. These suites are commonly referred to as 'calcalkaline', although the term has been used in numerous ways. Intrusive rocks range from gabbros and diorites to rocks as siliceous as granodiorites. Abundances of SiO_2 >~68% are rare, an important distinction from the more-silicic batholithic suites of continental margins. Magmatic rocks contain little or no K feldspar, and the absence of this mineral in clastic rocks of intra-oceanic arcs distinguishes them from continental-margin deposits.

The andesites of island arcs are too rich in LIL elements (e.g. K) to develop by fractionation of basalts or by partial melting of an unmodified mantle wedge between the subducted slab and the overlying arc (Fig. 4.7). Many models propose that initial melts formed in the mantle wedge under the influence of LIL-bearing fluids released from the descending oceanic crust. Most of the andesites exhibit abundant evidence of disequilibrium between crystals and groundmass (e.g. spongy plagioclase formed by partial remelting of sodic plagioclase incorporated in a basaltic or andesitic magma). The prevalence of such disequilibrium textures indicates that much of the compositional variability of the mature arc suites is a result of mixing of mantle-derived magma batches with earlier-formed volcanic rocks or partly crystallized magma batches.

The ophiolites of island arcs (and continental margins) are commonly regarded as a section across the mantle–crust transition in a spreading center (Fig. 4.8). In this model, the mantle consists largely of 'depleted'

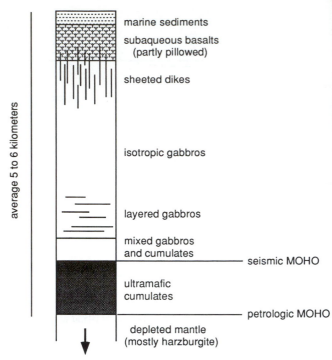

Fig. 4.8. Cross section through an 'ideal' ophiolite. Sheeted gabbroic dikes presumably formed because of extension at a ridge crest. (Ophiolites that do not contain these dikes may not represent oceanic crust formed by seafloor spreading.) The term 'isotropic' gabbros refers to rocks without well-defined layering. The difference between the seismic MOHO and petrologic MOHO results from the fact that the zone of ultramafic cumulates has a P-wave velocity of ~8 km/s (mantle velocity) but has formed by settling from magma injected into the crust (above the seismic MOHO).   □

ultramafic rocks (harzburgites, or orthopyroxene peridotites). Seafloor basalts are intruded by vertical 'sheeted dikes', which filled the void as the ridge underwent extension. Magmatic activity also produced large quantities of massive intrusions ranging from gabbro to minor trondhjemite (Na granite). The entire sequence is overlain by oceanic sediments (commonly red clays and cherts).

Several problems have arisen with respect to this 'classical' interpretation of ophiolite genesis. They include: 1) the fact that typical ophiolites on land are somewhat thinner than the crust of open oceans; 2) the absence of sheeted dikes from many ophiolites; 3) associated rocks, such as the siliceous volcanic suites that overlie many ophiolites – these rocks could not have formed along oceanic spreading centers; and 4) details of basalt composition that are incompatible

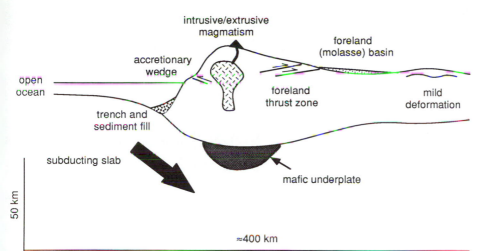

Fig. 4.9. Diagram of major features of continental-margin subduction zone. The foredeep trench forms by dragging of the lithosphere downward without adequate sedimentation to fill the basin. The exposed products of subduction commonly are intrusive and extrusive rocks with an andesitic to rhyolitic composition. Subduction may also produce mantle-derived basaltic to intermediate rocks that form a 'mafic underplate' at the base of the thickened crust. The foreland thrust zone brings the deformed orogen over the relatively undeformed foreland. The 'molasse basin' is a downwarp generated by the weight of the uplifted orogen.   □

with an origin at spreading centers. For these reasons, many ophiolites are now regarded as having formed through some set of processes in which oceanic crust has been modified by subduction-related magmatism, perhaps in a back-arc basin. Regardless of the exact mode of formation, ophiolites in orogenic belts can be used as indications of closure of an ocean that formerly occupied part of the belt (e.g see Section 5.1).

Many arcs contain 'remnant' (dormant) arcs parallel to the active arc but farther away from the subduction zone. These remnant arcs apparently formed during rifting within the active arc, which developed a small basin of active extension between the arc and an old arc no longer volcanically active. In some cases, however, the basin between the active and remnant arc may be old ocean crust left behind by some sudden jump of the subduction zone seaward.

The issue of back-arc basins is discussed more fully in Section 9.3, which summarizes the evolution of the western Pacific. Here we mention only that they generally have shallower depths than open oceans, have higher heat flows than oceanic crust of equal elevation, and commonly show only diffuse magnetic patterns. This combination of features probably indicates generation of the basins by a broadly distributed extension and synchronous magmatism, mostly over a zone of extension in the subducted slab. The magmatic products of these basins are very similar to those of open-ocean spreading ridges. The nature of 'back-arc' basins on land is discussed in Section 4.2.

Continental margins

A subduction zone in which the overlying plate consists of continental lithosphere (Fig. 4.9) rather than oceanic lithosphere shows several differences from intra-oceanic island arcs, including the following.

- Shallower trenches, presumably because of the large quantity of sediment supplied by the adjacent, uplifted, mountain range.
- Thicker crusts in the overlying slab. For example, depths to MOHO in some parts of the Andes are 70 km, contrasted with thicknesses of approximately one third as much in intra-oceanic arcs (see above). The exact reasons for this crustal thickening are problematic, with explanations ranging from addition of mantle-derived magmas to mechanical thickening of pre-existing crust by thrusting. We discuss this problem in Section 9.6.
- Sedimentary debris much richer in granitic components (quartz, K feldspar). The abundance of K feldspar commonly increases toward younger sediments as more of the granitic batholiths of the orogen are unroofed and eroded.
- Volcanic suites that range from minor basalt to abundant andesites and rhyolites. All rocks contain a higher abundance of LIL elements than rocks of equivalent SiO_2 content in intra-oceanic arcs, and the K_2O/SiO_2 ratio commonly increases landward from the subduction zone. The reasons for the variations in composition and differences from rocks in intra-oceanic arcs include partial melting of continental crust, the effect of continental crust on geotherms, and partial control of melting by depth to the subduction zone.
- Enormous amounts of tonalitic to granitic (calcalkaline?) intrusive rocks that form large batholiths along the strike of the orogen. These suites reflect the polarity of the subduction zone by showing increasing abundances of granitic rocks toward the continental interior.

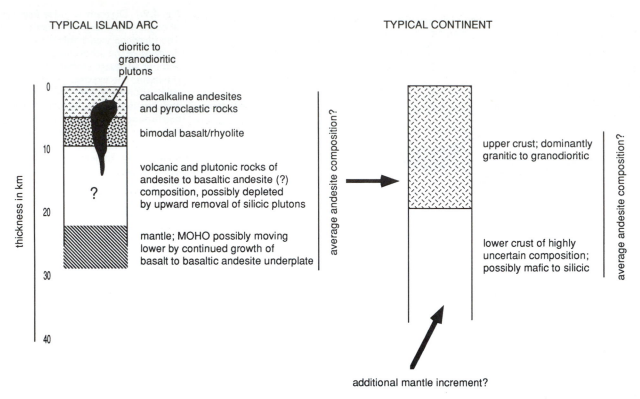

Fig. 4.10. Possible processes in the transformation of island arc to continent. The arc is the result of subduction-generated magmatism and is shown with an 'average' andesitic crust plus a possibly more mafic underplate at depth. The major problem in determining the composition of the continent is uncertainty about the composition of the lower crust. Depending on various assumptions, the two 'endmember' processes for development of the continent are: 1) internal fractionation of the island arc, with separation of the andesitic arc into a mafic lower crust and granitic upper crust; and 2) post-orogenic increment from the mantle-derived mafic underplate into the andesitic crust. If the increment consists of post-tectonic granites, then the average continental crust may be more 'granitic' than the original andesitic arc. □

The idealized continental-margin subduction zone shown in Fig. 4.9 contains a large area designated as 'mafic underplate'. A mafic underplate apparently represents the mafic/intermediate magmas generated by subduction, with the underplate forming because the magmas were unable to rise to the surface and had to crystallize in place. Virtually all models for the evolution of intermediate and silicic magmatic suites (andesites to rhyolites) from the mantle require a two-stage process because of the inability of quartz-bearing magmas to equilibrate with mantle peridotite. This process includes removal of basalt and/or basaltic andesite from the mantle, crystallization at the base of the crust, and further melting to form the silicic components of subduction-zone igneous suites.

Are continents formed by accretion of arcs?

Accumulation of island arcs into an accreted body of crust is clearly essential in the formation of continents. Is this accretion sufficient as well as necessary? That is, can continental crust develop solely from material accumulated in arcs, or is some additional component necessary?

Fig. 4.10 shows cross sections of the crusts of typical stable continental interiors and typical mature intra-oceanic island arcs. The average composition of the island arc is modeled as andesite, partly because that is the dominant rock exposed in most mature arcs and partly because sedimentary debris in arcs appears to have formed largely by erosion of andesites. The crust in the arc is thinner than in the continent, and the accretion of intra-oceanic arcs cannot produce a continent without further increment of mantle-derived material. Furthermore, because the arc does not contain the large volumes of silicic material that are exposed in the upper levels of continental crust, the arc must undergo some fractionation in order to form a continent.

The thickness of an arc crust may be increased as the arc collides with a continental margin because of the generation of magmas during that collision process. An arc crust plus a mafic underplate at a continental margin (see above) might have the same total thickness as the crust of a stable continent. Thus, the formation of continents by accretion of island arcs is volumetrically feasible provided that the arc is modified from its condition within an ocean basin.

Fractionation of an arc may involve several processes. One difficulty in suggesting processes is the lack of information about the nature of the lower continental crust (Section 2.3). If the lower crust is mafic, then an arc may fractionate very simply by moving of 'granitic' components upward and leaving a mafic residue behind as a lower crust. This fractionation could be accomplished largely by upward movement of magmas generated by partial melting of the lower arc and mafic continental-margin underplate. Such an origin has been proposed for the large volumes of post-orogenic granites that intrude orogenic belts at, or shortly following, the end of the deformational process. Magmatic transport of materials may be enhanced by metasomatic migration of LIL elements and volatiles upward, a process whose significance is difficult to evaluate quantitatively.

If the lower continental crust is not mafic but mostly silicic granulites, then simple intracrustal fractionation of an arc and underplate will not produce a continent. In this situation, the crust must be modified by additional increment of silicic material from the mantle. This type of increment may be associated with epeirogenic movements of 'stable' continental interiors long after the last vestiges of compressive orogeny have been recorded.

Varieties of orogenic belts

Orogenic belts have been classified in so many different ways by so many different investigators that we cannot even attempt a review of their efforts. This section, therefore, is simply a statement of the ways in which orogenic belts can differ from each other with respect to the direction of thrusting, the location of suture zones (if present), and the sources of the rocks involved in the orogenic process.

One terminology must be mentioned – the distinction between A-type and B-type subduction. The 'A' is derived from O. Ampferer, a geologist who proposed large-scale thrusting of crust beneath orogenically deformed areas. The 'B' is derived from H. Benioff, who proposed (in 1954) that oceanic lithosphere could be deeply underthrust beneath island arcs; this suggestion had originally been made by K. Wadati in 1935 and ignored by most of the geologic community. The dipping seismic zones are now commonly referred to as 'Wadati–Benioff zones.'

Based on these classifications, 'A-type' subduction occurs in areas where continental lithosphere overrides other continental lithosphere. Some overthrusting at shallow levels is required in most mountain belts by the presence of far-traveled thrust complexes overlying continental basement; examples are the Rocky Mountains (Section 9.4), the southern Appalachians (Section 7.3), and the eastern Andes (Section 9.6). Whether underthrusting of continental crust to levels of the upper mantle is possible, however, is questionable. Light continental crust is clearly difficult to subduct. Some models of Himalayan geology, however, propose that the Indian plate has been thrust more than 1000 km northward under the Asiatic mainland (largely in Tibet; Section 9.1).

'B-type' subduction is the common situation in which continental or oceanic lithosphere is underthrust by oceanic lithosphere. Whether this type of subduction, by itself, is capable of causing orogeny is controversial. Some investigators propose that major compressive deformation can occur only where the subducting slab contains island arcs, oceanic plateaus, or other buoyant crust. For example, as a spreading ridge and continental margin approach each other, continued movement would cause progressive subduction of younger, hotter, and less-dense oceanic lithosphere that would resist subduction and might cause compressive orogeny. Similarly, subduction of oceanic crust might continue until an ocean basin between two continents had been completely destroyed, at which time the collision of the continents would generate compressive orogeny. Most of the world's orogenic belts apparently have resulted from collision.

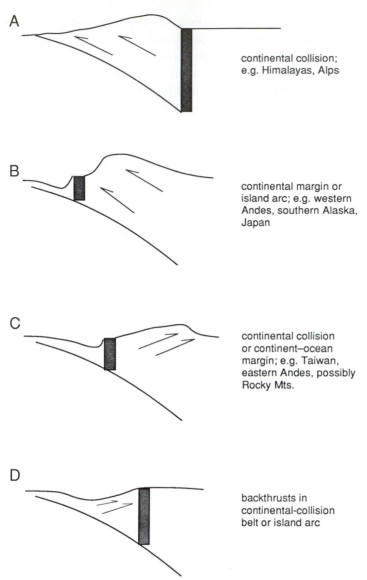

A
continental collision;
e.g. Himalayas, Alps

B
continental margin or
island arc; e.g. western
Andes, southern Alaska,
Japan

C
continental collision
or continent–ocean
margin; e.g. Taiwan,
eastern Andes, possibly
Rocky Mts.

D
backthrusts in
continental-collision
belt or island arc

Fig. 4.11.   Highly diagrammatic representation of the four geometrically possible relationships between subduction direction, thrust orientations and suture zones in orogenic belts. The sutures (containing oceanic crust) are shown by the dark rectangles. In diagram A, major thrusting occurs in an accretionary wedge deposited on the subducting plate and thrust over it by collision with the hinterland. In diagram B, thrusting occurs in material accreted on the margin of the over-riding plate (e.g. where an orogen occurs along a continental margin rather than between two collided continental blocks). In diagram C, thrusts are shown bringing the orogen over the foreland; this configuration may require transmission of compressional stress for considerable distance inland from the continental margin. Diagram D shows minor backthrusting in a thin wedge, possibly seaward of a suture zone along a continental margin.   □

This brief introduction provides the basis for distinguishing the varieties of orogenic belts shown in Fig. 4.11. These four types are the possible combinations of location of suture zones relative to the orogenic belt and direction of thrusting relative to the direction of subduction.

- The configuration shown in Fig. 4.11A is typical of the modern Alpine–Himalayan chain (Section 9.1). The mountain belt is constructed above the downgoing slab, with a suture zone separating the mountains from a comparatively (not completely) undeformed 'hinterland'. Faults are shown parallel to (synthetic with) the subducted slab. In most areas where this type of orogeny has occurred, sub-

duction has completely destroyed oceanic lithosphere and caused collision of the two continental blocks on either side of the former ocean (Neotethys in the Himalayas). In some areas, material from the hinterland has been thrust across the suture onto the mountain belt, bringing part of the suture zone into the compressive orogen (e.g. the Alps; Section 9.2). The uplifting orogen commonly causes downwarp of a large downwarp in the 'foreland' area, toward which the thrusts advance and override these foreland-basin deposits.

- The configuration shown in Fig. 4.11B differs from that in Fig. 4.11A in the location of the suture zone. Faults are synthetic in both diagrams, but in Fig. 4.11B they advance toward the suture instead of away from it. The deformed rocks are almost exclusively the upper plate plus material added to it by underplating and growth of an accretionary wedge. A prime example of this type of orogen is the western part of the Andes (Section 9.6) or southern Alaska (Section 9.4), where compression results from subduction of oceanic lithosphere beneath the continental margin.

- Fig. 4.11C shows antithetic faults developed in the upper plate of the subduction zone (in most locations, the base of these thrusts could be regarded as an A-type subduction zone). This type of orogen results from pressure caused by subduction, generally of oceanic lithosphere, and transmission of the stresses to some distance inland. The suture is simply the edge of the continent under which subduction occurs. An example of deformation near the subduction zone is Taiwan (Section 9.3). Examples of more distant transmission of stresses are the Rocky Mountains of North America (Section 9.4) and the eastern side of the Andes (Section 9.6).

- The configuration shown in Fig. 4.11D is included for completeness. In this situation, antithetic faults develop in material of the lower plate and move toward the suture zone. Backthrusting (see above) in some accretionary wedges follows this pattern but does not become the major structure of a mountain belt.

[**References** – Original discussions of the 'Moine thrust' are in Callaway (1883) and Lapworth (1883), and a more recent one is by Elliott and Johnson (1987). The edited issue of McKerrow (1987) provides general information about orogeny in Scotland. Widespread recognition of downgoing slabs resulted from the paper by Benioff (1954). Information concerning the geometry of thrusts, accretionary wedges and related features is provided by the volumes edited by Mitra and Wojtal (1988) and McClay and Price (1981) and papers by Davis, Suppe and Dahlen (1983) and Platt (1986). The discussion of melting in the mantle is from Yoder (1976). A discussion of melanges is in Cloos (1982). Island arcs are discussed by Sugimura and Uyeda (1973) and the volcanic rocks of arcs by Donnelly and Rogers (1980). Information on ophiolites can be obtained from Coleman (1977), Moores (1982) and Leitch (1984). Bally and Snelson (1980) review the general process of subsidence, and Ingersoll (1988) discusses types of basins. Andean volcanism and mechanisms of evolution are in the edited volume by Kay and Rapela (1990) and the paper by Sheffels (1990). Pearcy, DeBari and Sleep (1990) discuss the compositions of island arcs.]

4.2 Rifting

THE REYKJANES PENINSULA of southwestern Iceland is a windswept land of hummocky basalt and sparse vegetation. It is an on-land extension of the mid-Atlantic spreading ridge, responsible for the creation of the Atlantic Ocean. An undersea traveler could follow the mid-Atlantic ridge south from Iceland and, with some backtracking, connect with virtually (but not quite) the entire system of mid-ocean spreading ridges. This system extends for ~55 000 km and has an average spreading rate of ~6 cm/yr (total rate of separation of both plates from each other). Each year, the ridge system creates ~20 km^3 of new crust with an average thickness of ~6 km. It is the primary locus of extensional processes on the earth's surface.

Extension is manifested on the continents as well as the oceans. One result is the development of linear rift valleys or networks of intersecting rifts. Some of these rifts extend into the margins of continents, apparently as the 'failed arms' of triple junctions associated with oceanic opening along the other arms (Section 4.4). A second variety of continental extension, with uncertain relationships to the first, is the development of broad basins with widths of several hundred kilometers and numerous shallow-level extensional faults.

In this section, we first describe the nature and operation of the oceanic ridge system. This discussion is followed by an investigation of continental rifting, and the final topic is the evolution of the Red Sea and East African rift.

Oceanic spreading

The typical features of a mid-ocean spreading ridge are shown in map view and cross section in Fig. 4.12A–D. Slow-spreading ridges contain an axial valley, which is the site of active extension and tholeiitic (MORB) volcanism. Fast-spreading ridges do not contain the axial valley. The surface MORB is commonly succeeded downward by 'sheeted' diabase dikes, more massive intrusions of gabbro to diorite, and ultramafic rocks; (see discussion of ophiolites in Section 4.1).

The correlation between decreasing elevation and increasing age of crust away from a ridge crest is virtually identical for all of the world's ridges (Fig. 4.12A and B). This uniformity is caused by the fact that ridges are elevated isostatically because of a thermal anomaly that brings asthenosphere essentially to the surface at the ridge crest. The rate of elevation decrease, therefore, is directly controlled by the rate of cooling and follows an exponential law such that elevation is proportional to the square root of age over much of the ridge flank. The isostatically balanced elevation of exposed asthenosphere is ~2.5 km below sealevel.

The profile of ridge elevation vs distance from the crest is controlled not only by cooling rate but by the spreading rate, with crust of equivalent age farther from the ridge axis along fast spreading centers. If we

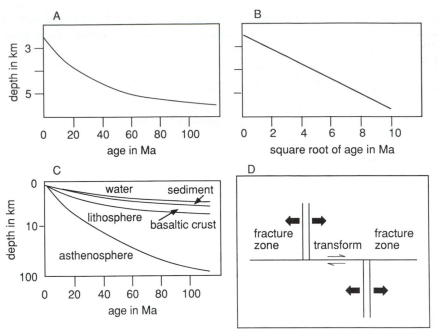

Fig. 4.12. A to C are cross sections of a typical mid-ocean ridge. Diagrams A and B show elevation changes plotted against age and against the square root of age, respectively. The linear relationship between elevation and the square root of age indicates subsidence by simple cooling of the oceanic lithosphere as it moves away from the spreading center. Diagram C shows the thickening of the lithosphere away from the ridge and the upper magmatic and sedimentary layers (note that the depth scale is not linear). Diagram D shows an oceanic transform generated by spreading on two offset ridge segments; the direction of movement on the transform is opposite to the apparent direction indicated by the offset of the segments (see also Fig. 4.21).   □

express spreading rates as 'half rates' (the rate of movement of one plate away from the ridge), the Atlantic has a spreading rate of 1 cm/yr to 2 cm/yr and the East Pacific rise about 5 cm/yr to 9 cm/yr. Rates remain comparatively constant throughout the spreading history of most ridge segments.

The origin of the forces responsible for ridge spreading is highly controversial. If ridges are regarded as 'active', produced by rising asthenosphere, then some of the lateral spreading force could be generated by injection of magmas into the spreading center. Conversely, a 'passive' ridge, or 'window to the mantle', could result from gravitational pulling of the slab away from the thermally elevated ridge (perhaps enhanced by descent into subduction zones). These 'push' and 'pull' forces may also be augmented by 'drag' of circulating asthenosphere away from the ridge beneath the oceanic lithosphere.

Modern spreading ridges are characterized by symmetrical magnetic stripes (Section 1.3), but oceans not bordered by subduction zones commonly show regions of 'magnetic quiet', (i.e. without reversals) near their outer margins. Particular examples are the Atlantic Ocean and the Red Sea. Much of the quiet area in the Atlantic Ocean has ages within the Cretaceous normal polarity interval, which continued without reversal from ~120 Ma to ~80 Ma. Other explanations

include: destruction of magnetization in the ocean crust by seafloor alteration; covering of the crust by sediments with diverse magnetic orientations deposited over time periods of many magnetic reversals; and an ocean floor consisting of thinned continental crust rather than oceanic crust.

A major feature of spreading ridges is their development of transform faults at intervals of approximately 70 km (Fig. 4.12D). These faults are perpendicular to the ridges, and their trace is part of a small circle whose center is the pole of rotation of the spreading plates. One of the earliest demonstrations of seafloor spreading was the use of seismic first-motion studies to show patterns of movement along fault segments between ridge crests opposite to the displacement of the ridge (Fig. 4.12D).

Varieties of continental rifts

Geologists have described different areas of continental rifting in ways that imply actual geologic differences between various types of rifts. One of our major efforts here is to ask whether these proposed differences represent real differences in the nature of the rifting process in different environments or whether they are the result of different perceptions of rifting by different geologists. We start this discussion with a sur-

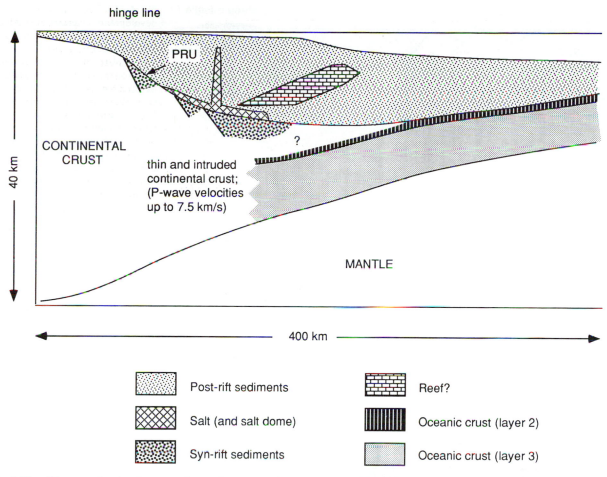

hinge line

PRU

CONTINENTAL
CRUST

thin and intruded
continental crust;
(P-wave velocities
up to 7.5 km/s)

40 km

?

MANTLE

400 km

	Post-rift sediments		Reef?
	Salt (and salt dome)		Oceanic crust (layer 2)
	Syn-rift sediments		Oceanic crust (layer 3)

Fig. 4.13. Diagram of major features of typical continental passive margin. The MOHO, separating the mantle and crust, increases from ~12 km depth in the open ocean to 35 km to 40 km depth under normal continental crust. Continental crust along the margin is thin and/or intruded by mafic magma (which increases its density). Some combination of thinning and/or increase in density causes thermal subsidence of the margin. Syn-rift sediments are formed in basins during thinning; post-rift sediments are formed along the continental margin as it subsides during drift away from a spreading ridge. The syn- and post-rift sediments are separated by a 'post-rift unconformity' (PRU). Evaporites (including salt) and reefs are common during the early stages of continental separation; clastic sediments are more common along older margins. The '?' signifies the continuing controversy over the type of contact (transition?) between thin continental crust and oceanic crust.

vey of the characteristics of *passive continental margins* that were formed by continental rifting and moved apart on the opposite sides of spreading oceans. The next topic is the development of *broad extensional areas*, with both widths and lengths measured in terms of hundreds of kilometers (e.g. the Basin and Range province of western North America). We finish with *linear systems of rifts* (e.g. East Africa).

Passive continental margins around spreading (Atlantic-type) ocean basins are the remnants of continental crust thinned during the rifting (Fig. 4.13). The amount of crustal thinning varies from one margin to

another and from place to place along margins. Commonly the original crustal thickness of 35 km to 40 km is reduced to 15 km to 20 km on the outer edge of the margin (edge of the continental shelf), but some margins show a very abrupt termination of thick continent against apparent oceanic lithosphere. The continental slope is generally regarded as a sediment wedge formed along the transition from continental to oceanic lithosphere.

The nature of the transition from continental to oceanic lithosphere is highly controversial. It is partly controlled by such processes as fracturing and

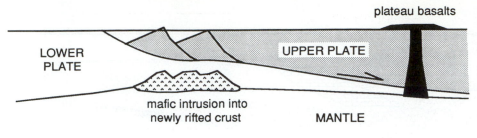

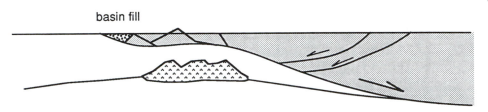

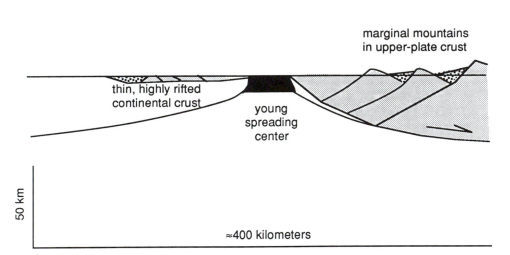

Fig. 4.14 Diagram of development of passive margin by rifting along major detachment. This type of rifting develops 'upper-plate' margins and 'lower-plate' margins. The upper-plate margins have a relatively thick crust that shows steep mountain fronts along normal faults and is truncated abruptly against oceanic lithosphere. The lower-plate margins consist of broad, thin, highly fractured, crust remaining after removal of the upper part.   □

foundering of continental lithosphere and intrusion of mafic (oceanic) magmas. The MOHO, at least as measured by seismic reflection studies, may be a secondary feature imposed on the lithologically complex transition zone. That is, the MOHO is not a simple lithologic change from ultramafic mantle to oceanic or continental crust but at least partly a zone of change in physical properties.

A characteristic structural feature of passive margins is a series of half graben (see below) parallel to the edge of the continent. Because most bounding faults on half graben dip toward the ocean, the opposing continental margins across an intervening ocean appear to be symmetrical. Thus, spreading oceans seem to have developed from full graben with boundary faults dipping toward each other from both sides. The appearance of symmetry on opposite sides of ocean basins may be incorrect, and some investigators have suggested that continental margins pull away from each other along detachments that sole into the lower crust or mantle, thereby forming opposing 'upper-plate' and 'lower-plate' margins (Fig. 4.14).

Half graben form on both plates, but the upper-plate margin has a thicker and more abrupt boundary against oceanic lithosphere than the lower-plate margin.

Active rifting (lithospheric separation) as continents begin to separate is replaced by thermal subsidence enhanced by sediment loading as the two continents drift apart. This change of process is commonly referred to as the 'rift-drift transition' and is marked by a 'post-rift unconformity' (or 'breakup unconformity'), which separates underlying rifted terrane and tilted sediments from overlying, generally undeformed, sediments of the passive margin. Most margins show a history approximately consistent with progressive cooling accompanied by only minor tectonic activity (that is, subsidence was proportional to the square root of time).

Many of the transform faults that now cut mid-ocean ridges (see above) have onland extensions and probably formed initially as strike-slip faults during the early phase of continental extension and rifting. Major faults of this type are commonly spaced at distances of hundreds of kilometers apart and delimit areas of different stretching and drifting histories (see opening of the Atlantic Ocean, Section 8.1). Both the major transforms and smaller, more closely spaced, ones are perpendicular to the extending margin, which may be a significant distinction from the oblique 'transfer' faults that are characteristic of linear intra-continental rift zones (e.g. East Africa; see below). This distinction may be one indication that the East African rift system is not a model for the formation of passive margins.

Broad extensional areas are characterized by extension of 100% to 300%, associated with significant crustal thinning, and formation of numerous horst-and-graben (and half-graben) structures parallel to each other. Two 'end-member' explanations for the structure of such areas are shown in Fig. 4.15. One is based on penetration of the crust (and upper mantle?) by deep-seated normal faults, above which the extensional terrane develops (a 'simple-shear' model). This model assumes that the lower crust behaves in a brittle fashion, at least on the time scales over which these faults move. The second model ('pure shear') presumes a transition from brittle behavior in the upper crust to ductile behavior in the lower crust. A ductile lower crust requires that faults in the upper crust 'sole' into the transition zone.

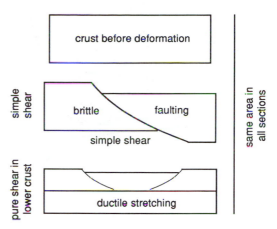

Fig. 4.15. Diagram showing difference between effects of simple and pure shear in extension of continental crust. Simple shear causes crustal thinning by offset of upper and lower plates along a deep fault. Pure shear causes thinning by a combination of ductile stretching and brittle fracturing over a broad area. □

All models for broad extensional regions must explain the presence of low-angle detachment faults. These detachments are commonly listric (concave upward), penetrate deeply enough to expose mid-crustal metamorphic 'core' complexes, and form the sole structures for overlying normal faults. Their importance has been recognized only within the past decade, partly because older geologists assumed that all normal faults had high dip angles as a result of control of shear angle by vertical principal stresses. Low-angle detachments also are now recognized in the upper parts of active orogens, where they overlie compressive structures at depth (see discussion of accretionary wedges above).

The 'type example' of a broad extensional basin is the Basin and Range of southwestern North America. This basin began its development during active subduction of oceanic lithosphere (Farallon plate; Section 9.4) under western North America in the late Mesozoic and early Cenozoic (Fig. 4.16). Extension in the northern part of the area plus andesitic/dacitic magmatism began during the early to middle Cenozoic. Most of the east–west extension, however, occurred during the late Cenozoic and was located south of the Mendocino fracture zone (MFZ) as it moved northward across the western margin of North America. The large effects of the MFZ resulted at least partly from its great length, which placed much younger, hotter and elevated lithosphere south of the MFZ against cooler and lower

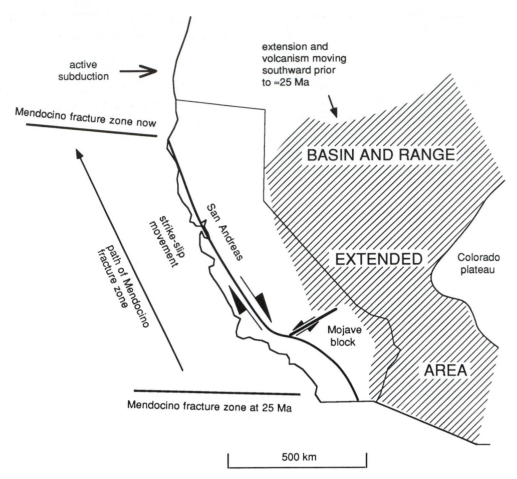

Fig. 4.16. Diagram of Basin and Range, showing major features of its evolution. Three dominant, and complexly related, processes control most of the evolution of the Basin and Range. 1) Subduction of oceanic crust under the margin of North America. This subduction occurs north of the extremely long Mendocino fracture zone (MFZ; Section 9.3). 2) Strike-slip movement along the San Andreas fault as the Pacific Ocean lithosphere moves northward relative to North America. Subduction does not occur south of the Mendocino fracture zone because the spreading ridge has intersected the continental margin. 3) Development of an east–west extended area (Basin and Range). Part of this extension occurred behind the prolongation of the Mendocino fracture zone as it moved northward beneath the continent. Part of the extension occurred by southward expansion of an extensional area that originally developed to the north of the fracture zone. The present Basin and Range formed by the merger of smaller areas of extension.   □

lithosphere to the north. This juxtaposition created an elevation difference of ~1 km, causing flexing of the overlying crust. Extension in the Basin and Range created an area two to four times wider than it was at the beginning.

Although lateral extension of more than 100% can be demonstrated in the Basin and Range, the present crust is not sufficiently thin (less than one half of its original thickness) to account for the extension. Depths to P-wave velocities of 8 km/s (mantle) are approximately 30 km throughout much of the extended area, with a lower crust containing numerous reflecting horizons and P-wave velocities >7 km/s. Thus, crustal volume must have increased during extension, presumably by intrusion of mantle-derived magmatic products.

Linear systems of rifts appear on maps as elongate 'rift valleys' with strike lengths of hundreds or more kilometers and widths of less than 50 km perpendicular to the extensional zone. In actuality, these valleys commonly consist of interlocking and alternating half graben ('graben' is plural in German). Modern examples are the East African rifts (see below) and the Rio Grande rift of western North America (Section 9.4).

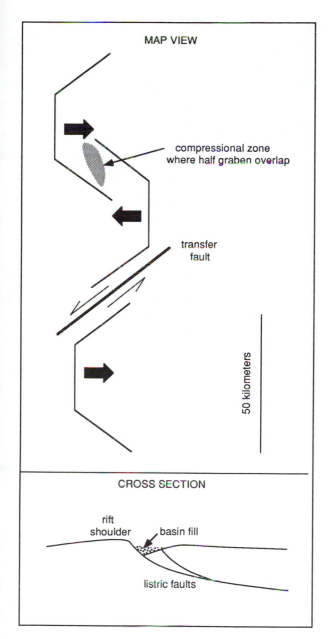

MAP VIEW

compressional zone
where half graben overlap

transfer
fault

50 kilometers

CROSS SECTION

rift
shoulder   basin fill

listric faults

Fig. 4.17.   Map view and cross section showing major features of linear rift belts. Three half graben are shown in the map view, with the black arrows showing the directions of downward movement of the upper blocks. Two of the half graben overlap, with a compressional zone caused by collision of the upper blocks as they move toward each other. The opposing directions of movement on two non-overlapping half graben are accommodated by movement along a 'transfer fault', which transfers the direction of movement on opposite sides.

The cross section shows an uplifted rift shoulder bordered by a basal listric (concave-upward) normal fault. A smaller fault joins the basal fault at depth. The sediment fill in the half graben is deep along the rift shoulder and grades to very thin where it overlaps the downgoing block.   □

The linear rift systems generally are described in quite different ways from the broad basins. Fig. 4.17 shows such characteristic features as: 1) alternating, locally overlapping, half graben; 2) 'transfer' faults, spaced at 50- to 100-km intervals, that separate areas of different movement of crustal blocks and that intersect the strike of the rifts at angles of approximately 60°; 3) total extension of less than 10% even after ten or more million years of rifting; and 4) narrowness of the rifted terrane, with few places where a cross section perpendicular to the rift zone intersects more than one rift basin. An important additional feature is the presence of elevated 'shoulders' around the margins of the rifts.

Linear rift systems can be classified in various ways. A genetic classification, which may be mostly semantic, is into 'active' and 'passive' rifts. An active rift is one that is caused by some active force, such as intrusion of magma to shallow levels. A passive rift is one in which separation of the opposite sides occurred because distribution of stresses elsewhere led to development of extensional stresses with components perpendicular to the resultant rift zone. The presence or absence of active volcanism is not a discriminant because magmas that fill the area left by crustal extension could be either the cause or the effect of the extension.

Magmatism in linear rifts is similar to that in the broad extensional areas. Intrusion of asthenosphere-derived basalt causes abundant melting of the continental crust, and the resulting magmas and solidified materials mix to form bimodal and other complex suites. The magmas of linear rifts tend to be very alkali rich, although some are similar to MORB (Section 2.3). No particular magmatic rock type is restricted to, or characteristic of, rift systems. P-wave velocities <8 km/s extend to depths of 40 km or more in the linear rift valleys, presumably indicating large thicknesses of hot, mafic rocks in a transition zone between lower crust and upper mantle (asthenosphere?).

The mechanism of uplift of rift shoulders is highly controversial. Older concepts postulated broad uplift (doming) of an area before rifting, resultant stretching of the continental crust, and subsequent rifting near the top of the dome. This model explains the rift shoulders as relics of the pre-rifting process, with the rift valley dropping isostatically toward its equilibrium position between the flanks. One problem with this explanation is the absence of coarse, conglomeratic,

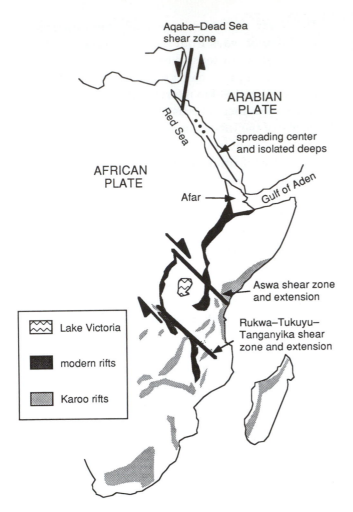

Fig. 4.18. Sketch map of Red Sea and East African rift system. The rift system radiates in three zones from the Afar triangle, which is an area of complex volcanism mostly surrounded by plateau basalts. 1) The Red Sea has opened by some combination of crustal stretching and seafloor spreading. A definable spreading center extends along approximately the southern two thirds of the sea and continues northward as isolated 'deeps'. The Aqaba–Dead Sea shear zone accommodates northward movement of the Arabian plate (Sections 8.1 and 9.1). 2) The Gulf of Aden, which is an extension of a spreading center in the Indian Ocean (Section 8.1). 3) The East African rift system. The Cenozoic rift system partly occupies the same area as older sequences, including the Permo-Triassic Karoo rifts. The rift system is offset along two major shear zones (Aswa and Rukwa–Tukuyu–Tanganyika). The movement associated with this shear couple forms two parallel rifts and an intervening stable platform, which is partly occupied by Lake Victoria. □

material in the initial sediments at the base of many rift sequences. Partly for this reason, recent models for the opening of rifts propose stretching of the lithosphere (including the crust), with initial tectonic downdrop-

ping of the stretched area, and no initial development of shoulders. In these models, uplift of rift flanks occurs for a variety of reasons; one is that asthenospheric heating around the rift raises the still-thick crust of the shoulders to an isostatic position at a higher elevation than the rift floor.

The differences shown by the three varieties of continental rifting pose serious questions as to whether they represent the same extensional process, but we cannot be sure how many of the differences are real and how many represent different perceptions of different investigators. Passive continental margins commonly have been thinned to approximately one half of original crustal thickness and are characterized by transfer faults perpendicular to the margins and connecting with oceanic fracture (transform) zones. The presence of low-angle, master, detachments along these margins is controversial. Detachments are characteristic of broad extensional basins, where total extension may be several hundred percent. These broad basins commonly are associated with preceding or synchronous subduction, whereas modern spreading oceans (Atlantic, Indian, Red Sea) initially developed in continental interiors completely unaffected by subduction. Linear rift valleys are characterized by small amounts of extension, transfer faults at an oblique angle to the extension direction, and have not been shown to be associated with low-angle detachments.

We illustrate some of these issues further by discussion of the Red Sea and East African rift system.

Red Sea and East African rift system

Over a distance of nearly 5000 km, the world's largest active rift system separates the sweltering coastal-plain deserts on either side of the Red Sea, lifts a rift shoulder up to 3500 m high in the Arabian peninsula, cuts the basaltic uplands of Ethiopia, and passes southward through the savannahs and forests of eastern Africa (Figs 4.18 and 4.19). The intracontinental rift system south of the Red Sea is the latest of several (perhaps as many as six) systems formed during repeated, or nearly continuous, rifting episodes throughout much of the Mesozoic and Cenozoic.

We start a discussion of the modern Red Sea and East African rift system in the Afar region, one of the world's most famous triple junctions. The Afar is a tri-

Fig. 4.19. Scarp with a relief of ~2 km along eastern margin of Red Sea extensional zone near Taif, Saudi Arabia.   □

angle of crust that could not have existed prior to separation of Arabia and Africa because it occupies the spot that would be filled by the southern tip of the Arabian peninsula in any reconstruction of a pre-drift configuration (Fig. 4.18). The triangle consists primarily of a diverse assemblage of volcanic rocks that began forming at about 30 Ma, presumably representing the initial surface activity of a plume. Horsts of crystalline African basement occur at several places, apparently the result of small-scale rifting from the adjoining plateaus. Long-continued plume activity centered around Afar has also caused the high elevations (locally >4000 m) in adjacent areas of Ethiopia and Yemen.

The three rifts that emanate from the Afar triangle have very different histories (Fig. 4.18). In conventional terminology, the East African rift system is a 'failed' or 'unsuccessful' arm, and the Gulf of Aden and Red Sea have been successful in the development of ocean basins. The simplest of the three arms is the east–west-trending Gulf of Aden, formed by an prolongation of the Carlsberg Ridge of the Indian Ocean between Africa and Arabia. The origins of the Red Sea and East African rifts are more controversial and, probably, more complex.

The opening of the Red Sea represents a relative northeastward movement of the Arabian peninsula away from its former position against Africa (Fig. 4.18). Extension between the African and Arabian plates began at about 30 Ma, but precursor activity (volcanism, deposition of coarse clastics in small basins, etc.) may have begun in the late Cretaceous. The movement pattern across the Red Sea is both a rotation of Arabia away from Africa and a northward movement of Arabia along the Aqaba–Dead Sea transform. Movement along the transform began at about 20 Ma; it was associated with changes in opening patterns in the Indian Ocean (Section 8.1) and formation of the Turkish syntaxis by crushing of orogenic terranes between Eurasia and the northward-colliding Arabia (Section 9.1).

The central part of the southern Red Sea is clearly oceanic lithosphere, with magnetic stripes dating from ~5 Ma (Figs 4.18 and 4.20). The tip of this oceanic wedge appears to have been migrating northward, away from the triple junction. North of the oceanic tip, and on line with it, are several deep holes in the Red Sea crust that are filled with hot, metal-bearing, brines and apparently represent the earliest stages of oceanic ridge formation in the overlying, thin, crust. The elevation of zero-age crust at the Red Sea spreading center is only slightly shallower than the crests of other ridges in the world's major oceans. Thus, although the oceanic lithosphere in the Red Sea forms a trough rather than ridge, the trough results mostly from the high elevation of the surrounding oceanic shelves

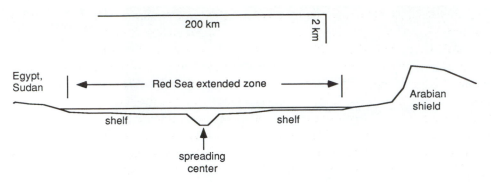

Fig. 4.20. Cross section of southern Red Sea. The margins of the ocean basin are asymmetrical, with a steep uplift on the Arabian side and a more subdued topography on the African side. The extended area occupied by the Red Sea has broad shelves proposed to represent either thin continental crust or elevated oceanic crust. The southern part of the Red Sea contains a spreading center that occupies a trough between the elevated shelves.   □

rather than from a low elevation of the spreading center.

Topographic features of the Red Sea and surrounding continents provide some information on the evolution of the area. The margins of the deep areas of the Red Sea are very abrupt, and coastal areas of the surrounding continents appear to be underlain by normal thicknesses of Precambrian sialic crust (Fig. 4.20). Thus, extension of the Red Sea has occurred primarily within the Red Sea basin rather than over a more diffuse area of the Nubian–Arabian shield. The Arabian and African sides of the Red Sea are topographically asymmetrical, with a high, steep scarp on the Arabian peninsula and only low hills of Precambrian crust on the African side. Some investigators have attributed this asymmetry to development of the ocean above an eastward-dipping detachment fault that underlies the entire Red Sea, but the existence of such a fault has never been confirmed.

Proposals for the origin of the deep parts of the Red Sea range between two 'endmembers'. One is that the Red Sea is underlain largely by thinned continental crust, with successful creation of oceanic lithosphere occurring first in the south and spreading northward. Refitting of the sides of the Red Sea indicates that their present separation would have required thinning of the continental crust to about one third of its normal value, with consequent rapid downwarp and sediment filling. A thinned continental crust also is consistent with the observation that the Red Sea contains magnetic stripes only within the spreading axis in the southern part of the basin and not within the marginal

areas (magnetic anomalies in the margins are larger and do not have the regular pattern found around spreading ridges).

The second proposal for the origin of the Red Sea is that the major part of the ocean is underlain by oceanic crust, even in the relatively shallow marginal plains. This proposal is supported by some aspects of the distribution of magmatic rocks and by the close fit of the present ocean margins against each other, implying lack of stretching when they formed. If the Red Sea basin is underlain by lithosphere formed at a spreading ridge, some process must have obscured the magnetic stripe pattern.

The northern end of the Red Sea terminates against the Aqaba–Dead Sea transform (Fig. 4.18). This transform has isolated the Gulf of Suez from active spreading since ~20 Ma, permitting purely thermal subsidence in the central Gulf since that time. The transform also localizes deep basins, both within the Gulf of Aqaba and at various places along the onland part of the fault; the principal basin on land is the Dead Sea, whose elevation of ~400 m below sea level makes it the world's lowest land area. The deep basins along the transform have been attributed to small-scale rifting or to the development of pull-aparts (Section 4.4).

Southward from the Afar triple junction, the rift system penetrates an otherwise-stable African continent. The major area of extension is between two large transforms, Aswa and Rukwa–Tukuyu–Tanganyika (Fig. 4.18). Between these zones are two parallel rift branches, an eastern one that is largely filled by volcanic activity (the Gregory rift), and a western one in which volcanic activity is limited to a few, highly alkaline, volcanic centers. The reason for these 'wet' and 'dry' branches is unknown. The rift valleys almost completely bypass the Archean Tanzanian craton, with its shallow Lake Victoria, and are localized within belts of Proterozoic deformation on its sides (Chapter 5). The rift system crosses two broad areas of domal uplift where modern rift shoulders are high, one in Ethiopia (a late-Cenozoic basalt plateau, Section 8.2) and in one Kenya. Volcanism in these domes began in the middle Tertiary (perhaps slightly older), but development of the modern rift valleys is a younger phenomenon.

The East African rift system provides the archetypical example of rifts consisting of alternating and overlapping half graben (see above). Transfer faults with an average spacing of 70 km separate half graben of different polarities. Thus, most large 'rift valleys', including the major lakes of the western branch of the system, consist of four to eight individual half graben with slightly different faulting and subsidence histories. As measured by movement on bounding faults, total extension across most of the rift valleys is less than 10%. This low value of separation seems remarkable considering that extension in the area has been active for more than 20 million years, enough time for most of the opening of the Red Sea.

[**References** – The evolution of basins and related thermal effects are discussed by McKenzie (1978) and Sclater, Jaupart and Galson (1980). Information on spreading ridges is partly from Karson and Elthon (1987). Passive margins are discussed in volumes edited by Vogt and Tucholke (1986), Sheridan and Grow (1988) and Tankard and Balkwill (1989) and the paper by Rosendahl et al. (1992). The Basin and Range is discussed in the volume edited by Lipman and Glazner (1991) and papers by Glazner and Bartley (1984) and Catchings and Mooney (1991). The East African rifts and the general nature of continental rifting is in a volume edited by Rosendahl, Rogers and Rach (1989). Makris, Mohr and Rihm (1991) provide an edited issue on the Red Sea.]

4.3 Transform margins

WHEN we draw simple diagrams, as in the two preceding sections, we imply movement of plates at nearly right angles to the plate boundaries (either accreting or consuming). In most situations, however, plates approach subduction zones at angles other than orthogonal ('transpression'), and many accreting margins exhibit non-orthogonal extension ('transtension'). Approximately 50% of the world's plate margins show movement vectors of plates at angles more than 30° from the orthogonal to the margin (less than a 60° angle between the movement vector and the subduction zone or rift). In this section, we review these lateral movements between plates and their interactions with other types of plate movements.

The lateral movements are along faults that have a highly diverse terminology. The term 'transform' is used both in reference to the faults that offset oceanic spreading centers (Fig. 4.12) and as a general term for large faults that separate lithospheric plates. The terms 'transcurrent' and 'strike-slip' are also used for various types of faults with lateral movements, also with no unanimity as to their appropriateness. We will avoid entanglement in these issues and simply refer to zones of 'lateral movement'.

Major intracontinental fractures do not form simple shear planes but large zones of displacement locally many kilometers wide. An ancient example is the Proterozoic Snowbird suture of the Canadian shield, which shows combined compressional and lateral movements throughout a zone 20 or more kilometers wide (Section 5.2). The modern San Andreas system includes numerous en echelon faults distributed over a width of up to 150 km across the fault trend; the complexity of the fault patterns produces local stress orientations that cause displacements on some small faults to be opposite to the principal direction of offset.

Large intracontinental fault systems appear to develop in general zones of weakness rather than to form sharp breaks across regions of resistant lithosphere. The most-studied fault system, the San Andreas, demonstrates this weakness by its lack of high heat flow. If movement on the fault were strongly resisted, then it would have to use a high proportion of its stored strain energy to overcome the resistance, and the released frictional heat should generate high heat flow anomalies along the fault trace. Fragmentary evidence indicates that the absence of these anomalies on the San Andreas is characteristic of other major zones of lateral movement. Possibly these zones develop predominantly in lithosphere that has been weakened by some previous events.

The relationship between oceanic transforms and major areas of lateral movement on continents is complicated, and we discuss two possible relationships.

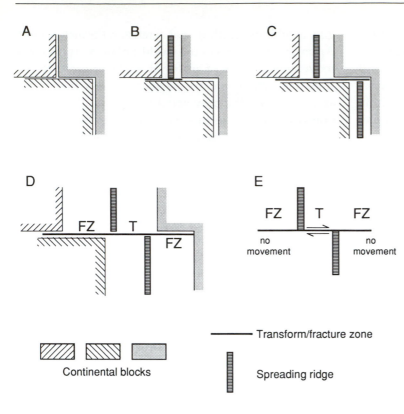

Fig. 4.21. Diagrammatic evolution of oceanic transform from continental fracture zone. A, initial condition. B and C, strike-slip movement between continents accompanies development of spreading centers. D, after continental separation is complete, the fracture is a transform (T) between spreading ridges and a fracture zone (FZ) beyond each ridge. E, if spreading rates are equal on both ridge segments, no movement occurs on the fracture zones after continental separation. (Note: this diagram is a simplified representation of the separation of North America, South America and Africa. The actual separation was complicated by development of the Caribbean and the establishment of a spreading center between the Bahama–Guinea and Romanche transforms, which separates the North Atlantic and South Atlantic ridges. See Section 8.1 for further information.)   ☐

——————— Transform/fracture zone

Continental blocks

Spreading ridge

One is the ideal oceanic transform (Fig. 4.12), in which the only lateral movement is in the area between the two offset spreading ridges. The fracture zones extending to either side of the ridge area separate two blocks of oceanic lithosphere moving at the same rate and in the same direction. Hence, the fracture zones show no lateral offset but have a vertical relief resulting from the higher elevation of young crust than of older crust. A typical example is the Mendocino fracture zone, which has had major effects on the Basin and Range area (Sections 4.2 and 9.4).

A second type of interaction between oceanic and continental offsets is shown diagrammatically in Fig. 4.21. Two spreading ridges, possibly initiated at different times, cause separation of the single continent into three smaller ones. The consequent strike-slip faulting along part of the new continental margin becomes inactive as complete continental separation occurs. In the resulting ocean, the original fault becomes partly a transform between the ridges and partly two fracture zones (probably inactive) extending from each ridge to the adjacent continent. The strike-slip continental margins are straighter and show more abrupt terminations against oceanic crust than many margins formed by extensional processes.

Some major complexities of lateral offsets are shown in Fig. 4.22A–F. Parts A–C depict situations that can develop along otherwise-simple faults. Many faults contain bends, and the stresses generated there depend on the relationship between the geometry of the bend and the movement on the fault. Fig. 4.22A shows a 'releasing bend', at which movement opens a space between the blocks on either side of the fault. This type of movement is one method for the generation of 'pull-apart' basins, which commonly undergo very rapid subsidence because of the sudden removal of thick segments of the crust. A second geometrical interpretation of the development of pull-aparts is shown in Fig. 4.22B, where the basin opens within a broad fault zone. The opposite configuration to a releasing bend is the 'restraining bend', where blocks on either side of the fault are crushed together in the zone of the bend (Fig. 4.22C).

Fig. 4.22D–F shows relationships along zones where lateral movement and orthogonal movement are both generated by external forces. In many situations, the motions may be described best as taking place within a broad zone of weakness rather than by attempting to isolate individual offsets along specific structures. Fig. 4.22D shows transtension along a closely spaced com-

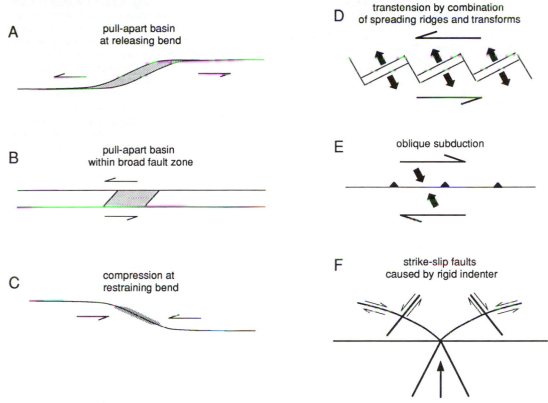

Fig. 4.22. Features along strike-slip faults and transform margins. (A) Development of pull-apart basin along releasing bend in strike-slip fault. (B) Development of pull-apart basin in broad zone of lateral movement. (C) Compression at restraining bend of strike-slip fault. (D) Transtensional area established by combination of spreading centers and connecting transforms. (E) Transpressional area established by combination of subduction and strike-slip movement (including oblique subduction). (F) Strike-slip faults established by compression caused by a 'rigid indenter' colliding with a continental margin.   □

bination of spreading centers and transform faults, which occurs along several mid-ocean ridges (Section 8.1). Fig. 4.22E diagrammatically illustrates subduction at an oblique angle. During active subduction, some orogenic belts show reversals of the directions of lateral movement (e.g. see discussion of the North Atlantic orogenic system in Section 7.3). Fig. 4.22F depicts movement directions along strike-slip faults generated by the collision of a block ('rigid indenter') with another block. An excellent example of this process is the development of strike-slip faults in eastern Asia as a result of the collision of India with the Eurasian mainland (Section 9.1). Many investigators refer to movement of blocks away from the rigid indenter as a form of 'tectonic escape'.

[**References** – Strike-slip faults are summarized by Lemiszki and Brown (1988) and Sylvester (1988) and also covered by a volume edited by Biddle and Christie-Blick (1985). Woodcock (1986) dis-

cusses plate margins that show transform movement. The San Andreas fault is summarized in a volume edited by Wallace (1990).]

4.4 Plumes

IN its original concept, the ideal plume rises abruptly and rapidly from a source near the mantle–core boundary (Fig. 4.23). It has a diameter of approximately 100 km to 300 km and produces a 'hotspot' of magmatic activity that endures for 100 million years or longer. Its initial activity is marked by voluminous eruptions of basalts, producing continental and oceanic basalt plateaus at the 'head' of the plume before activity tapers off along the 'tail' of the plume track. The ideal plume occupies a fixed position relative to the earth's interior, thus generating hotspot

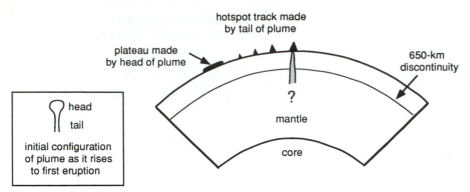

Fig. 4.23. Diagram of ideal plume. The inset shows the enlarged head of the plume, which generates large outpourings of basalt as it initially breaches the earth's surface. Small-scale magmatism along the tail of the plume is responsible for the hotspot track as the lithosphere moves over it. The 650-km discontinuity is a major seismic discontinuity. The '?' indicates that some investigators propose that plumes develop below the 650-km discontinuity, whereas others propose that the plumes must exist only above the discontinuity. ☐

tracks of former volcanoes that ride away from the plume location on plates passing overhead. The fixed location of hotspots permits us to determine absolute motions of the earth's plates. The ideal plume samples the deep mantle, below the zone of asthenospheric depletion (Section 2.3), thus causing oceanic islands and comparable intracontinental volcanic centers to contain rocks relatively rich in LIL elements. The ideal plume may not exist – in this section, we discuss some of its problems.

One difficulty with the concept of an ideal plume is the question of the immobility of its position below the moving lithosphere. The question involves two issues. One is the relative positions of plumes with respect to each other, which is equivalent to the question of whether plumes move within the mantle. The second issue is whether the entire mantle moves with respect to the spin axis of the earth (causing 'true polar wander').

The best evidence for unvarying locations is the relative position of plumes with respect to each other. Most investigators have proposed that plate motions during the past few tens of millions of years demonstrate the fixed positions of groups of hotspots. For example, our diagrams of the evolution of the Pacific Ocean (Section 8.1) are consistent with relative invariance of the locations of the Galapagos, Hawaii and Yellowstone hotspots. Despite these indications of invariance, not all hotspot tracks show perfect sequences of ages of volcanism along the track, suggesting that plumes undergo at least small-scale movements in comparison with plate motions.

The second issue of plume movement, true polar wander resulting from a shift of the mantle, is more difficult to analyze than the first. It depends on comparison of plate motions determined from hotspot tracks and plate motions calculated from paleomagnetic measurements. Differences between these two sets of data result from an absolute movement of the mantle, which contains the plumes, relative to the earth's magnetic field, presumably generated by the core. Present data indicate that the mantle 'rolled' ~20° counterclockwise during the period of 110 Ma to 40 Ma and ~12° clockwise from 40 Ma to the present. Both rotations occurred about an axis emerging near the equator in the Indian Ocean. The shift at 40 Ma coincides with significant plate reorganization on the earth's surface (Sections 8.1 and 9.1), which could have changed the rotational properties of the earth.

A larger difficulty with the idealized concept of plumes than their absolute movements is their source and the process by which they rise through the mantle. A plume is obviously the result of localized high temperatures. The question is the extent to which these high temperatures are caused by: 1) a thermal anomaly, in which isotherms rise by conductive heating and without mass transfer of a body of material into overlying parts of the mantle; and 2) a plume of low-density mantle that moves upward, perhaps displacing rocks of different composition. As we discussed in Chapter 2, mass transfer must occur within the deep mantle. It also must occur locally near the earth's surface, where magmas are intruded and erupted into the hotspot track. The magmas, however, may represent melting at relatively shallow depths (a few hundred kilometers or less) of mantle that was activated by a thermal anomaly; that is, no component of the deep mantle may have risen into the source region of the magmas that we see at the surface.

The problem of convective transfer of mass and heat is closely related to the question of the nature of a seismic discontinuity in the mantle at ~650 km depth (variable over a depth range of approximately 100 km). If this discontinuity results from a phase change, then its position is controlled solely by temperature and pressure acting on materials with the composition of average mantle. A phase boundary at ~650 km is consistent with concepts that plumes can rise through the entire mantle (core to crust) without disrupting the discontinuity. This type of convection effectively 'stirs' the mantle.

If the ~650-km discontinuity, and possibly other mantle discontinuities, represent compositional changes, then mantle stirring cannot occur across the boundary. In that situation, plumes reaching the surface must originate within the upper ~650 km of the mantle. They could be localized either by a conductive thermal anomaly at a depth of ~650 km or by a plume rising through the lower mantle and generating a second plume in the upper mantle above the area of high temperatures at the discontinuity.

Attempts to determine the source, including depth of melting, of hotspot magmas yield uncertain results. Hotspot magmatism has been proposed to be responsible not only for chains of oceanic and continental volcanoes but also for many oceanic plateaus and for continental plateau basalts. All magmatic products postulated to have been generated by plumes have been derived from sources that are enriched in LIL elements relative to the asthenospheric source of MORB (Section 2.3). This enrichment is shown partly by the comparatively high abundances of LIL elements in erupted rocks, including some highly alkaline rock types, and mostly by the initial isotope ratios of Pb, Sr and Nd in the magmatic products, which indicate isotopic evolution in source regions containing high abundances of LIL over long periods of time. High LIL concentrations support such diverse concepts as an LIL-enriched lower mantle and an upper mantle that shows MORB-type depletion only in the uppermost lid of the asthenosphere.

Part of the difficulty in the interpretation of plume magmas is that, although all are LIL enriched, they exhibit wide compositional diversity. This diversity is shown both in time at one place and from place to place along the course of the hotspot track. Some investigators suggest that rising plumes entrain part of their wall rocks, thus causing magmas to be generated from at least two different types of source material. Such a process could explain the synchroneity of eruption of both alkali basalts and basalts with a more LIL-poor (tholeiitic) composition. In the Archean, the two eruptive suites might be shallow-level basalt magmas and deep-level komatiites. Flow patterns in rising plumes may generate two parallel columns of deep-source material surrounded by material melted at higher levels, forming bilaterally symmetrical plumes that explain the generation of parallel sets of islands in some hotspot tracks.

Our final issue is the process or processes that control the number and location of plumes. At present, a few dozen plumes are active on the earth; the number that we cite depends highly on various assumptions as to what constitutes a plume. Plumes appear to be concentrated in areas of global high gravity as measured by highs on the 'slab residual geoid'. (The slab residual geoid is the geoid from which the effects of subducted slabs, which increase gravity, have been removed.) Within these areas, and on a geographic (latitude–longitude) grid of the earth, however, plume locations are random. If plumes are derived from the lowest 100 km to 300 km of the mantle, (the D'' layer), then presumably some variability in or near the core–mantle boundary determines the precise location of the thermal anomaly. No evidence currently exists for any regularity in the pattern of core–mantle variations.

[**References** – An early recognition of the importance of hotspots is in Morgan (1972). The distribution of hotspots is discussed by Stefanick and Jurdy (1984) and Richards and Hager (1988). The concept of heads and tails of plumes is summarized by Richards, Duncan and Courtillot (1989). Archean hotspots and mantle magmatism are in Campbell, Griffiths and Hill (1989). A summary of basalt plateaus, hotspots, plumes, and true polar wander is in Alt, Sears and Hyndman (1988), Davies (1980), Sleep (1990), Duncan and Richards (1991) and Griffiths and Campbell (1991). Decker, Wright and Stauffer (1987) discuss Hawaii. Reunion Island is discussed in Fisk *et al.* (1989) and a volume edited by McBirney (1989).]

4.5 Summary

THIS CHAPTER has concentrated primarily on earth processes associated with the existence of a rigid lithosphere that overlies a more mobile asthenosphere. Lithospheric rigidity permits description of shallow-level processes in terms of 'plate tectonics', which recognizes an outer earth consisting of stable plates bounded by three ideal types of plate margins: 1) subducting margin, in which a descending lithospheric slab (generally oceanic) is consumed back into the mantle: 2) accreting margin, generally along a mid-ocean ridge, from which plates move away; and 3) a strike-slip (or transform) margin along which plates move laterally past each other. Most margins are some combination of subduction or accretion with strike-slip movement; recent investigations continue to reveal lateral movements of several hundred, to more than 1000, kilometers along most orogenic belts.

Subduction has three principal effects. One is compression, leading to the characteristic thrust faults, folds, cleavage and foliation of orogenic belts. The lateral shortening represented by these structures is associated with the development of thick crusts in mountain belts, but the thickening may be caused partly by addition of mantle-derived magmatic rocks. Compression is commonly generated by collision of terranes, and many orogenic belts consist of deformed suture zones between stable continental blocks.

The second major effect of subduction is magmatism. The descending slab places wet sediment and (commonly hydrated) oceanic crust into a region of pressures and temperatures in which dehydration and partial melting occur. The wedge of mantle overlying the slab may be partly mobilized by volatiles escaping from the descending sediments and crust. Thus, the magmatic suites contain components from subducted oceanic lithosphere, overlying mantle and commonly from crust of the arc or mountain range incorporated into the rising magmas. Early magmatic suites are commonly bimodal (basalt–rhyolite), but later ones are typically 'calcalkaline' (andesite-dominated).

The third consequence of subduction is the development of basins of sediment accumulation. Foredeep trenches result from the drag of the subducting plate. Fore-arc basins develop when the large wedge of sediments and volcanic rocks developed above the subducted slab becomes too thick to be stable and partly collapses. Foreland basins occur on the opposite side of an orogen from the subduction zone and result from the downbuckling of the crust by the weight of the accumulating orogen. Both the fore-arc and foreland basin generally contain less volcanic material than the foredeep trench.

The principal site of extension in the earth is along the system of mid-ocean ridges that extends for ~55 000 km and generates ~20 km³ of new oceanic crust each year. These ridges are related to the mantle convection system of the earth, and much of the earth's heat flow is dissipated through the ridges. The forces responsible for ridge spreading, however, are controversial, probably containing components of pushing by newly inserted magmas along the ridge, pulling by the weight of subducting slabs, and dragging by circulation of convecting mantle beneath the oceanic plate. The major product of ridge volcanism is mid-ocean-ridge basalt (MORB), apparently derived from partial melts of LIL-depleted asthenosphere.

Extension in continents can be caused by forces within the extending area, perhaps above plumes or other rising limbs of convection cells, or by forces outside the extending area that require continental separation. These forces generate a variety of extensional terranes, and we have somewhat arbitrarily classified them into three categories. One type of continental extension results in long linear belts across which a small amount of separation has occurred. These belts are generally characterized by sets of half graben separated by transfer faults (accommodation zones) oblique to the extensional system. Some continental areas exhibit overlapping sets of the extensional belts formed at different times through intervals of several hundred million years.

A second variety of continental extension is a broad basin across which total separation may be 100% or more of the original width. These basins contain parallel sets of half graben (or full graben) and intervening horsts, with the movement on each fault contributing to the overall amount of separation. Shallow-level extension in basins of this type is apparently controlled partly by the development of large, low-angle, detachment faults, which act as master faults for an overlying block containing numerous listric faults. A controversial issue in the development of these basins is whether the detachments extend into the mantle or disappear into the top of a ductile lower crust. Broad extensional basins occur in areas of active or recently active subduction, and low-angle detachments are also found in the upper parts of compressional orogens.

A third variety of extension, with uncertain relationships to the other two, develops passive (rifted) continental margins. These margins are thinned during the rift phase of continental separation and then cool and subside as drifting begins. The rate of subsidence is determined largely by the amount of thinning that the margin has undergone, and the accumulated sediments overlie an unconformity on the synrift sediments and volcanic rocks. Rifting may be controlled by low-angle detachments, with 'upper-plate' and 'lower-plate' margins forming opposite each other, although the existence of detachment faults has not been proved. A possibly significant difference between rifted margins and linear belts of intracontinental rifting is that transfer faults in the rifted margins are commonly orthogonal to the margin and, thus, may connect with transform and fracture zones of adjacent mid-ocean ridges.

Strike-slip faulting (as along major transforms) commonly produces straight plate margins. Zones of pure strike-slip movement can produce 'pull-apart' basins. A combination of strike-slip movement and near-orthogonal rifting is more likely and causes sharp irregularities in the continental margins, particularly along the extensions of large transforms that cut the mid-ocean ridges.

The concept of plumes is not strictly associated with the concept of rigid lithospheric plates. Plates moving across a stationary plume, however, produce the familiar chains of volcanic centers that have been so important in developing the idea of plate tectonics. Plumes tap some part of the mantle below the outer, depleted, asthenosphere, but their precise origin is unclear. Most plumes appear to have occupied stationary positions with respect to each other during their entire measured period of existence (100 m.y. or more). Irregularities in age distributions of volcanism along some plume tracks, however, indicate small-scale movement in the mantle. Plume magmatism is most voluminous at the start of activity (head of the plume) and decreases toward younger events (tail of the plume).

References

Alt, D., Sears, J. M. & Hyndman, D. W. (1988). Terrestrial maria: the origins of large basalt plateaus, hotspot tracks and spreading ridges. *Journal of Geology*, **96**, 647–62.

Bally, A. W. & Snelson, S. (1980). Realms of subsidence. In *Facts and Principles of World Petroleum Occurrence*, ed. A. D.Miall, pp. 9–94. Canadian Society of Petroleum Geologists Memoir 6.

Benioff, H. (1954). Orogenesis and deep crustal structure – additional evidence from seismology. *Geological Society of America Bulletin*, **65**, 385–400.

Biddle, K. T. & Christie-Blick, N., eds. (1985). *Strike-Slip Deformation, Basin Formation, and Sedimentation*. Society of Economic Paleontologists and Mineralogists Special Publication 37, 386 pp.

Callaway, C. (1883). The age of the Newer Gneissic Rocks of the Northern Highlands. *Quarterly Journal of the Geological Society of London*, **39**, 355–414.

Campbell, I. H., Griffiths, R. W. & Hill, R. I. (1989). Melting in an Archaean mantle plume: heads it's basalts, tails it's komatiites. *Nature*, **339**, 697–9.

Catchings, R. D. & Mooney, W.D (1991). Basin and Range crustal and upper mantle structure, northwest to central Nevada. *Journal of Geophysical Research*, **96**, 6247–67.

Cloos, M. (1982). Flow melanges: numerical modeling and geologic constraints on their origin in the Franciscan subduction complex, California. *Geological Society of America Bulletin*, **93**, 330–45.

Coleman, R. G. (1977). *Ophiolites*. Berlin: Springer Verlag, 229 pp.

Davies, G. F. (1980). Thermal histories of convective earth models and constraints on radiogenic heat production in the Earth. *Journal of Geophysical Research*, **85**, 2517–30.

Davis, D., Suppe, J. & Dahlen, F. A. (1983). Mechanics of fold-and-thrust belts and accretionary wedges. *Journal of Geophysical Research*, **88**, 1153–72.

Decker, R. W., Wright, T. L. & Stauffer, P. H. (1987). *Volcanism in Hawaii*. U.S. Geological Survey Professional Paper 1350, 2 volumes, 1667 pp.

Donnelly, T. W. & Rogers, J. J. W. (1980). Igneous series in island arcs – the northeastern Caribbean compared with worldwide island-arc assemblages. *Bulletin Volcanogique*, **43**, 347–82.

Duncan, R. A. & Richards, M. A. (1991). Hotspots, mantle plumes, flood basalts, and true polar wander. *Reviews of Geophysics*, **29**, 31–50.

Elliott, D. & Johnson, M. R. W. (1987). Structural evolution in the northern part of the Moine thrust belt, NW Scotland. *Transactions of the Royal Society of Edinburgh, Earth Sciences*, **71**, 69–96.

Fisk, M. R., Duncan, R.A., Baxter, A. N., Grenough, J. D., Hargraves, R. B. & Tatsumi, Y. (1989). Reunion hotspot magma chemistry over the past 65 m.y.: results from Leg 115 of the Ocean Drilling Program. *Geology*, **17**, 934–7.

Glazner, A. F. & Bartley, J. M. (1984). Timing and tectonic setting of Tertiary low-angle normal faulting and associated magmatism in the southwestern United States. *Tectonics*, **3**, 385–96.

Griffiths, R. W. & Campbell, I. H. (1991). On the dynamics of long-lived plume conduits in the convecting mantle. *Earth and Planetary Science Letters*, **103**, 214–27.

Hess, H. H. (1962). History of ocean basins. In *Petrologic Studies – A Volume to Honor A.F. Buddington*, ed. A. E. J. Engel, H. L. James & B. F. Leonard, pp. 599–620. New York: Geological Society of America.

Hills, E. S. (1940). *Outlines of Structural Geology*. London: Methuen and Co., 172 pp.

Ingersoll, R. V. (1988). Tectonics of sedimentary basins. *Geological Society of America Bulletin*, **100**, 1704–19.

Karson, J. A. & Elthon, D. (1987). Evidence for variations in magma production along oceanic spreading centers: a critical appraisal. *Geology*, **15**, 127–31.

Kay, S. M. & Rapela, C. W., eds. (1990). *Plutonism from Antarctica to Alaska*. Geological Society of America Special Publication 241, 263 pp.

Lapworth, C. (1883). The secret of the Highlands. *Geological Magazine*, Decade II, vol. **10**, 120–8.

Leitch, E. C. (1984). Island-arc elements and arc-related ophiolites. *Tectonophysics*, **106**, 177–203.

Lemiszki, P. J. & Brown, L. D. (1988). Variable crustal structure of strike-slip fault zones as observed on deep seismic reflection profiles. *Geological Society of America Bulletin*, **100**, 665–76.

Lipman, P. W. & Glazner, A. F., eds. (1991). Mid-Tertiary Cordilleran magmatism: plate convergence vs. intraplate processes. Special section of *Journal of Geophysical Research*, **96**, 13,193–735.

Makris, J., Mohr, P. & Rihm, R., eds. (1991). Red Sea: birth and early history of a new oceanic basin. *Tectonophysics*, **198**, 149–468.

McBirney, A. R., ed. (1989). Piton de la Fournaise volcano, Reunion Island. Special Issue of *Journal of Volcanology and Geothermal Research*, **36**, 1–232.

McClay, K. R. & Price, N. J., eds. (1981). *Thrust and Nappe Tectonics*. Geological Society of London Special Publication 9, 539 pp.

McKenzie, D. (1978). Some remarks on the development of sedimentary basins. *Earth and Planetary Science Letters*, **40**, 25–32.

McKenzie, D. & Morgan, W. J. (1969). Evolution of triple junctions. *Nature*, **224**, 125–33.

McKerrow, W. S., ed. (1987). The southern uplands controversy. *Geological Society of London Journal*, **144**, 735–838.

Mitra, G, & Wojtal, S., eds. (1988). *Geometries and Mechanisms of Thrusting, with Special Reference to the Appalachians*. Geological Society of America Special Paper 222, 236 pp.

Moores, E. M. (1982). Origin and emplacement of ophiolites. *Reviews of Geophysics and Space Physics*, **20**, 735–60.

Morgan, W. J. (1972). Plate motions and deep mantle convection. In *Studies in Earth and Space Sciences*, ed. R. Shagam, R. B. Hargraves, W. J. Morgan, F. B. van Houten, C. A. Burk, H. D. Holland & L. C. Hollister, pp. 7–22. Geological Society of America Memoir 132.

Pearcy, L. G., DeBari, S. M. & Sleep, N. H. (1990). Mass balance calculations for two sections of island-arc crust and implications for the formation of continents. *Earth and Planetary Science Letters*, **96**, 427–42.

Platt, J. P. (1986). Dynamics of orogenic wedges and the uplift of high-pressure metamorphic rocks. *Geological Society of America Bulletin*, **97**, 1037–53.

Richards, M. A., Duncan, R. A. & Courtillot, V. E. (1989). Flood basalts and hot-spot tracks: plume heads and tails. *Science*, **246**, 103–7.

Richards, M. A. & Hager, B. H. (1988). Dynamically supported geoid highs over hotspots: observation and theory. *Journal of Geophysical Research*, **93**, 7690–708.

Rosendahl, B. R., Meyers, J., Groschel, H. & Scott, D. (1992). Nature of the transition from continental to oceanic crust and the meaning of reflection Moho. *Geology*, **20**, 721–4.

Rosendahl, B. R., Rogers, J. J. W. & Rach, N. M., eds. (1989). African Rifting: Special Issue of *Journal of African Earth Sciences*, **8**, 137–629.

Sclater, J. G., Jaupart, C. & Galson, D. (1980). The heat flow through oceanic and continental crust and the heat loss of the earth. *Reviews of Geophysics and Space Physics*, **18**, 269–311.

Sheffels, B. M. (1990). Lower bound on the amount of crustal shortening in the central Bolivian Andes. *Geology*, **18**, 812–15.

Sheridan, R. E. & Grow, J. A., eds. (1988). *The Atlantic Continental Margin: Geology of North America, Volume I-2, U.S.* Boulder, Colorado: Geological Society of America, 610 pp.

Sleep, N. H. (1990). Hotspots and mantle plumes: some phenomenology. *Journal of Geophysical Research*, **95**, 6715–36.

Stefanick, M. & Jurdy, D. M. (1984). The distribution of hot spots. *Journal of Geophysical Research*, **89**, 9919–25.

Stewart, J. A. (1990). *Drifting Continents & Colliding Paradigms*. Bloomington, Indiana: Indiana University Press, 285 pp.

Sugimura, A. & Uyeda, S. (1973). *Island Arcs – Japan and its Environs*. Amsterdam: Elsevier, Amsterdam, 247 pp.

Sylvester, A. G. (1988). Strike-slip faults. *Geological Society of America Bulletin*, **100**, 1666–703.

Tankard, A. J. & Balkwill, H. R., eds. (1989). *Extensional Tectonics and Stratigraphy of the North Atlantic Margins*. American Association of Petroleum Geologists Memoir 46, 641 pp.

Vogt, P. R, & Tucholke, B. E., eds. (1986). *The Western North Atlantic Region: Geology of North America, Volume M*. Geological Society of America, 696 pp.

Wallace, R. E., ed. (1990). *The San Andreas Fault System, California*. U.S. Geological Survey Professional Paper 1515, 283 pp.

Wegener, A. (1912). Die Entstehung der Kontinente. *Geologische Rundschau*, **3**, 276–92.

Woodcock, N. H. (1986). The role of strike-slip fault systems at plate boundaries. *Philosophical Transactions of the Royal Society of London*, A **317**, 13–29.

Yoder, H. S., Jr. (1976). *Generation of Basaltic Magma*. Washington, D.C.: National Academy of Sciences, 265 pp.

5

Lower-Cambrian Wajid Sandstone lying unconformably on late-Proterozoic rocks of the Arabian shield, Saudi Arabia. The covering of the shield by fluvial/shallow-water sediments only slightly younger than the igneous and metamorphic rocks of the shield indicates very rapid uplift, erosion, and subsidence of the newly formed shield. □

5

THE PROTEROZOIC

5.0 Introduction

THE PROTEROZOIC is a difficult era to study. It is too long, lasting at least 2000 m.y. It is too variable, representing a time when the earth changed from a 'strange' Archean world to a 'normal' Phanerozoic one. We approach it cautiously.

To begin the discussion of the Proterozoic, we ask when it began, but we find no firm answer. The most commonly cited absolute age for the Archean/ Proterozoic boundary is ~2500 Ma. Typical 'Proterozoic' processes, however, began in some parts of the earth by at least 3000 Ma and possibly earlier (Section 3.0). These processes include development of intra-cratonic basins and reasonably clear evidence for the existence of rigid lithosphere ('plate tectonics'). In this chapter, we discuss some areas that were stable before 2500 Ma and regard them as 'tectonically Proterozoic'.

The conventional end of the Proterozoic is not difficult to define, but thus far it has defied understanding. At numerous places in the world, we can walk stratigraphically upward through sequences of sediments that contain no fossils, or only trace fossils, until suddenly we encounter skeletal remains. This abrupt appearance of organisms capable of secreting hard parts is the universally accepted end of the Proterozoic and beginning of the Phanerozoic, and we discuss it and the organic changes associated with it in considerable detail in Sections 6.1 and 6.2.

Within its bounds, the Proterozoic contains at least two events that suggest separation into periods or even separate eras. The most significant time may have been at approximately 2000 to 1900 Ma. This time records: the generation of large amounts of new continental crust; the end of a period of incorporation of 2500- to 2000-Ma crust so thoroughly into Archean cratons that most maps still show these areas as Archean; the end of deposition of large volumes of banded iron formation; and possibly the generation of an oxidizing atmosphere (all of these observations are discussed more thoroughly in the rest of the chapter). A second time of major change during the Proterozoic is roughly from 1300 to 1100 Ma, admittedly a long time in which to find some definable period or era boundary (almost the entire Mesozoic plus Cenozoic would fit within it). The principal events at this time were compressional orogenies, which appear to have occurred in most of the now-preserved continental fragments during this period.

We investigate the history of this confusing Proterozoic in a number of ways. First, we discuss whether compressive Proterozoic orogenies have the same origin and structural style as Phanerozoic ones (Section 5.1). The issue is the (possibly semantic) difference between orogenies caused by closure of ocean basins, (inter-cratonic belts) and those caused by compression within a previously formed craton (intra-cratonic belts). Most Phanerozoic orogenic belts are the result of collision of continental blocks or, in situations such as the Andes, the continued subduction of oceanic crust under a continental margin. Oceanic closure has not been demonstrated in many Proterozoic orogenic belts, however, leading to the possibility of compressive stresses generated within presumably stable continental lithosphere. The discussion leads to an application of these principles to specific orogenic belts (Section 5.2).

Our discussion of Proterozoic orogenic belts and other aspects of Proterozoic continental geology is based on a four-fold grouping of continental assemblages (the same grouping used in Chapter 3). We can be reasonably certain that two of these assemblages have been coherent supercontinental blocks since the middle Proterozoic (or possibly earlier): 1) a combined North America, Greenland, northern Scotland, Baltic shield, and possibly the remainder of the Russian platform; 2) a combined India, Australia, Antarctica, and Madagascar. A third assemblage is a combined South America, Africa, and Arabia. Many investigators have regarded South America and Africa/Arabia as a supercontinent since the middle Proterozoic, but recent work has suggested that it was assembled in the late Proterozoic or early Paleozoic (Section 7.0). The fourth assemblage is eastern Asia, including cratonic blocks in Siberia and China that clearly were assembled during the late Paleozoic and possibly as late as the Mesozoic.

After orogeny, the next topics in this chapter are continental rifting (Section 5.3) and the related issue of anorogenic magmatism (Section 5.4). The Proterozoic apparently was the first time in the history of the earth in which continental lithosphere was rigid enough to undergo rifting of the type that has characterized the Phanerozoic. These rift networks have been mapped in all continental areas and range in age from the earliest Proterozoic to the end of the era. The spacing of the rifts is affected both by the strength of the lithosphere and the size of the convection cells at its base, and the evidence in Section 5.3 suggests that possibly the tendency of rifts to be more closely spaced in the early

Proterozoic than later in the era indicates that convection cells became broader as the earth cooled.

Our discussion of Proterozoic anorogenic magmatism (Section 5.4) covers three suites of rocks: 1) sets of diabase dikes that extend for hundreds of kilometers across previously stabilized Archean and early-Proterozoic cratons; 2) layered mafic/ultramafic complexes, such as the Bushveld, injected as voluminous amounts of magma into upper levels of the crust; and 3) anorthosites possibly related to the layered complexes but occurring as massive bodies, commonly associated with rapakivi granites. The anorthosite/rapakivi suite is areally restricted to the North American and Baltic shields (including the Russian platform), despite the presence of large areas of stable continental crust elsewhere in the world.

Meanwhile, back on the surface, sedimentary and volcanic accumulations record the multiple effects of sediment provenance, the distribution and composition of oceans, and the composition of the atmosphere (Sections 5.5 and 5.6). With some exceptions, Proterozoic sediments are similar to Phanerozoic ones. Accumulation of oxygen in the atmosphere at ~2000 Ma is suggested by the sudden decrease in abundance of banded iron formation after that time, and also (possibly) by changes in compositional profiles through Proterozoic soils. An increase in the abundance of carbonate rocks toward the later Proterozoic probably results from an increasing oceanic pH (more basic) and/or an increase in biologic activity.

The surface of the Proterozoic earth was dominated organically by bacteria (Section 5.7). Apparently they occupied terrestrial and aquatic niches, anaerobic and aerobic environments, and probably environments with a wide range of oceanic and atmospheric composition and temperature. Whether metazoan (higher animal and plant) life was significant during the Proterozoic is uncertain, but the bacteria were assuredly busy. Bacteria are prokaryotic (without a cell nucleus), but eukaryotic cells (with a nucleus) almost certainly evolved during the Proterozoic, although perhaps only as algae and fungi.

We continue the chapter on the Proterozoic with a discussion of the compositions of atmosphere and oceans (Section 5.8). After the establishment of an oxidizing atmosphere, the major issue was the concentration of O_2. As shown in Section 6.2, the partial pressure of O_2 may have reached only a few percent of its pre-

INTRA-CRATONIC OROGENIC BELT (?)

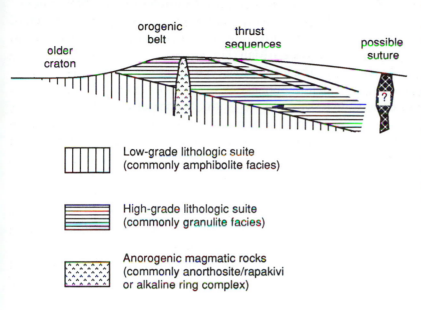

older craton

orogenic belt

thrust sequences

possible suture

?

Low-grade lithologic suite
(commonly amphibolite facies)

High-grade lithologic suite
(commonly granulite facies)

Anorogenic magmatic rocks
(commonly anorthosite/rapakivi
or alkaline ring complex)

Fig. 5.1. Diagrammatic representation of possible differences between intra-cratonic and inter-cratonic orogenic belts. Intra-cratonic orogenic belts develop without the closure of an ocean basin. Thus, they do not contain ophiolites or other deep-ocean materials and commonly are represented by thrusting of one continental terrane over another continental terrane. If compression in intra-cratonic belts is generated by continent collision resulting from ocean closure, the suture is outside of the intra-cratonic belt itself. Conversely, an inter-cratonic orogenic belt contains preserved oceanic sediments, basaltic crust and/or ophiolites. The problem of generating compressive stress without ocean closure is a major drawback to the concept of intra-cratonic orogeny. ☐

INTER-CRATONIC OROGENIC BELT (?)

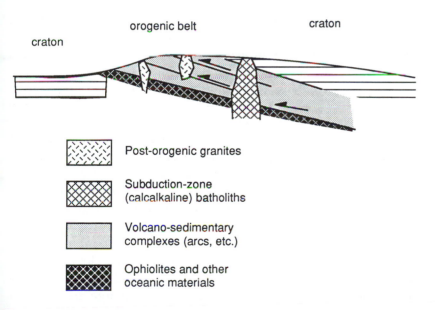

craton

orogenic belt

craton

Post-orogenic granites

Subduction-zone
(calcalkaline) batholiths

Volcano-sedimentary
complexes (arcs, etc.)

Ophiolites and other
oceanic materials

sent level by the end of the Proterozoic, but reducing gases such as ammonia and methane probably had already declined to very small amounts, if present at all, throughout much of the Proterozoic. Similarly, composition of sea water had probably become comparable to that of modern oceans at least by the middle of the Proterozoic.

The final section (Section 5.9) is a brief summary of the major events of the Proterozoic. We emphasize the significance of activities at ~2000 Ma, when the earth apparently underwent internal and external reorganization, and at the end of the Proterozoic, when the remarkable transition to skeletal fauna occurred.

5.1 The nature of Proterozoic orogenies

ONE of the enduring controversies regarding Proterozoic orogenic belts is the possibility that many of the deformed zones developed 'within' a continent rather than by subduction and destruction of oceanic lithosphere, which commonly involves collision of continental blocks (Fig. 5.1). In theory, the distinction between intra- and inter-continental orogenic belts also permits us to distinguish zones of reactivation and deformation of old crust from zones of marginal growth and expansion of continents. The issues, however, are complicated by availability of geologic data, by perceptions and biases of investigators, and by semantics. The ensuing discussion is a complex intermingling of controversial tectonic principles and information about individual areas. Here are some road signs. Section 5.1 discusses the general principles of the distinction between inter- and intra-continental orogenic belts and areas of rapid outgrowth of continental margins (Figs 5.1and 5.2). Some of the supporting illustrative material for Section 5.1 is in Section 5.2, which discusses Proterozoic orogenic belts in four different areas (see Section 5.0). Maps of these areas and illustrations of several of the individual belts are shown in Figs 5.3 to 5.6.

Inter- and intra-cratonic orogenic belts

The distinction between varieties of orogenic belts is shown diagrammatically in Fig. 5.1. The lower part of the diagram illustrates the typical features of presumed collisional zones, characterized by a thick wedge of oceanic crust that includes: ophiolites and other oceanic sedimentary/volcanic assemblages; island-arc volcanic suites; and sedimentary deposits of the arc, trench, foredeep, and associated basins. The volcano–sedimentary sequence contains 'calcalkaline' batholiths, and the entire orogenic belt is commonly intruded by small granite plutons formed immediately after the cessation of orogeny (Section 4.1). Both the batholiths and the postorogenic granites exhibit isotopic ratios prohibiting derivation by melting of upper continental crust or sediments derived by erosion of continental cratons. Many belts are bordered on both sides by older continental blocks, with the block above

the subducted slab thrust over the deformed oceanic/arc complex. Examples of this 'inter-cratonic' belt are the middle-Proterozoic Singhbhum orogen of eastern India (Fig. 5.5) and the late-Proterozoic orogen of western Mali (Fig. 5.4).

The alternative type of orogen is an 'intra-cratonic' deformational belt (upper part of Fig. 5.1). The characteristic feature is an absence of ophiolitic and other oceanic and island-arc assemblages. Continental rocks are thrust over other continental rocks, with the main thrust zone separating suites of greatly different metamorphic grade. Commonly the over-riding suites are in granulite facies, although lower-grade assemblages occur in some belts. In at least one belt (the Eastern Ghats of eastern India; Fig. 5.5), the margin between granulite- and amphibolite-facies rocks is within the underthrust block. Zones of far-travelled thrusts are common within the overriding block. Subduction-zone magmatism is absent, and the extensive post-orogenic granites intruded into compressed oceanic rock assemblages are not developed. Examples of this 'intra-cratonic' orogeny include the early-Proterozoic Limpopo belt of southern Africa (Fig. 5.4), the middle-Proterozoic Eastern Ghats belt of India (Fig. 5.5), and possibly the late-Proterozoic Grenville belt of North America (Fig. 5.3).

The diagrams shown in Fig. 5.1 are subject to a variety of interpretations. Here are some of the major problems. First, does the phrase 'intra-cratonic orogeny' have any meaning? Although evidence of ocean closure may not be found within the orogenic belt itself, the lateral compression necessary for development of the belt must be generated somewhere. Thus, the shoving of one continental assemblage over another one may simply be the result of ocean closure outside of the belt itself. The difficulty in Proterozoic belts is that some orogens have very large and apparently stable cratons on both sides, requiring ocean closure at great distances from the deformed zone itself. If these marginal cratons were actually fused before the deformation of the orogenic belt, and if compression was caused by subduction of oceanic crust, then compressive stress must have been transmitted over very large distances. Semantically, can such orogenic belts be regarded as 'intra-cratonic'?

A second problem in Fig. 5.1 is the possibility that evidence for the closure of oceanic basins is actually preserved in the alleged 'intra-cratonic' belts. Many of these belts contain small, discontinuous, layers of amphibolitic or serpentinitic rocks that may represent metamorphosed former oceanic crust or graywacke assemblages. Do these suites demonstrate the closure of ocean basins? Are they simply metamorphosed continental basalts or flysch-facies sediments formed on thin continental crust? Recent discovery of possible oceanic assemblages and sutures has resulted in the recent reclassification of many 'intra-cratonic' belts into the more familiar zones of oceanic consumption between continental blocks. A primary example of this change in interpretation is the conversion of the former 'Churchill' province of Canada, which had been considered as a zone of reactivation of Archean crust, into an area of closure of ocean basin between the separate Rae and Hearn provinces (Fig. 5.3).

A third problem in the categorization of Proterozoic orogenic belts is possible differences in interpretation caused by different levels of exposure in the various belts. In particular, closure between continental blocks may eliminate preserved ophiolites and arc assemblages at depth, essentially 'squeezing' them out and upward and showing only a thrust relationship between the cratons at deep levels within the orogen. Two arguments may be used to show that the differences proposed in Fig. 5.1 are not simply functions of depth of exposure. One is the absence of calcalkaline batholithic suites and post-orogenic granite plutons in areas that do not preserve evidence of oceanic consumption. If present within the orogen, these suites should be preserved at the levels of exposure in the postulated 'intra-cratonic' orogens. A second argument is the concentration of anorthosite/rapakivi suites and alkaline ring complexes only in areas that do not show oceanic closure; their preservation should also be independent of depth of exposure.

A fourth problem in interpreting Proterozoic orogeny is the difficulty of generating compressive deformation in 'intra-cratonic' belts. Such a process requires the development of a detachment surface separating upper-level rocks, which undergo compression, from lower-level rocks that are 'subducted' and destroyed. Thus, major shortening across the orogen requires destruction of large quantities of lower continental crust, a process whose feasibility is uncertain.

A potential method for distinguishing belts of oceanic closure from those formed by reactivation within a craton is the relationship between possibly correlative features on either side of the orogen. Oceanic opening and closure should, at least theoretically, prevent tracing of features across the orogen from one marginal block to the other. Conversely, pure crustal reactivation within a craton might permit such a correlation. A principal area for the attempt to correlate old orogenic belts across younger ones is Africa (Fig. 5.4). For example, the Irumide orogen intersects the Ubendian orogen without any offset of the older belt, an observation that has been used to indicate that the Irumide belt is merely an area of crustal reworking rather than a closure between blocks.

Continental-margin orogenic belts

Several broad areas of continental crust have been regarded as zones of marginal growth of the continents during the Proterozoic. These zones are 'inter-cratonic' in the sense that they developed along the margin between continental and oceanic lithosphere, although they did not result from collision of two cratons. The ones most commonly placed in this category are the 1900- to 1400-Ma terranes of the southern mid-continent of the United States (Fig. 5.3), the Svecofennian belt of the Baltic shield (Fig. 5.3) and the Nubian–Arabian shield (Fig. 5.4). Fig. 5.2 is a composite of relationships in these three areas.

The information obtained from our three examples of marginal continental growth is highly variable in quality and quantity. The Nubian–Arabian shield is split by the Red Sea. It is well exposed in desert areas along the coastal plain of the Arabian peninsula, on the Arabian plateau, and in the Red Sea Hills of Egypt and Sudan; much of the shield, however, has received only limited investigation. The Svecofennian belt is partly covered by the Pleistocene glacial deposits, and where they are absent the dense northern forests provide a further challenge to investigation. The southern US provides even more difficulties, for the Precambrian is exposed only toward the west (Yavapai and Mazatzal terranes), where it is involved in intricate Phanerozoic events. In the south-center of the continent, nearly the entire granite–rhyolite terrane is buried under Phanerozoic platform sediments, and inferences must be made from sparse basement samples brought up by drilling.

The basic structure of a belt of marginal growth of

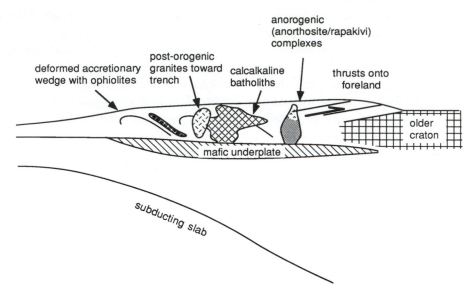

Fig. 5.2. Diagrammatic representation of continental-margin accretionary belt. Symbols are as shown in Fig. 5.1 plus post-orogenic granite (conventional granite symbol) and anorthosite/rapakivi complex (rapakivi granite above mafic rocks). The diagram shows the outward construction of an old craton during subduction of oceanic lithosphere. The accretionary wedge is partly thrust onto the craton but accretes primarily toward the ocean. Magmatic rocks formed during subduction are primarily calcalkaline batholiths. Younger magmatic rocks are largely post-orogenic granites possibly formed by partial melting of the mafic underplate. Crustal stabilization proceeds 'outward' from the craton, and anorogenic magmatism (anorthosite/rapakivi complex) occurs in crust that was stabilized early in the accretionary process.

new continental crust is shown diagrammatically in Fig. 5.2. The major components of crust exposed at the present surface are thick sequences of continental- and arc-derived (immature) sediments interlayered with volcanic flow and pyroclastic units. The volcanic suites tend to be bimodal in SiO_2 contents rather than andesite-dominated (calcalkaline; Section 4.1). The volcano–sedimentary assemblages are interspersed with ophiolitic and other rocks from ocean floors or back-arc basins along zones that can be regarded as local sutures (Figs 5.3 and 5.4). The immaturity of the sandstones and the bimodality of the volcanic rocks signifies accumulation of much of the belt on oceanic lithosphere.

The volcano–sedimentary suites are invariably intruded by calcalkaline batholiths. These batholiths range from the relatively granite-poor suites of intra-oceanic island arcs to the granite-dominated assemblages of Andean-type continental margins (Section 4.1). Most batholithic rocks show little or no evidence of interaction with old continental crust, and presumably they represent addition of new sialic material to the growing continents. The oldest batholiths appear

to be only slightly younger than their wallrocks. In the Svecofennian belt, batholiths become younger toward the southwest as the margin grows in the same direction and the subduction zone moves farther away from the Archean core of the Baltic shield. Subduction appears to have been directed downward beneath the craton in all three of the shields discussed here.

An important characteristic of the marginal belts is the abundance of undeformed plutons that intruded the areas both post-orogenically (within a few tens of millions of years after the end of compression) and anorogenically (one hundred million years or more after stabilization). The post-orogenic granites are particularly numerous in the Nubian–Arabian shield. In the central US, some post-orogenic suites are associated with ring complexes, and undeformed to slightly deformed sequences of rhyolite and interbedded quartzite were deposited locally on the newly cratonized terrane (the buried 'granite–rhyolite' terrane of the midcontinent). Anorogenic granites are associated with anorthosites and/or alkaline rocks and commonly were intruded one hundred or more million years after the post-orogenic granites. In both the cen-

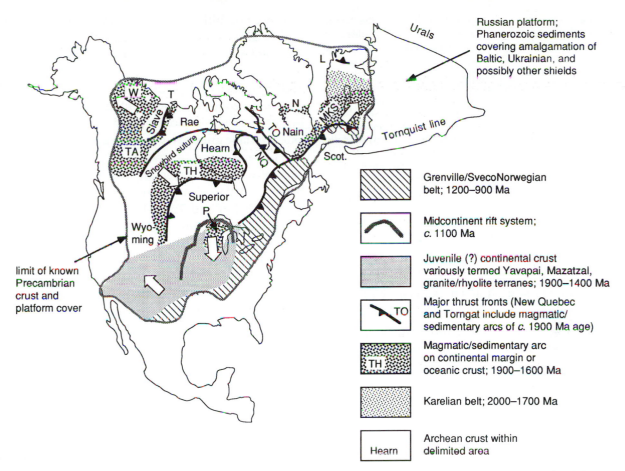

Fig. 5.3. Proterozoic mobile belts in the North Atlantic craton (North America, Greenland, northernmost Scotland and the Baltic shield and possibly farther south in Russia). This configuration of continents and continental fragments probably was attained in the middle Proterozoic.

Many of the belts appear to be magmatic/sedimentary arcs accreted onto a continental margin and associated with frontal thrust systems; the map symbol shows where broad areas of these juvenile (mantle-derived) arcs accumulated. White arrows show the inferred directions of subduction of the descending slabs. Symbols for the various arcs are: L, Lapland granulites; MKS, Makkovian–Ketilidian–Svecofennian; N, Nagssugtoqidian; NQ, New Quebec; P, Penokean; T, Thelon; TA, Taltsin; TH, Trans-Hudson; TO, Torngat; W, Wopmay.

Separate patterns are used for: 1) the broad area of juvenile (?) crust, that apparently represents exceptional crustal growth during the middle Proterozoic in the southern part of the mid-continent ('granite–rhyolite terrane' and associated suites); 2) the Grenville belt of middle- to late-Proterozoic age; and 3) the Karelian belt of early- to middle-Proterozoic age.   ☐

tral US and the Svecofennian belt, anorogenic magmatism occurred closer to the Archean craton at approximately the same time as post-orogenic granites were forming in younger crust away from the craton. The implication is that a 'wave' of continental stabilization and thickening continued after deformation ceased and moved outward from the craton through time.

The ages of orogenic activity are different in the three areas combined for Fig. 5.2. The oldest is the Svecofennian belt, in which subduction occurred in the range of 1900 to 1700 Ma and anorogenic (anorthosite/rapakivi) magmatism at ~1600 Ma. The central and southwestern US encompass a broad range of ages, with island-arc formation during ~1900 to ~1700 Ma and anorogenic magmatism and other events from ~1600 to ~1400 Ma. The Nubian–Arabian shield is the youngest, consisting of arc and batholithic suites younger than ~1000 Ma to ~700 Ma, post-orogenic activity beginning close to 600 Ma, and anorogenic intrusions occurring throughout much of the Phanerozoic.

[References – The discussion of the Proterozoic of the North American and Baltic areas is based on material in: volumes edited by Moore, Davidson and Baer (1986), Gaal and Gorbatschev (1987) and Gower, Rivers and Ryan (1991) and papers by Lindh (1987) on the Baltic shield and Karlstrom and Bowring (1988) on the southwestern United States. Information on the Nubian–Arabian shield is in Rogers *et al.* (1978), Johnson, Scheibner and Smith (1987), Stoeser and Stacey (1988), Rogers (1991) and Stern and Dawoud (1991). Discussion of specific orogenic belts is provided for the Damaran belt in the volume edited by Miller (1983), on the Limpopo belt in the volumes edited by Van Biljon and Legg (1983) and Van Reenen *et al.* (1992), and on the Damaran belt by Kukla and Stanistreet (1991). Bickford (1988) gives a summary of continental evolution.]

5.2 Areas of Proterozoic orogeny

AS DISCUSSED in Section 5.0, the history of continents during the Proterozoic can be described in terms of four different sialic blocks (North Atlantic, South America/Africa/Arabia, India/Australia/Antarctica, and eastern Asia). We will outline their histories separately and summarize the information at the end of the section in a more general discussion that compares Proterozoic orogenies worldwide.

North Atlantic craton

Proterozoic rocks that apparently evolved in one coherent craton are now separated in North America, Greenland, northernmost Scotland, and the Baltic shield (and possibly farther south into the Russian platform). Fig. 5.3 reconstructs this craton as it existed near the end of the Proterozoic and shows the various orogenic belts that evolved during the Proterozoic. With the exception of depicting a general area of exposure of the Ukrainian shield, no effort has been made to reconstruct the Precambrian basement beneath the Russian platform.

The tectonics shown in Fig. 5.3 are quite different from those that would have been depicted on a map drawn only ten years ago. At that time, the major part of the Canadian shield was divided into the Superior and Slave Archean cratons, the late-Proterozoic Grenville belt, and a broad 'Churchill province' between the Superior and Slave cratons. The Churchill province was widely regarded as an area of reworking of Archean continental crust (an intra-cratonic orogeny) during the Hudsonian event, ~1800 million years ago.

The Churchill province is now generally divided into at least two major provinces, the Rae and Hearn. They are separated by the Snowbird 'suture' zone, a belt of mylonites and other evidence of both lateral and compressive deformation up to 25 km wide. The Snowbird zone does not contain ophiolites or other evidence of ocean closure, and the terranes on each side are sialic gneisses of the Rae and Hearn provinces. The other margins (away from the Snowbird zone) of each province, however, are collisional zones with the Slave and Superior provinces respectively. These zones (Thelon and Trans-Hudson) were sites of consumption of oceanic lithosphere during the period ~1900 to ~1600 Ma, and that age is regarded as the age of suturing of the Rae and Hearn provinces.

Most of the orogenic belts shown in Fig. 5.3 developed approximately during the period of 1900 to 1600 Ma. In North America, all belts of this age are now regarded as locations of closure of cratons and consumption of oceanic lithosphere, with compressional deformation largely restricted to the period of ~1900 to ~1800 Ma and younger ages represented by development of post- and anorogenic magmatic rocks. The similarity of these ages throughout North America implies suturing of the Slave, Rae, Hearn (and Wyoming?), Superior and Nain provinces by ~1700 Ma. No inference can be made about the distances between these Archean blocks prior to that time.

The Grenville belt has undergone numerous changes in interpretation during the past few decades. Many of the early plate-tectonic interpretations regarded it as a microcontinent that collided with cratonic North America, forming the southeast-dipping Grenville front (thrust) that brings granulite-facies rocks up over amphibolite-facies rocks. This viewpoint foundered on the observations that numerous units of the craton can be traced across the Grenville front from the adjoining Superior province. Modern interpretations of the Grenville province require: 1) a 'parautochthonous' (slightly moved) terrane up to 150 km wide southeast of the front, in which older rocks of the Canadian shield were metamorphosed beginning at ~1100 Ma and thrust upward over the craton; and 2) highly allochthonous terranes southeast of the parautochthon containing continental-margin suites. Prior to the metamorphic event, the Grenville crust was the site of platformal or rift-related sedimentation and volcanism and the intrusion of anorthosite/

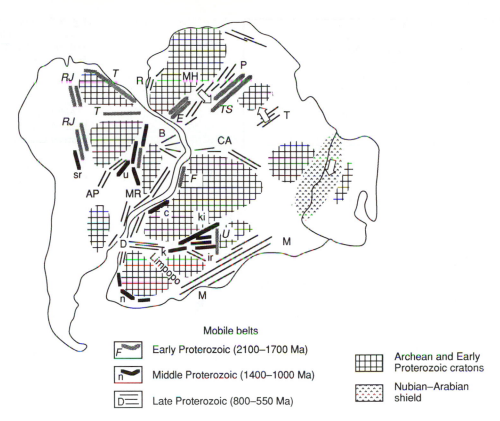

Fig. 5.4. Proterozoic mobile belts in South America, Africa, and Arabia. The continents are shown in the assembly used for Fig. 3.2. This configuration, however, may not have been attained until the end of the Proterozoic or early Paleozoic (see Section 7.0). The white arrows show proposed downward directions of subduction of oceanic crust during the late-Proterozoic (Pan-African, or Braziliano) event. Most of the late-Proterozoic belts are not shown with subduction directions because of the unresolved problem of inter-cratonic vs intra-cratonic orogeny. The only distinction attempted in this figure is between belts of different apparent ages of deformation. The late-Proterozoic Nubian–Arabian shield is shown with a separate pattern in order to emphasize its separation as juvenile, arc-derived, crust. The Limpopo belt was formed at ~2700 to ~2650 Ma.

Symbols are as follows. Early Proterozoic: E, Eburnian; F, Francevillian; RJ, Rio Negro-Juruena; T, Transamazonian; TS, Transsaharan; U, Ubendian. Middle Proterozoic: c, West Congo; ir, Irumide; k, Katangan; ki, Kibaran; n, Namaqua–Natal; sr, Sunsas–Rondonian; u, Uruacuano. Late Proterozoic: AP, Araguaia–Paraguay; B, Borborema; CA, Central African; D, Damaran; M, Mozambique; MH, Mali–Hoggar; MR, Mantiqueira–Ribeira; P, Pharusian (West Hoggar); R, Rokelide; T, Tibesti.

rapakivi–granite suites (Section 5.4). The anorthosites are anorogenic with regard to older rocks of the belt but precede the Grenville metamorphism so closely that they are presumably related to thermal effects associated with crustal thickening. The Grenville and SvecoNorwegian belts are probably correlative

The Makkovik orogenic belt in northeastern North America is correlated with the Ketilidian of Greenland and the Svecofennian of southern Scandinavia. The concept of the Svecofennian as an area of outgrowth along a continental margin is discussed in Section 5.1. The Svecofennian orogeny overlaps, but is somewhat younger than, deformation of platformal sediments in the Karelian orogenic belt (Fig. 5.3), which may be regarded as the (miogeoclinal) cratonward extension of the Svecofennian orogen, with a former continental margin at the join between the Karelian and Svecofennian areas.

South America/Africa/Arabia

The present plates of South America, Africa and Arabia were once joined along zones largely occupied by orogenic belts of Pan-African age (end Proterozoic; ~600 Ma; Fig. 5.4). Fig. 5.4 shows three definable periods of orogeny within an assembled South America/Africa/Arabia. Because of the strong Pan-African overprint and the scarcity of data in many

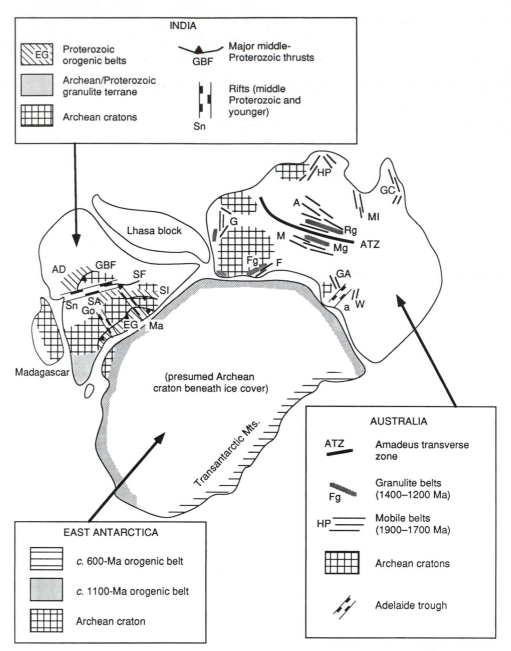

Fig. 5.5. Proterozoic mobile belts and rifts in India, Australia, and Antarctica. The continental assembly is probably one achieved during the middle to late Proterozoic. Symbols are as follows.

India. Proterozoic orogenic belts: AD, Aravalli–Delhi; EG, Eastern Ghats and bordering Eastern Ghats front; SA, Satpura; SI, Singhbhum and bordering Singhbhum (Copper Belt) thrust. Other major thrusts: GBF, Great Boundary fault; SF, Son fault. Rifts: Go, Godavari; Ma, Mahanadi; Sn, Narmada–Son; .

Australia. Mobile belts: A, Arunta–Granites–Tanami–Warra munga; F, Fraser; G, Gascoyne (part of Capricorn); GA, Gawler; GC, Georgetown–Coen; HP, Halls Creek–Pine Creek; M, Musgrave; Ml, Mount Isa; W, Willyama (Broken Hill). Granulite belts: Fg, Fraser; Mg, Musgrave (Woodroffe zone); Rg, Redbank. Rift: a, Adelaidean rift and orogen.   □

areas, however, the correlations for early- and middle-Proterozoic events are not reliable, and additional information may well indicate a near continuity of deformation at some place in the area shown throughout most of the Proterozoic.

The Pan-African (largely 700 to 600 Ma) is one of the most extensive orogenic episodes in the history of the earth. It is displayed in Africa and South America, in the Nubian–Arabian shield, in fragments of continental crust that accreted both to North America and Europe during the Paleozoic (Section 7.3), and possibly in such apparently isolated areas as the Lake Baikal region of eastern Asia (see below). It coincides with, or immediately precedes, the transition from Precambrian to Cambrian, and the major late-Proterozoic period of glaciation may have been related to worldwide orogeny (Section 5.7).

The Pan-African areas of Africa may be regarded as the 'type area' for the discussion of intra- and inter-cratonic orogeny (Figs 5.1 and 5.7). Subduction and other evidence of oceanic closure is prominent throughout much of North Africa and the Nubian–Arabian shield, but elsewhere the situation is unclear. Most Pan-African belts are compressed between two or more cratons, and one can walk for tens or hundreds of miles across such areas encountering only silicic rocks that appear to have been pre-existing granitic gneiss or sediments deposited on it. Isotopic data commonly confirm a crustal precursor for both metamorphic and igneous rocks. Conversely, apparent oceanic lithosphere occurs along some thin zones, and the possibility of oceanic closure cannot be dismissed.

India/Australia/Antarctica

If viewed from space, the continents of India, Australia and Antarctica would show virtually no similarity. The Indian peninsula is green to brown, ranging from dense forests to semi-arid and desert regions. Australia appears green only around the coasts and in the mountainous region of the Paleozoic orogenic belts on the east (Section 7.5). The 'red center' of Australia is an enormous expanse of semi-arid flatlands, where 'mountain ranges' may have a relief of only a few hundred feet and outcrops appear sparsely through the alluvium. East Antarctica, naturally, is mostly white, with investigations conducted only with great difficulty around the coast and along the belt of Transantarctic Mountains that separates the Archean craton of the east from accreted terranes in the west (Section 9.6).

Despite these differences, strong evidence suggests that these continents were fused together at some time during the Proterozoic (Fig. 5.5). The most compelling evidence is the similarity of cooling ages of ~1100 Ma in the Eastern Ghats of India and the adjacent late-Proterozoic belt of Antarctica (Rayner complex). This join is also consistent with a continuity between the Archean complex on the tip of Antarctica (Napier complex) and the adjacent Archean craton of southern India. The evidence for fusion of Australia to India/Antarctica prior to the Paleozoic is less clear. It consists mainly of possible correlation of the Fraser belt of Australia into the Antarctic and possible correlation of late-Proterozoic activity along the Transantarctic Mountains with activity preceding the major Phanerozoic deformation of eastern Australia.

The orogenic belts of India are shown simply as 'Proterozoic' in Fig. 5.5 because of uncertainties and wide ranges in their ages. For example, the Aravalli–Delhi belt of northwestern India contains sedimentary sequences that were deposited throughout much of the early and middle Proterozoic and deformed certainly in the middle Proterozoic and possibly much later. Other belts appear to have been active mostly in the middle Proterozoic, although ages (possibly cooling or reactivation) of ~1000 Ma have been measured. The ages of the major thrusts are presumed to be similar to the ages of deformation in the adjacent orogenic belts, but they are determined largely by their effects on middle-Proterozoic platform suites and are not well constrained. The Eastern Ghats is tectonically similar to the Grenville belt, including intra-continental movement on the bounding thrust and uplift of granulite-facies rocks previously deposited in a platformal setting.

Australia appears to show a very different pattern of Proterozoic orogenies than is found in India. Virtually all of the central and western part of the continent not covered by Phanerozoic sediments was affected by compressive deformation in the time range of ~1900 to ~1700 Ma (Fig. 5.5). Assuming that these orogenic belts connect beneath the younger cover, a likelihood based on geophysical data, then most of the Australian continent was traversed by compressive orogens that separated older cratons into a patchwork of isolated blocks at this time. Many investigators regard them as embryonic rifts that opened far enough to undergo mafic underplating from the mantle, accumulated thick sequences of rift and platformal sediments, and then closed back upon themselves. A younger Proterozoic event, starting at ~1400 Ma, reworked the older belts, caused granulite metamorphism at depth, and uplifted the high-grade terranes along thrust faults within the older Proterozoic areas.

The major tectonic join in Australia is the Amadeus transverse zone (Fig. 5.5). This zone underlies the Amadeus sedimentary basin, which separates the Musgrave and Arunta blocks, and probably extends for considerable distance on either side. It appears to have formed by a combination of strike-slip and thrust faulting and may be analogous to the Snowbird suture of the Canadian shield (Fig. 5.3).

The coasts of eastern Antarctica expose orogenic belts that surround the probably fused set of Archean terranes that underlie the great Antarctic icecap (Fig. 5.5). Orogenic activity was largely in the age range of 1100 Ma, but a younger, roughly 600-Ma, belt has been found

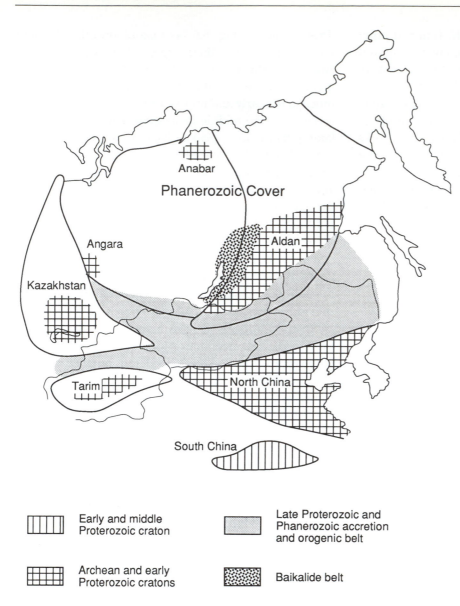

Fig. 5.6. Proterozoic events in eastern Asia. Geologic province boundaries and national borders are the same as on Fig. 7.5. The South China Proterozoic craton is shown detached from the rest of eastern Asia as an indication of its evolution in a remote area. Archean and early-Proterozoic cratons are shown over broad areas of outcrop, although the North China block is partly covered by Phanerozoic sediments, and Archean crust is mostly inferred under sediments of the Tarim basin. The Baikalide belt is apparently late Proterozoic and has been regarded as 'Pan-African' by many investigators.   ☐

Early and middle Proterozoic craton

Late Proterozoic and Phanerozoic accretion and orogenic belt

Archean and early Proterozoic cratons

Baikalide belt

to be the earliest sign of orogenic activity along parts of the dominantly Paleozoic Transantarctic Mountains.

Eastern Asia

Fig. 5.6 is a much simpler map than those of other areas covered in this section. Some of the reasons are human-related – the Siberian winters, the inability of geologists from elsewhere in the world to read Russian or Chinese. Some of the reasons are related to exposures – for example, extensive Phanerozoic cover obscures the tectonic relationships among the three major shields of Siberia (Aldan, Anabar, and Angara). Some of the reasons, however, may be that the

Proterozoic tectonic history of the area really is simpler than those of other continental blocks.

Archean and early-Proterozoic crust is exposed in six areas of northeastern Asia: the Anabar and Angara shields, which are commonly assumed to be connected into a single paleoplate at depth; the Aldan shield and Kazakh crystalline nucleus, in both of which apparent Archean is scattered through intensely deformed younger rocks; and the North China block (Sino-Korean craton) and Tarim cratons. The suturing of North China and Tarim against Siberia in the Paleozoic is discussed in Section 7.2, and our indication of comparable Proterozoic histories for North China and Siberia is almost certainly unreasonable. The South China

Fig. 5.7 Anatectic granites in metasedimentary/metavolcanic suites of the Pan-African Damaran orogen, Namibia. □

(Yangtze) craton appears to contain no crust older than early Proterozoic; apparently it evolved elsewhere and was sutured against the rest of Asia at a much later time.

Based on the limited data available, the cratons shown in Fig. 5.6 (excluding South China) have very similar histories. They record Archean events partly, to largely, overprinted by early-Proterozoic events in the general time range of ~2300 to ~1900 Ma (to ~1700 Ma locally). By the middle Proterozoic, compressive deformation ceased, rifting began, and deposition of platform and rift sequences became widespread. This comparative stability appears to have lasted into, or through, the latest Proterozoic (Vendian in Russia; Sinian in China). The transition from Precambrian to Cambrian is well displayed in undeformed sequences both in China and Russia (Section 6.1), indicating continued quiet sedimentation into the Phanerozoic.

The Baikalide belt to the east and north of Lake Baikal contains thick sequences of highly deformed sedimentary and volcanic rocks intruded by orogenic (calcalkaline) batholiths (Fig. 5.6). It probably formed by closure of the Aldan craton against a continental block formed by fusion of the Angara and Anabar (plus other?) Archean cratons. The Baikalides have long been regarded as a late-Proterozoic assemblage, and the term has been widely applied to lithologic suites accumulated along active continental margins elsewhere in Eurasia (e.g. early phases of accumulation in the Urals).

Comparison of Proterozoic histories of different continents

Having discussed the Proterozoic development of individual continental blocks, we now summarize the tectonic history of all continental areas together.

Individual orogenic belts generally have not been located in the age range of 2500 to 2000 Ma. This early Proterozoic was a time of consolidation of Archean cratons, with intricate reworking of older rocks, incorporation of new magmatic products, and a mobility of elements that is reflected in resetting of older isotopic systems. The activity in most cratons appears to have been pervasive rather than organized in preserved linear belts. For many cratons, tectonic 'stability' did not arrive until near the end of the early Proterozoic.

The period from 1900 to 1700 Ma seems to have been a time of major crustal evolution. Belts of this age are found in North America/Greenland/Baltica, where they are commonly regarded as inter-continental sutures that cause addition of new material to the crust, and in Australia, where they may represent compression of materials deposited in a complex intra-continental rift system. Belts of the same age occur in other areas but are not as well known as in North America.

The middle Proterozoic, in the range of ~1500 Ma, is

proposed as a major time of compressive orogeny only in India, and there the dating is very meager. Outside of India, the cratons now present in eastern Asia seem to have become stable by ~1500 Ma, the orogeny that formed granulite belts in Australia did not begin until slightly later, and the Kibaran and related belts of Africa and South America are also younger. The later, more anorogenic, stages of southward continental growth in North America overlap the 1500-Ma time, but compression had largely ceased before then. Possibly the middle-Proterozoic dates in India are incorrect, in which case we could propose a middle-Proterozoic hiatus in worldwide continental orogenic activity. Possibly they are correct, and India is simply different from other continental blocks.

The period of about 1300 to 1100 Ma was a time of compression throughout many of the world's continents. Major orogeny began in the Grenville–SvecoNorwegian belts at this time, a 'Kibaran-age' orogeny is widespread throughout South America and Africa, and similar ages are common around the outer edge of Antarctica. Reactivation and/or cooling ages in the Eastern Ghats of India permit correlation of India with Antarctica. The extent of compression in Australia at this time is unclear. Possibly a collisional assembly of most of the world's continental blocks occurred at ~1000 Ma, but more intercontinental correlation is needed before we can be certain.

The last major event assigned to the Proterozoic is the Pan-African orogeny, which continued in some areas into the early Paleozoic. The entire continent of Africa and much of the exposed craton of South America were affected. The Pan-African, however, is not recorded in North America, Greenland, or Baltica except in exotic fragments (possibly from Africa). Similarly, Pan-African ages have been found only sparsely in southern India (the Granulite terrane), and the only possible Pan-African belt in Australia and Antarctica is a precursor to later Paleozoic activity. The Baikalides have been proposed to be a Pan-African orogeny, but other interpretations are likely.

We conclude our discussion of Proterozoic orogenies with two questions. First, are the apparent differences in orogenic histories among the various continental blocks real differences or artifacts based on inadequate data and interpretation? We cannot answer this question without additional geochronologic data and need not discuss it further. Second, are the proposed differences real or imaginary between (1) orogenic belts that form sutures and new crust between old blocks and (2) those belts that rework older crust? Many of the differences in interpretation seem to be related to the home of the geologist making the interpretation, with northern-hemisphere investigators favoring suturing and southern-hemisphere geologists favoring crustal reactivation. Solution to this problem apparently requires understanding of the work done in many different places (hardly a surprising thought).

[References – The discussion of North America/Greenland/Baltica is based partly on volumes edited by Tobi and Touret (1985), Park and Tarney (1987), Lewry and Staufer (1990) and Gower, Rivers and Ryan (1991). Hoffman (1988) provides a general summary of North America. Information for specific areas is from Kalsbeek, Pidgeon and Taylor (1987) for Greenland, Bowring and Karlstrom (1990) for the southwestern United States, and Emslie and Hunt (1990) for the Grenville province.

Information for South America/Africa/Arabia is from a volume on the southern hemisphere edited by Hunter (1981) and papers in a volume edited by Coward and Ries (1986) on collision tectonics. Edited volumes on Africa/Arabia include Fabre (1983) for West Africa and Bowden and Kinnaird (1987) for a broader coverage. Specific papers on Africa are Pin and Poidevin (1987), Key et al. (1989), and Ledru et al. (1989). Cahen et al. (1984) summarize the geochronology of Africa. Information for South America is in a volume edited by Mabesoone, de Brito Neves and Sial (1981) and a paper by Texeira et al. (1989) plus a paper by Porada (1989) on the linkage of Africa and Brazil.

The discussion of India/Australia/Antarctica is based on: Grew and Manton (1986) and the book by Naqvi and Rogers (1987) for India; on edited volumes by Oliver, James and Jago (1983) and Thomson, Crame and Thomson (1991) for Antarctica; and on papers in Wyborn and Etheridge (1988) for Australia.

Information on Eastern Asia is from: books on Russia and Siberia by Nalivkin (1960, 1973) and Khain (1985) and a paper on the Baikalides by Butov, Zanvilevich and Litvinovskiy (1974); and papers on China by H. Wang and Qiao (1984), Zhang, Liou and Coleman (1984), and Jahn, Zhou and Li (1990).]

5.3 Rifting and related processes

THE CONSOLIDATION of continental cratons during the Proterozoic provided large blocks brittle enough to rift. The process involves various mechanisms of rifting (Section 4.2), variation in type of rifts

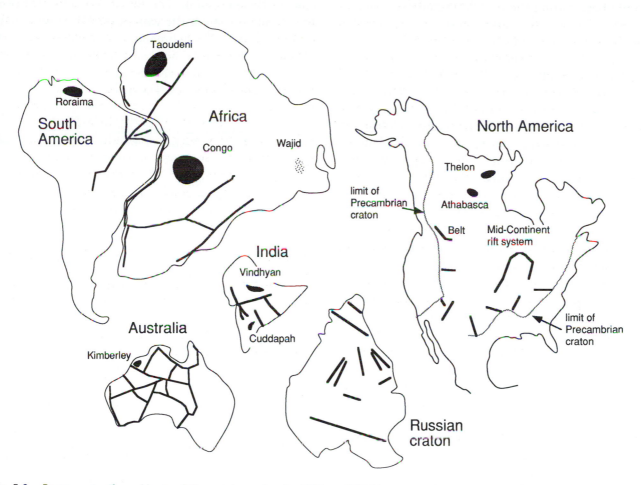

Fig. 5.8. Proterozoic rifts and basins. Rifts are shown for the 1900- to 1700-Ma range in Australia, ~1500-Ma range for India, ~1000-Ma and younger range for North America, and late Proterozoic for South America/Africa and the Russian craton. The difference in rift patterns among the various continents may be related to age of rifting or to some other controlling factor.

Areas designated as 'basins' are ones in which linear underlying rifts have not been demonstrated, although they may be present.

Only rifts and basins referenced in the text are shown here. Other rifts and basins are shown in Figs 5.11 to 5.14; the Tarim basin, referred to in the text, is shown on Fig. 5.6.

from place to place, and possible variations in type of rifting with time.

Proposed major rift sequences in several cratons are shown in Fig. 5.8, together with estimated ages at which the various networks were established; (the various blocks of eastern Asia are not shown because of limited availability of information). All cratonic blocks shown in the figure are at the same scale. In addition to the rifts shown, sets of smaller rifts occur in many areas and can be inferred in others (discussion below).

The oldest rift sequence shown is the ~1900-Ma rifts of Australia, which are shown by gravity highs that testify to the presence of underplated mafic rocks

beneath a thin continental crust. The age is derived from the presumption that unsuccessful rifting (i.e. no oceanic lithosphere was formed) preceded the major intra-cratonic compressive orogeny that affected much of Australia at about 1900 Ma to 1700 Ma (Section 5.2). If the diagram shown for Australia in Fig. 5.8 is valid, then a very closely spaced network of rifts developed during the early Proterozoic within a broad, stable, continental craton. Possibly the close spacing is the result of small convection cells that characterized the high heat flow of this period of time, but such a conclusion must be regarded as highly speculative.

Fig. 5.8 also shows a network of rifts formed in India

at some time during the middle Proterozoic. Two of the rifts follow previous zones of compression (Narmada–Son and Mahanadi; Fig. 5.5), but the Godavari rift appears simply to be a half graben cutting Archean crust. The Indian rifts have remained episodically active for the past 1500 million years, serving also as the locus for Karoo (late-Paleozoic) sedimentation and still showing mild seismic activity. The reason for this long-continued activity, without significant separation across the rift valleys, is unclear but is probably related to some very unusual thermal structure beneath the Indian subcontinent (Section 3.3).

Two possibly distinct types of rifts are shown for North America. The Mid-Continent rift system contains thick sequences of basalts (dated at ~1100 Ma) and sediments that form a pronounced gravity high along the entire western arm of the rift despite burial beneath flat-lying Phanerozoic sediments. The accumulation of basalts may indicate that the western arm of the rift system was beginning to open to an ocean at ~1100 Ma but was interrupted, possibly by the beginning of the Grenville compressional event. The eastern arm of the Mid-Continent rift system underlies the Michigan basin, and several other rifts in eastern North America are also associated with overlying basins that extend far beyond the borders of the rift itself (Fig. 7.2A). In some basins, the rift is not the locus of thickest deposition but merely accentuated the subsidence process over a broad area.

All North American rifts except the Mid-Continent rift system have similar characteristics. They all extend into the craton margin as aulacogens, they are not associated with precursor orogenic belts, and they localize depositional basins much larger than the rifts themselves. The aulacogens apparently have a range of ages from middle to late Proterozoic.

As in North America, rifts of presumed late-Proterozoic age (Riphean, ~800 Ma) on the Russian craton have had significant effects on the formation of basins for the accumulation of Phanerozoic sediments (Fig. 5.8; Section 7.1). All of the rifts shown appear to have formed at the same time and represent a post-stabilization fracturing.

The rift patterns shown in Fig. 5.8 for South America and Africa represent the possibility that Pan-African (Brasiliano in South America) compressional events are intra-cratonic (Section 5.1). Thus, the pattern is drawn to follow the Pan-African belts. The implied history is one of unsuccessful rifting, sedimentation, and final compression. The Brasiliano belts also contain numerous small rift valleys that preserve minor sequences of clastic sediments within the framework of the broader belts. These small rifts may represent adjustments after compression and, thus, be younger than the belts in which they occur.

[**References** – Specific studies of rifting include Ross (1963) for the Belt Series, J. Stewart (1972) for initial rifting along the western margin of North America, Klevtsova (1979) for the Russian platform, Aleinikov *et al.* (1980) for Russia and Siberia, Keller *et al.* (1983) for central North America, van Schmus and Hinze (1985) for the mid-continent rift system of North America, Naqvi and Rogers (1987) for India, and Porada (1989) for Brazil. Papers on the evolution of platform sediments and platformal basins are Bronner *et al.* (1980) for the Taoudeni basin of West Africa, the volume edited by Cameron (1983) for the Athabasca basin of Canada, Greenberg and Brown (1984) for the northern mid-continent of the United States, and Rogers *et al.* (1984) for a general statement of the issues. The volume edited by Hinze and Braile (1988) provides general information on the North American craton.]

5.4 Proterozoic anorogenic magmatism

THE RIGIDITY and antiquity of Proterozoic continental crust provided opportunities for injection and eruption of magmas not directly related to subduction or its aftermath. Such igneous activity is commonly referred to as 'anorogenic' and results from rising plumes or another (commonly unknown) cause of local heating. Anorogenic melts are derived from the asthenosphere, lithospheric mantle, crust, or some combination of source regions. We will review them in three categories: mafic dike swarms, layered intrusions, and anorthosite/rapakivi complexes.

Mafic dike swarms

Although Archean mafic dikes are abundant, they are commonly deformed and do not extend over great distances. In the Proterozoic, however, individual dikes have lengths of 100 km or more, and dike swarms with consistent orientation patterns have lengths up to 2000 km. The dikes cross pre-existing geologic structures without significant offset, indicating establishment of a stable lithosphere before dike injection. The radiating, parallel, and other patterns indicating a single stress field must signify a crustal rigidity that permitted fracturing over such a broad area plus some mechanism of generating consistent stresses within it.

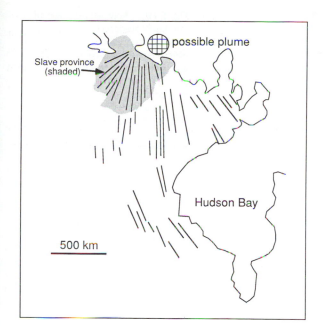

Fig. 5.9. Diagrammatic map of the ~1200-Ma Mackenzie dike swarm, northern Canada. Note the pattern radiating from a point (possibly a plume) that may have been associated with rifting of the northern margin of the Canadian shield. (Simplified from Fahrig, 1987.)   □

Individual dikes vary greatly in size. The longest ones are probably in the Mackenzie swarm (see below). Other very long dikes include the 'Great Dyke' of the Zimbabwe craton (a product of multiple intrusions), the Binneringie dike of the Yilgarn craton and the 'Great Abitibi dike' of the southern Superior province of Canada, all of which have lengths between 500 km and 600 km. Large dikes have widths commonly of several hundred meters, and the Great Dyke of Zimbabwe is up to 11 km wide.

The largest set of dikes known is the Mackenzie swarm of the Canadian shield (Fig. 5.9). It extends over a distance of 1500 km, apparently radiating from some point now in the Arctic Ocean just north of the Canadian coast. Many of the dikes are composite, consisting of several batches of closely related basaltic magmas. The dikes contain at least two separate basaltic batches that cannot be related by fractionation and must represent at least two mantle source regions with compositional differences. The radiating pattern suggests a plume origin, but that possibility has not been demonstrated.

Mechanisms of dike intrusion are controversial. Lateral spreading from a single magma source for hundreds of kilometers seems impossible, but some investigators have proposed that dikes become less mafic and more 'evolved' along strike away from their apparent center of injection. Such compositional variation implies lateral injection. Conversely, the presence of more than one magma batch along a single dike, plus differences between dikes, presumably indicates tapping of different mantle source volumes for the different batches that filled the dike. Individual dikes can be used as pathways by magmas formed at greatly different periods of time.

Ages of dike injection span the Proterozoic. The Great Dyke of Zimbabwe and satellite dikes not much smaller have ages greater than 2400 Ma. The Binneringie and other dikes in the Yilgarn craton were emplaced at about the same time as the Great Dyke. The Mackenzie swarm has an age of ~1200 Ma, and the Abitibi and related dikes ~1100 Ma. Many areas contain intersecting dike sets formed at different times.

Layered complexes

Layered mafic/ultramafic complexes of various sizes do not constitute a major part of the earth's crust, but they are common throughout the world's Archean and Proterozoic terranes. Well-studied Proterozoic examples include the Bushveld complex of South Africa, the Duluth complex of the United States and the Sudbury complex of Canada. Several suites in the complexes have been attributed to meteorite impact (e.g. Vredefort, in the Bushveld complex of South Africa, and Sudbury, Canada).

The long-time, 'classic' example of a layered mafic intrusion is the ~2100-Ma Bushveld complex of the Kaapvaal craton of southern Africa, which contains a greater variety of ore metals and total amount of ore than any other rock suite in the world. The Bushveld suite consists primarily of orthopyroxene- and olivine-gabbros plus anorthositic rocks in a sequence up to 9 km thick. The magmas intruded sediments of the Transvaal Supergroup, one of the cratonic basins developed on the craton beginning in the late Archean (Section 5.5). The area of intrusion is multi-lobed and has an average diameter of approximately 350 km. The complex clearly represents one or more pulses of mantle-derived basaltic magma that underwent fractionation before, during, and after injection. The textures and layered pattern, however, contain many features

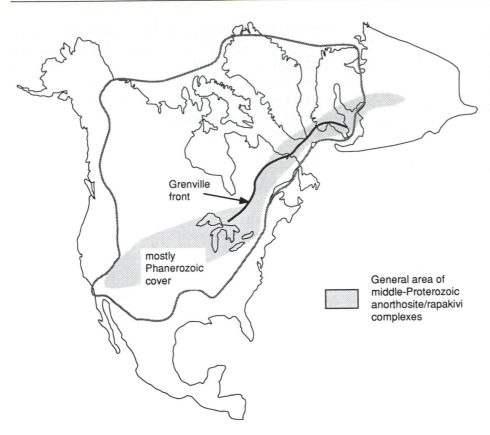

Fig. 5.10. Area of intrusion of middle-Proterozoic anorthosite/rapakivi complexes in North America, Greenland, the Baltic shield and the Russian platform (Fig. 5.3). The limit of known Precambrian crust and the Grenville front are from Fig. 5.3.

The development of middle-Proterozoic anorthosite/rapakivi complexes along the belt shown here, and apparently nowhere else on earth, is a remarkable episode in earth history. It probably is a response to the passage of this supercontinental block over some mantle anomaly that generated the long belt of mafic intrusions, but a convincing explanation is not available.   □

Grenville front

mostly Phanerozoic cover

General area of middle-Proterozoic anorthosite/rapakivi complexes

that must be explained by processes other than traditional crystal settling and accumulation.

In addition to the layered rocks of the Bushveld complex, one part of the Bushveld outcrop area is occupied by the Vredefort dome, an uplifted area about 40 km in diameter. The core of the Vredefort dome exposes Archean basement with equilibration pressures indicating as much as 30 km of uplift. The core is rimmed by uplifted and deformed older sediments and intruded by granitic and alkalic rocks, presumably crustal melts. Because of these features and the abundance of shock-related structures, many investigators have regarded the Vredefort structure as an impact site (Section 2.1). The dome, however, has an age only a few tens of millions of years younger than the Bushveld complex, and other investigators propose a link between the extraordinary Bushveld magmatism and 'internal' (crypto-explosion?) processes that formed the Vredefort dome.

Anorthosite/rapakivi complexes

Bodies of anorthosite with diameters in the tens of kilometers occur throughout the world. Many of them

are not layered and do not obviously have an origin within a layered mafic/ultramafic intrusion. They have, therefore, been regarded as forming a suite of rocks ('massif anorthosites') separate from the layered complexes, but the petrology of the relationship is controversial. We will simply describe their occurrence and some aspects of the tectonic conditions under which they formed. Archean equivalents (?) are discussed in Section 3.5.

Most information about the anorthosite/rapakivi–granite association results from studies of the dominantly anorthositic massifs of the Grenville province (Fig. 5.3) and the dominantly rapakivi granites of the Baltic shield (Fig. 5.3). The term 'rapakivi' means 'rotten rock', but it has been applied broadly to granites in which white sodic plagioclase surrounds pink K-feldspar megacrysts to make the characteristic ovoid texture. Cogenetic relationships between the anorthosites and granites are controversial, with one proposal being that the granite magmas formed by melting of continental crust by anorthosite intrusions.

In addition to the anorthosite/rapakivi massifs, both rock types occur in other localities. Most massif

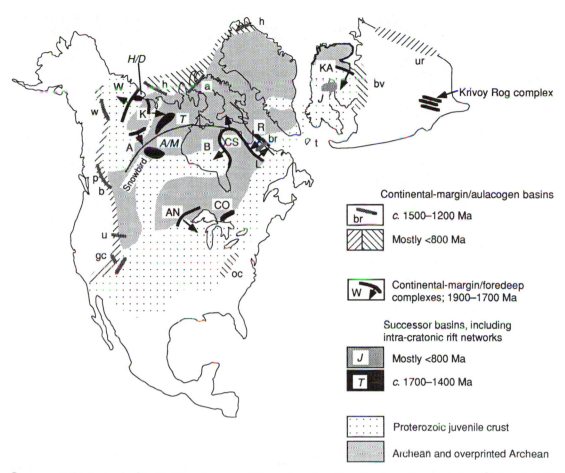

Fig. 5.11. Proterozoic basins in the North Atlantic craton (North America, Greenland, northernmost Scotland, and the Baltic shield and possibly farther south in Russia). (The tectonic relationships of the area are shown in Fig. 5.3.) Basins are divided into four categories: 1) basins developed along undeformed continental margins, including aulacogens that extend deeply into the continent; 2) foredeep basins along deformed continental margins, including both the deep-water (eugeoclinal) and platformal (miogeoclinal) areas; arrows show the direction of thickening of these magmatic/sedimentary complexes; 3) basins formed on unstable crust immediately following orogeny (successor basins); and 4) intra-cratonic basins that may or may not be associated with underlying rifts.

Symbols are as follows. Continental-margin and aulacogen suites (category 1 above): a, Amundsen basin; b, Belt; br, Bruce River; bv, Vendian around Baltic shield; gc, Grand Canyon; h, general Helikian (c. Middle Proterozoic) suites; oc, Ocoee; p, Purcell; t, Torridonian; u, Uinta; ur, Uralian margin of Russian platform; w, Wernecke. Continental-margin/foredeep complexes of ~1900 Ma to ~1700 Ma (category 2 above): A, Athapuscow; AN, Animikie (and basinal materials deformed in Penokean orogeny); B, Belcher; CO, Cobalt and other Huronian suites; CS, Cape Smith; K, Kilohigok; KA, Karelian; L, Labrador trough (Torngat); R, Ramah and other suites of Labrador; W, Wopmay. Successor and other intra-cratonic basins (categories 3 and 4 above): A/M, Athabasca and Martin; H/D, Hornby–Dismal Lake; J, Jotnian; T, Thelon.

anorthosites do not have associated granite. Granite with rapakivi texture occurs in numerous anorogenic intrusions, including 'red rocks' fractionated from mafic layered complexes. The complete association of anorthosite and rapakivi is actually restricted to an area transecting the joined Laurentian and Baltic shields (Fig. 5.10). Within this area, anorthosite and/or rapakivi intrusions were emplaced anorogenically in the general time range of 1700 to 1300 Ma.

Because the massifs of the Grenville belt were metamorphosed, many to granulite facies, at ~1100 Ma, some investigators have regarded them as related to the orogenic process that caused the Grenville deformation. This concept receives some support from observations elsewhere (e.g. southern India; Section 5.2) that anorthosites were emplaced during granulite-facies metamorphism. The Laurentian–Baltic suites, however, appear to have crystallized during an anoro-

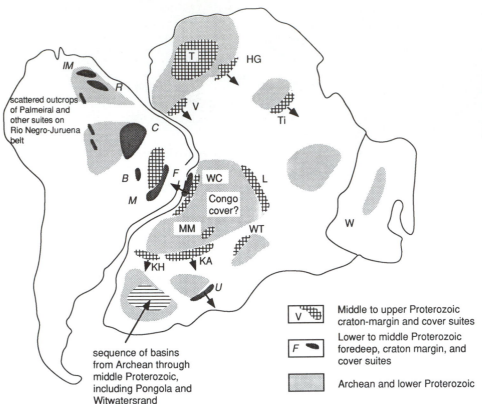

Fig. 5.12. Proterozoic basins in South America, Africa and Arabia. (The tectonic relationships of this area are shown in Fig. 5.4.) Basins are classified primarily by age rather than by inferred origin. The arrows show the direction of thickening of apparent continental-margin complexes.

Symbols are as follows. Lower to Middle Proterozoic: *B*, Bambui; *C*, Carajas; *F*, Francevillian; *IM*, Imataca; *M*, Minas Gerais; *R*, Roraima; *U*, Umkondo. Middle to Upper Proterozoic: *HG*, Hoggar; *KA*, Katangan; *KH*, Khomas (Damaran orogen); *L*, Lindian; *MM*, Mbuji–Mayi; *T*, Taoudeni; *Ti*, Tibesti; *V*, Volta; *W*, Wajid; *WC*, West Congolian; *WT*, West Tanzanian. Arrows show direction of thickening of sedimentary suites off of craton platforms. Archean through Middle Proterozoic basins of southern Africa are considerably superimposed and not depicted separately. □

scattered outcrops of Palmeiral and other suites on Rio Negro-Juruena belt

sequence of basins from Archean through middle Proterozoic, including Pongola and Witwatersrand

Middle to upper Proterozoic craton-margin and cover suites

Lower to middle Proterozoic foredeep, craton margin, and cover suites

Archean and lower Proterozoic

genic period following early Proterozoic orogenies and only locally been incorporated in a later orogenic belt.

[**References** – Mafic dike swarms are discussed in the edited volumes of Halls and Fahrig (1987) and Parker, Rickwood and Tucker (1990); the Mackenzie swarm is shown by Fahrig (1987). The discussion of the Bushveld and Vredefort suites is mostly from papers in Nicolaysen and Reimold (1990). Information on anorthosites and rapakivi complexes is from a volume edited by Lindsley and Simmons (1990), a monograph by Ramo (1991), and papers by Emslie (1978) and Emslie and Hunt (1990).]

5.5 Stratigraphy in Proterozoic basins

PROTEROZOIC basin formation and rifting represent major processes not found in the tectonically defined Archean (Section 3.0). The stabilization of sizable blocks of continental crust provided 'basement' for the accumulation of volcano–sedimentary sequences in platforms and basins of various types. Indeed, the ability to serve as a basement may be the best definition of a 'stable craton'. The basins range in age from ~3000 Ma (chronologically Archean) to latest Proterozoic, gradational into the Cambrian.

Locations of Proterozoic volcano–sedimentary accumulations are shown in Figs 5.11 to 5.14. Each map covers the same assemblage of present continents as the corresponding maps in Section 5.2, namely: North America/Greenland/Baltica (Fig. 5.11); South America/Africa/Arabia (Fig. 5.12); India/Australia/Antarctica (Fig. 5.13); and eastern Asia (Fig. 5.14). Representative stratigraphic sections from eleven of the basins are shown in Fig. 5.15. The degree of detail shown on Figs 5.11 to 5.14 varies greatly from area to area. For example, the amount of information available for the Canadian shield is far greater than can be shown on a summary map.

Varieties and examples of basins

In theory, we can distinguish four types of basins, but in reality the assignment of a Proterozoic volcano–sedimentary accumulation to one category or another is very difficult. The *first* type of basin is represented by

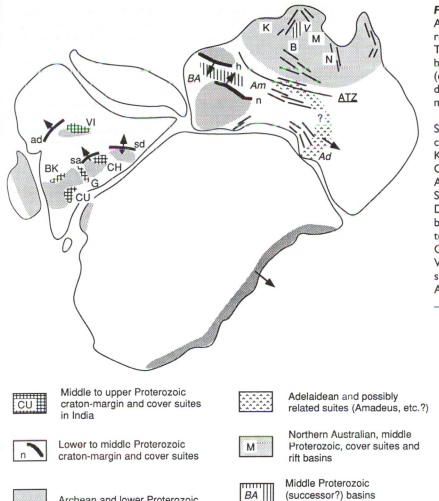

Fig. 5.13. Proterozoic basins in India, Australia, and Antarctica. (The tectonic relationships of this area are shown in Fig. 5.5.) The only attempted tectonic subdivision of the basins is shown in the legend for the figure (compare Fig. 5.11). The arrows show the direction of thickening of apparent continental-margin complexes.

ATZ is the Amadeus transverse zone. Symbols are as follows. Northern Australian cover suites and rift basins: B, Birrundudu; K, Kimberley; M, MacArthur; N, South Nicholson. Other Lower to Middle Proterozoic: ad, Aravalli–Delhi; h, Hamersley; n, Nabberu; sa, Sausar (Satpura orogen); sd, Singhbhum–Dhalbhum. Middle Proterozoic (successor?) basins: BA, Bangemall; V, South Victoria. Middle to Upper Proterozoic: BK, Bhima–Kaladgi; CH, Chhattisgarh; CU, Cuddapah; G, Godavari; VI, Vindhyan. Adelaidean and possibly related suites: Ad, Adelaide basin and Sturt shelf; Am, Amadeus. □

Middle to upper Proterozoic craton-margin and cover suites in India	Adelaidean and possibly related suites (Amadeus, etc.?)
Lower to middle Proterozoic craton-margin and cover suites	Northern Australian, middle Proterozoic, cover suites and rift basins
Archean and lower Proterozoic	Middle Proterozoic (successor?) basins in Australia

deposition along a comparatively undeformed continental margin, possibly recently rifted and commonly accompanied by aulacogens extending into the craton. Examples of continental-margin suites include the late-Proterozoic accumulations around most of the North American and Siberian cratons (Figs 5.11 and 5.14). In North America, aulacogens extending into the continent have localized some basins with late-Proterozoic to Phanerozoic sediments (e.g. Belt and Illinois basins). Similar basins occur in the buried Russian craton, although (for some mysterious reason) the rifts extend farther into the craton than they do in North America.

A *second* type of Proterozoic basin is associated with orogenic belts and includes thick, deformed, eugeoclinal accumulations plus foredeeps and miogeoclinal basins on adjoining cratonic platforms. These belts are shown on their respective illustrations with arrows pointing toward the thicker accumulations, presumably areas originally underlain by oceanic lithosphere.

The *third* type of Proterozoic basin includes those that are successors to orogenic belts. Crust underlying these basins has stabilization ages only slightly older than the oldest sediment in the basin. Typical examples are the Athabasca and Thelon basins of Canada (Fig. 5.11), the Roraima basin of the Guiana craton (Fig. 5.12), and the Wajid basin of the Nubian–Arabian shield (Fig. 5.12). Basins in this category appear to have developed by subsidence of crust stabilized only a few millions or tens of millions of years before the start of basin downwarp and may simply represent an epeirogenic response to vertical instability in a new block of continental crust.

The *fourth* type of Proterozoic basin occurs within continental platforms, commonly localized by rifting

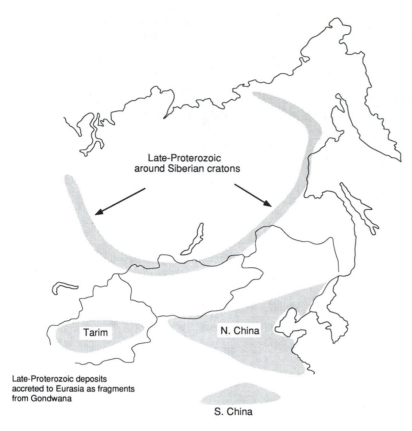

Tarim

N. China

Late-Proterozoic
around Siberian cratons

Late-Proterozoic deposits
accreted to Eurasia as fragments
from Gondwana

S. China

General areas of late-Proterozoic
(Vendian and Sinian) sedimentation

Fig. 5.14. Proterozoic basins in eastern Asia. (Tectonic relationships of this area are shown in Fig. 5.6.) The pattern shows areas of deposits generally classified as 'Vendian' or 'Sinian'. These areas include South China and other cratonic fragments accreted to Eurasia following the Proterozoic. The term 'Sinian' is used throughout China regardless of the fact that South China was presumably remote from North China and Tarim during late-Proterozoic time. Early and Middle Proterozoic deposits occur throughout much of the Siberian, Tarim and North China blocks but are not distinguished here; South China apparently was not a stable craton until the middle Proterozoic.   □

in crust considerably older than the initial deposits of the basins. Examples include the Taoudeni basin of the West African craton, the Witwatersrand basin of the Kaapvaal craton (both in Fig. 5.12), and the Vindhyan basin of India (Fig. 5.13). These basins consist of sediments deposited on basements that have variable age relationships to the age of the oldest sediments. In some places, this crystalline basement is apparently separated from the basinal deposits by thin platform sediments of uniform thickness across much of the craton. The basins may require embryonic rifting in order to develop, but commonly there is no stratigraphic or geophysical evidence to support such an origin.

Variation in place and time

Differences among Figs 5.11 to 5.14 in the categories used to depict Proterozoic basins are probably more apparent than real. For example, numerous orogenic basin complexes of ~1900 Ma have been shown in North America but not on other maps. This concentration of 1900-Ma, eugeoclinal, basins in one area almost certainly represents two factors. One is the much greater availability of geochronologic information for the Canadian shield than for other shields. The second is the tendency for geologists in North America to perceive geologic processes of all ages in terms of Phanerozoic plate tectonics, including continental collisions, in contrast to the more 'intra-cratonic' orientation of geologists from other areas (Section 5.2).

Considering these problems of perception and observation, we can propose, with any confidence at all, only one major difference between different continental areas with regard to the types of basins formed on them. This difference is the direct result of the old age of stabilization (at ~ 3000 Ma) in the Kaapvaal cra-

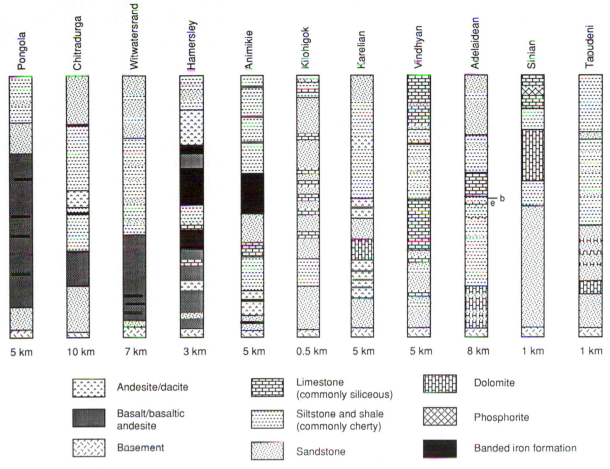

Fig. 5.15. Examples of Proterozoic stratigraphic sequences. Diagrammatic sections are shown for the basins listed below. These sections represent the general lithology and sequence and do not reproduce the exact section at any one place. Figures below each section show general thickness of each suite above crystalline basement. Basins are listed in approximate age sequence (oldest to left).

Pongola (on Kaapvaal craton of southern Africa); deposition initiated at ~3000 Ma.

Chitradurga belt of the Dharwar supracrustal sequences of Western Dharwar craton of southern India; deposition between 3000 Ma and 2500 Ma.

Witwatersrand (on Kaapvaal craton of southern Africa); deposition initiated at ~2600 Ma.

Hamersley (on margin of Pilbara craton of Australia); deposition initiated at ~2100 Ma.

Animikie (on southern part of Superior craton of North America); possibly a foreland basin for the Penokean orogenic belt; deposition initiated at ~2000 Ma.

Kilohigok (on margin of Slave province of North America); possibly a foreland basin; deposition initiated at ~1900 Ma.

Karelian (on southwestern margin of Baltic shield); apparently the cratonic cover associated with basinal accumulation of eugeoclinal deposits of the Svecofennian orogenic belt; deposition initiated at ~1900 Ma.

Vindhyan (on southern margin of Bundelkhand craton of India); deposition initiated possibly at ~1400 Ma.

Adelaidean (on eastern margin of Gawler craton of southern Australia); the section shows initiation of rifting after deposition under platform conditions; b denotes the location of a breakup unconformity; e shows location of Ediacaran fossils; deposition initiated at ~1000 Ma.

Sinian (on South China craton); deposition initiated at ~800 Ma.

Taoudeni basin (on West African craton); deposition initiated at ~700 Ma. □

ton (Fig. 3.2) and Western Dharwar craton (Figs 3.3 and 3.7; Section 3.3). Basins underlain by continental crust began to develop in the chronological Archean (>2500 Ma) in these two areas but have not been found

elsewhere (including the even older Pilbara craton; Section 3.3). In the Kaapvaal craton, a series of basins developed in overlapping locations for a period of more than one billion years, providing an extraordinary opportunity to investigate the time-dependent variation in depositional sequences.

The absence of any demonstrable, comprehensive, difference in the Proterozoic basins of different cratons suggests that we can investigate the possibility of world-wide, time-dependent variations in the nature of volcano–sedimentary accumulations. (Differences between the compositions of Archean and Proterozoic sediments are discussed in Section 2.3.) The sections shown in Fig. 5.15 have been chosen as representative examples of their respective ages, partly in an effort to illustrate these variations through time.

The sections depicted in Fig. 5.15 cover the range of basin types discussed above (details and times of initiation are shown in the caption). *Rifted continental margins* are represented by the Adelaide trough, with pre-rift, syn-rift, and post-rift sediments. The syn-rift sediments contain the type section of the Ediacaran fauna in shallow-water (tidal?) deposits (Section 6.2). *Eugeocline/miogeocline/foredeep complexes* include the Hamersley, Animikie and Karelian suites. The section for the Hamersley sequence is in the deeper parts of the basin, and the Animikie and Karelian suites sections represent the miogeoclinal facies. The thin Kilohigok suite formed by downward buckling of the Slave craton during collision with an eastern terrane and should be included in the same category. *Successor basins* mostly contain thick sequences of coarse clastic rocks. They are represented in Fig. 5.15, however, by the ~3000-Ma, volcanic-rich, Pongola basin and possibly by the Chitradurga belt, which may be partly a rift basin. *Basins formed on old crust* (by rifting?) include the Witwatersrand, Vindhyan, Sinian and Taoudeni, showing an age span of approximately 2000 million years.

The most apparent variations in rock suites with time are described below. Inferences about atmospheric and oceanic composition that can be drawn from the composition and lithology of the rock suites are discussed in Section 5.6.

- Except in the Kaapvaal and Western Dharwar cratons, no basins that retain recognizable stratigraphy are older than ~2100 Ma. Apparently, most continental crust was not sufficiently stable to support coherent basins until that time.
- Volcanic (and volcaniclastic) assemblages are far more common in older basins. This abundance is particularly evident in the very old Pongola basin, where deposition began at ~3000 Ma on an unconformity over crystalline

rocks only slightly older. The high abundance of volcanic material in the older basins may represent some combination of thin lithosphere, hot and unstable crust, and rapid upwelling of magmatic sources.
- Banded iron formations have been deposited throughout almost all of geologic time, but most of the world's accumulations were formed during a few hundred million years in the early Proterozoic, prior to ~2000 Ma (Section 5.6).

A predominant environmental, as opposed to tectonic, control over the deposition of iron formation is shown by comparison of the two principal basins containing iron formations shown in Fig. 5.15. The Animikie basin, of the southern Canadian shield, is probably a miogeoclinal or foredeep basin associated with the evolution of the Penokean orogeny (Section 5.2), which deformed a typically thick, eugeoclinal, volcano–sedimentary suite. Iron formation in this region is concentrated largely in the miogeocline. By contrast, the Hamersley basin of western Australia shows major accumulation of iron formation in the thickest (eugeoclinal?) area, with more 'normal' volcanic, clastic, and some carbonate rocks in the miogeocline.

- Carbonate rocks have formed throughout virtually all of geologic time, but their abundance appears to increase from the early to the late Proterozoic. They are very abundant in middle-Proterozoic and younger suites, including the Vindhyan and Sinian platform covers. The carbonate rocks of the early- to middle-Proterozoic Karelian miogeocline are described primarily as dolomites, and some investigators have suggested that the world-wide dolomite/limestone ratio decreases toward younger time. This proposal, however, is not definite, particularly when we note the abundance of old limestones (e.g. Kilohigok) and young dolomites (e.g. Sinian).
- Because of the limited selection of basins in Fig. 5.15, the distribution of rocks such as phosphorite and evaporite is not adequately represented. The chronologic distribution of evaporites is controversial, with the oldest undoubted examples in the late-Proterozoic craton cover of northern Australia (Bitter Springs Formation of the Amadeus basin). Phosphorites are shown in Fig. 5.15 only in the Sinian, where they are important members of suites that cross the Precambrian/Cambrian boundary (Section 6.1). Phosphate rocks, however, occur throughout the Proterozoic, with some particularly old examples in the Aravalli suite of northwestern India.

[**References** – General discussions of Proterozoic sedimentary basins are in the special issues edited by Bonhomme (1982) and

Condie and Sun (1990), in Hunter (1981) for the southern hemisphere, and in Cahen *et al.* (1984) for Africa. Banded iron formations are discussed in volumes edited by James and Sims (1973) and Trendall and Morris (1983). Information about specific areas is provided for the following.

- South Africa, including the Pongola and Witwatersrand Groups, by the edited volumes of Eriksson, Callaghan and Zawada (1991) and Tankard *et al.* (1982); Grobler, van der Westhuizen and Tordiffe (1989) discuss the Sodium Group.
- West Africa; the Taoudeni basin is discussed by Bertrand-Sarfati and Moussine-Pouchkin (1983).
- India, by Naqvi and Rogers (1987); information on the Dharwar sequence is given by Mukhopadhyay, Baral and Ghosh (1981) and Chadwick *et al.* (1989), and on the Vindhyan sequence by Kale and Phansalkar (1991).
- Australia, by papers on the Hamersley sequence in Trendall and Morris (1983), the Adelaidean suite by von der Borch (1980), and the late Proterozoic by Plumb (1985).
- North America, by the volume edited by Campbell (1981) for Canada, plus specific discussion of circum-North American suites by Young (1981), the Kilohigok suite by Grotzinger and McCormick (1988), and rocks in the Great Lakes region by the volume edited by Medaris (1983).
- The Baltic shield, by papers in the volume edited by Gaal and Gorbatschev (1987) and the study of Karelian rocks by Laayoki (1986, 1990).
- China, by H. Wang and Qiao (1984) and Z. Wang, Cheng and Hongzhen (1986); information on the Sinian is provided by Chen *et al.* (1981).
- Russia and Siberia, by the general work of Nalivkin (1960) and specific papers on the Riphean and Vendian by Chumakov and Semikhatov (1981) and Korkutis (1981).
- South America, by papers on the Guiana shield by Gibbs and Barron (1983), on the late-Proterozoic of South America by de Brito Neves and Cordani (1991), on the Imataca complex by Onstott, Hall and York (1989), and on the Carajas and other iron formations by Hoppe, Schobbenhaus and Walde (1987), Olszewski *et al.* (1989), and Machado *et al.* (1991).]

5.6 Varieties of Proterozoic supracrustal rocks

ARMED with a knowledge of stratigraphy from the preceding section, we now turn to an investigation of the individual rock types that characterize the various deposits. Our attention is oriented toward lithologies that provide information about the special history of the Proterozoic rather than to a general survey of all rock types. For example, graywacke–silt–shale (flysch) sequences are probably the most common supracrustal Proterozoic rock, but they occur in the same orogenic environments as in the Phanerozoic and are not further discussed. The rocks investigated are (in sequence): volcanic rocks; soils and detrital minerals; banded iron formations (BIF) and related Mn deposits;

shales; limestones and dolomites; evaporites; and glaciogenic deposits. Inferences concerning Proterozoic oceans and atmospheres drawn from these rocks are discussed in Section 5.8.

Volcanic rocks

The magmatic products of middle- to late-Proterozoic orogenic belts are virtually identical to those of modern subduction zones (calcalkaline suites?), and it is in the intra-cratonic basins that we must look for specific information about the condition of Proterozoic crust and the nature of the Proterozoic mantle. Volcanism was particularly intense in the oldest basins (chronologically Archean) and became rare by the middle Proterozoic.

The questions posed by intra-cratonic volcanism in early Proterozoic basins are: how early in earth history did subduction become an active process?; and, were early Proterozoic basins floored by continental crust similar to that of present continents? These issues are exemplified by the Pongola basin (Figs 5.12 and 5.15), initiated at ~3000 Ma and apparently the world's oldest. The Pongola basin contains several kilometers of basalts and basaltic andesites with some more silicic varieties, all primarily in the lower part of the section. Eruptive environments appear to have varied from subaqueous to subaerial.

Tectonic inferences can be drawn from the compositions of Pongola rocks. Two conclusions, not necessarily mutually exclusive, have emerged from these data. One is that the volcanic suites represent mantle-derived magmas, similar to modern intracontinental basalts and basaltic andesites, that were contaminated by continental crust. This concept is supported by radiogenic isotope ratios higher than would be possible for magmas with wholly mantle signatures. The inference is that the continental 'basement' to the Pongola supracrustal suite was thick enough and silicic enough to permit significant contamination of rising magmas. Thus, the apparently rapid accumulation of volcanic rocks was controlled at least as much by rapid dissipation of heat from a hot earth as by lesser thickness of crust three billion years ago.

The other point of view concerning Pongola volcanic rocks is that some of their trace-element contents are similar to those of modern subduction-zone magmas and, thus, may demonstrate the existence of subduction within the chronological Archean. If these geochemical indicators apply to rocks of ~3000-Ma age as well as they do to modern rocks, then the conclusion is probably valid. The difficulty is that we cannot be sure that mantle compositions and mineralogy and the thermal gradients associated with subduction zones were sufficiently similar in the late Archean and early Proterozoic to permit direct compositional comparison with modern volcanic suites.

Soil zones, regolith and detrital minerals

Many Proterozoic basin sequences lie above unconformities on the basement, and some basins contain major depositional breaks within the supracrustal sequences. Paleosols are present along some of these horizons, and where the paleosols were not eroded before deposition or eruption of the overlying rocks, they provide information on conditions of weathering. The paleosols range in age from pre-Pongola (~3000 Ma) to the present.

Before describing and interpreting Proterozoic paleosols, we must consider several cautionary statements. One is that, particularly in the older rocks, the paleosols have been metamorphosed and/or metasomatized. The degree to which these processes altered the original soil compositions are commonly unknown. A second warning is the necessity of realizing that Precambrian soil formation occurred without the effects of surface organic activity that dominate so much modern weathering. Microbial activity (Section 5.7) was probably present on the Proterozoic surface, but there seems little possibility that it could have had the chemical effect of modern vascular plants or the physical effect of organisms (e.g. worms) moving through the soil.

Paleosols of all ages tend to concentrate Al and Ti at increasing abundances upward in the soil profile. This process appears to have been relatively independent of atmospheric composition and, in its extreme development, accounts for the abundance of Proterozoic bauxites. Similarly, the cations Ca, Mg, and Na tend to be lost from all soils (see below). Different Proterozoic paleosols show both loss and retention of K, largely dependent on the formation of illite (smectite), but K contents are very likely to be affected by later metaso-

matism and probably are unreliable indicators of paleo-environments.

The loss of Ca, Mg, and Na has been used to infer atmospheric CO_2 concentrations. All three elements are released from silicate minerals largely by corrosion by H_2CO_3 (many investigators refer to the process as 'titration' by carbonic acid). In the Precambrian, and probably at any time prior to the development of land plants in the Late Silurian, atmospheric CO_2 was almost certainly higher than present concentrations. Thus, we might expect Proterozoic soils to show much more leaching of titratable cations than modern ones. Unfortunately, evidence for this process is inconsistent, with some soils exhibiting great depletion and others showing compositional profiles similar to present ones.

Because of its importance in determining oxidation states of the atmosphere, the element of greatest interest in the study of Precambrian paleosols is Fe, which is more abundant in paleosols younger than ~2000 Ma than in older ones. Carbonic acid attacks Fe-bearing silicates, forming dissolved ions and possibly depositing siderite ($FeCO_3$). The Fe is leached from the soil unless it is oxidized into some variety of ferric oxide and/or hydroxide and concentrated in the weathering profile. The critical environmental parameter for determining the behavior of Fe is not just O_2 concentration but, more likely, the atmospheric ratio of O_2/CO_2. Limited data indicate that atmospheric O_2 concentrations only a few percent of present atmospheric levels (PAL) are capable of oxidizing and retaining Fe in soils provided that atmospheric CO_2 was not more than about 10 PAL. Thus, we may infer a significant increase in atmospheric O_2/CO_2 at ~2000 Ma.

Although the two valence states of Fe make it ideal for assessing redox conditions, other elements may be used for the same purpose. In particular, the Th/U ratio is a sensitive index. Th has only one valence state and is unaffected by oxidation or reduction under any reasonable conditions at the earth's surface, whereas U occurs in both the reduced (tetravalent) and oxidized (hexavalent) form. Thus, soils formed under oxidizing conditions retain Th/U ratios much higher than those formed under reducing conditions. Information on Th/U ratios in Proterozoic soils is much more limited than for Fe, but the available Th/U data are consistent with increase in atmospheric O_2 at some time in the early Proterozoic.

One of the earliest attempts to infer atmospheric O_2 levels was based on the occurrence of detrital uraninite and associated minerals in late-Archean/early-Proterozoic quartz sandstones. Because the mineralogy of clastic rocks is determined by numerous factors, such as atmospheric and oceanic compositions, rate of transport and burial, and post-depositional processes, interpretation of detrital mineral abundances has always been highly controversial. Thus, many investigators have challenged early conclusions that the highly oxidizable uraninite (and pyrite) in quartzites from South Africa and other locations indicated a lack of O_2 in the early Proterozoic atmosphere. In modern sediments, however, such oxidizable minerals are preserved only where erosion, transport, and sedimentation rates have been extremely rapid (e.g. debris shed from the Himalayas and buried in the Indian Ocean). Conversely, the early Proterozoic occurrences of oxidizable minerals are in clean, shallow-water to fluvial, sandstones that were deposited slowly and exposed to the atmosphere for considerable periods of time. Thus, we can probably conclude that they confirm a lack of atmospheric oxidation before ~2000 Ma.

Banded iron formations (BIF) and manganese deposits

Banded iron formations (BIF) have long been one of geology's major enigmas. Because none are forming today, and only a few within the past two billion years, we cannot examine a modern model. The problem would not be acute if they were some insignificant rock type, but they are abundant through the early Precambrian and also constitute almost all of the world's economic deposits of Fe. The deposition of BIF, evaporites, red beds, carbonate rocks etc. are all intimately related to abundances and ratios of O_2 and CO_2 in the atmosphere and oceans (Section 5.8.)

BIF are a (generally) laminated sequence of Fe minerals and chert (as some form of SiO_2). Laminations are on a scale of millimeters and have been regarded as varves, although that statement is not proven. The Fe minerals consist of oxides, silicates, carbonates, and sulfides, commonly with more than one mineral in each rock. The oxide suite is most common; many investigators regard magnetite–quartz rocks as primary, with most hematite–quartz rocks formed by later oxidation. Contacts between Fe-oxide bands and chert bands are mostly sharp, implying deposition of the Fe minerals and SiO_2 minerals at non-overlapping times.

Many BIF suites contain micro-organisms preserved in the chert. Their exact nature has aroused intense debate, although most investigators regard them as some form of prokaryotic bacteria (Section 5.7). Because some (primitive?) bacteria can obtain energy both aerobically and anaerobically, they may have been ideally suited to life in oceans that precipitated silica and iron minerals. Indeed, their release of O_2 may have caused the precipitation of Fe minerals from an otherwise reducing ocean. Stromatolitic mats consisting of chert and Fe minerals, possibly controlled by cyanobacteria (blue-green algae), are common in BIF.

The accumulation of BIF requires several environmental factors. Waters draining source terranes cannot be subjected to oxidation because they would precipitate their Fe before reaching a depositional basin. Conversely, deposition within a basin may be the direct result of oxidation, either mediated by organisms or possibly caused by photolytic conversion of Fe^{2+} to Fe^{3+} and release of H_2. An alternative explanation for deposition of Fe minerals is simple evaporation of water in a restricted basin, with oxidation caused by the action of the limited O_2 available in the atmosphere on the precipitated ferrous minerals. Regardless of the exact process of deposition, the basin of accumulation must have permitted only restricted water movement and cannot have had a large clastic input.

The general model for deposition of BIF, with numerous controversial refinements, is as follows. A terrane of mafic/intermediate igneous rocks (and metamorphic equivalents) was eroded. Both Fe and Si were placed into solution under conditions of very low O_2/CO_2 ratios in both atmosphere and oceans. The dissolved ions were transported to a restricted basin, either marine or fresh water. In the basin, SiO_2 was precipitated by evaporative oversaturation. Fe minerals, mostly magnetite, were precipitated either by oxidation, mediated by bacteria and/or light or by evaporation. Later diagenesis/metamorphism produced Fe silicates, hematite, and various mineralogical complexities. This process appears to have become impossible about 2000 million years ago because of increase in atmospheric (and oceanic) O_2.

A

5 cm

Fig. 5.16. Archean bacteria. (Courtesy J. W. Schopf; see Schopf, 1992.) (A) Postulated ~3500-Ma (early Archean) stromatolite of the Warrawoona Group, Pilbara craton, Western Australia. The stromatolite is controversial and regarded as inorganic by some investigators. (B) Filamentous bacterium- or cyanobacterium-like (prokaryotic) microfossils (*Primaevifilium amoenum* Schopf) in petrographic thin sections of 3400- to 3500-Ma Apex Chert from the Pilbara craton, Western Australia. As of 1992, they are the oldest undoubted fossils known. D_1 and D_2 are photograph and drawing, respectively, of specimen D from Schopf.   □

B

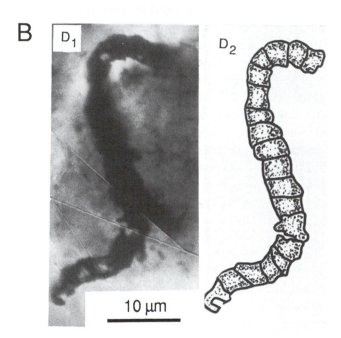

D_1 D_2

10 μm

The development of Mn-bearing sedimentary rocks poses problems somewhat different from those associated with BIF. Although the Mn was presumably transported to its depositional site in a reduced state (Mn^{2+}) and then precipitated in oxidized form, the Mn^{2+} ion is not as readily oxidized as Fe^{2+}. Therefore, Mn is mobile even in an oxidizing atmosphere, and the major Mn deposits of the world are Phanerozoic (including present Mn nodules on the ocean floor). Some BIF contain significant amounts of Mn, but most Mn deposits are not particularly Fe rich.

Shales

The compositions and mineralogy of shales are controlled primarily by their source regions. Some long-term variations, particularly between the Archean and Proterozoic, are discussed in Section 2.3. The only issue that we should mention here is the greater abundance of Ca in early-Proterozoic shales than in those deposited in the later Proterozoic and Phanerozoic. This increased Ca content correlates well with the apparently low abundance of carbonate rocks before ~2000 Ma. Presumably, Ca removed from igneous/metamorphic source rocks exceeded the capacity of the oceans to precipitate calcium-bearing minerals (calcite, dolomite, gypsum) and was retained in shales. The exact site of the Ca in shales has not been investigated in most occurrences; possibly it is retained in incompletely weathered feldspars.

Limestones and dolomites

The presence of calcsilicates in the oldest supracrustal rocks known (Isua, Greenland; Section 3.4) indicates precipitation of Ca and/or Mg carbonate from very early stages of earth history. Comparatively thick limestone banks are preserved in rocks of late-Archean (>2500 Ma) basins. Nevertheless, limestones and dolomites are relatively scarce members of supracrustal sequences until the middle Proterozoic.

The deposition of any limestones before the advent of calcite-secreting organisms at the base of the Cambrian poses numerous problems. Only two mechanisms of deposition were available. One was inorganic precipitation caused by supersaturation of sea water, probably by evaporation. Modern sea water appears to be slightly supersaturated in $CaCO_3$, presumably enhancing the stability of shelled organisms, but inorganic precipitation in modern oceans is either rare or impossible (depending on which carbonate petrologist is polled). Proterozoic suites, however, show rapid accumulation of carbonate muds, textures indicative of sudden crystallization of aragonite (apparently dendritically), and an abundance of calcite trapped in stromatolites. Possibly Proterozoic oceans were more supersaturated in $CaCO_3$ than modern ones, and small perturbations in environmental conditions (temperature, CO_2 pressure, etc.) led to inorganic precipitation.

A second important mechanism of carbonate accumulation in the Proterozoic may have been stromatolites (Fig. 5.16). During this time, stromatolites acquired an enormous variety of morphologic forms and appear to have occupied many of the ecological niches now inhabited by a diverse assemblage of modern calcareous organisms. The stromatolites were incapable of secreting calcite or aragonite by direct biological action, but they withdrew CO_2 from the sea water for photosynthesis and thereby could have promoted $CaCO_3$ precipitation.

The increase in abundance of limestones at ~2000 Ma may have been the result either of tectonic or environmental factors. A tectonic explanation is that limestones can form abundantly only on shallow-water platforms, thus requiring the evolution of large, stable cratons. Prior to ~2000 Ma, this stability was achieved only in southern Africa and southwestern India (Section 5.5). An environmental explanation can be based on the observation that ~2000 Ma is a time of major increase in atmospheric O_2, (see above and below). This increase may have been caused by a rapid increase in the abundance of cyanobacteria (blue-green algae) that could secrete oxygen photosynthetically, thus consuming CO_2 from the oceans and causing precipitation of calcium carbonate (Section 5.7).

The abundance of dolomite in Precambrian carbonates is controversial. Some investigators have proposed that the dolomite/calcite ratio is higher in older rocks, but others propose that dolomite may not have become an important component of carbonate rocks until the middle and later Proterozoic, at the same time as evaporites became abundant. If dolomite was not a major component of carbonate rocks in the early Proterozoic, the disposition of Mg weathered from source rocks and delivered to the oceans is unknown.

Evaporites

Precipitation of gypsum, and its resulting anhydrite, has been inferred in rocks as old as the middle Archean. The occurrences are controversial, however, because no actual sulfate minerals remain, and their former presence commonly is indicated by pseudomorphs now largely occupied by silica.

The oldest complete sequence of evaporites thus far discovered is in the Bitter Springs Formation of the Amadeus basin of Australia (Fig. 5.13), probably deposited shortly after ~1000 Ma. These deposits are about 300 m thick, with the basal section consisting largely of carbonaceous dolomite and the upper section primarily a dolomite–gypsum breccia. The deposit apparently formed by evaporation in a barred, mostly anoxic, basin. Stromatolites formed a fringing reef, contributed to the reduced circulation, and aided in the trapping of sediment. The similarity of the Bitter Springs and other proposed Proterozoic evaporites to Phanerozoic ones has been used to conclude that late-Proterozoic sea water had roughly the same ratios of cations as in the present oceans.

Glaciogenic deposits

The problem with reconstructing the glacial history of the Proterozoic is, as in younger ages, the recognition of glacial deposits. The characteristic features of glacial tills are: 1) they are diamictic (i.e. unsorted and possibly bimodal or multimodal in sizes); 2) they contain distantly derived rocks and minerals (as erratics); 3) they contain clasts that have been abraded and possibly striated; and 4) they rest on erosional surfaces, locally showing striations and/or other glacial grooving. Unfortunately, sediments deposited by mud flows, sub-aqueous slumps etc., can have similar properties, and their distinction from glacial tills has not always been made successfully.

Using the preceding criteria, we can identify one or more glacial periods in the late Proterozoic and possibly earlier ones. Glaciation in the latest Proterozoic seems to have affected virtually all of the world's continental areas. This glaciation is referred to as Varangian or Varanganian, from deposits in Norway, and occurred between about 700 Ma and 600 Ma. Worldwide glaciation, with accompanying sealevel depression, may have been partly responsible for the environmental stresses that led to the development of organisms with skeletal parts (Section 6.2), although other environmental factors may have been much more important.

Glaciation prior to the Varangian is difficult to determine. Two or three pulses in different parts of the earth have been proposed for the period from ~1000 Ma and ~700 Ma. The degree of correlation between events in widely spaced areas is undetermined, and thus we do not know whether they were global or only local. No major evidence exists for glaciation between approximately 2000 Ma and 1000 Ma. Some apparent tillites in North America and Baltica (and possibly correlative elsewhere), however, suggest major glaciation in the time shortly before 2000 Ma.

[**References** – General references include summaries of the atmosphere and oceans in the volume edited by Nagy *et al.* (1983), the paper of Maynard, Ritger and Sutton (1991), and discussions of the effects of organic activity by Garrels (1987b) and De Duve (1991). Information on volcanic rocks in the Pongola Group is from Hegner, Kroner and Hofmann (1984), Armstrong, Wilson and Hunter (1986), and Crow *et al.* (1989); Grobler *et al.* (1989) discuss the very old Sodium Group. The discussion of soil formation is based on the volume edited by Retallack (1986) and papers by Schau and Henderson (1983), G-Farrow and Mossman (1988), Holland, Feakes and Zbinden (1989), Palmer, Phillips and McCarthy (1989), Rainbird, Nesbitt and Donaldson (1990), and Wiggering and Beukes (1990). The discussion of banded iron formations is from the edited volume of Trendall and Morris (1983) and the paper of Garrels (1987a). Manganese deposition is summarized by Roy (1981, 1988). The composition of Canadian shales is discussed by Cameron and Garrels (1980). The discussion of stromatolites is based mostly on Southgate (1989) and Grotzinger (1990). A Proterozoic evaporite sequence is discussed by A. Stewart (1979). Proterozoic glaciation is summarized in the volumes edited by Hambrey and Harland (1981) and

Deynoux (1985) and the papers of Ojakangas (1988) and Young (1988).

5.7 Life in the Proterozoic

BACTERIA are not as dramatic as dinosaurs. They did, however, rule the earth for a far longer time than the 'Age of Reptiles', and in many respects the Proterozoic may be called 'The Age of Bacteria'. Because bacteria are constructed with complex organic molecules, they required some 'pre-biotic' synthesis before they could evolve, a topic covered in Section 3.6. Bacteria were obviously present in the Archean, but we discuss both their Archean and Proterozoic history in this section. First, we need to establish some fundamental properties of bacteria and their relationships to other organisms.

Properties of bacteria

Bacteria are single-celled organisms. They are prokaryotic, signifying the absence of a nucleus and the distribution of organs that regulate the life of the organism throughout the entire cell. (In eukaryotic organisms, the regulators are clustered into a nucleus and in various organelles, such as the mitochondria.) Reproduction is asexual except in unusual situations, in which genetic material may even be transferred between different species. Despite this apparent simplicity, bacteria show many elements of physiological similarity to higher organisms. The similarities include the use of DNA and RNA for the coding of reproductive information, construction of proteins based on essentially the same amino acids as metazoa, and the use of adenosine phosphates (principally ATP) for storage and transfer of energy produced by metabolism (Section 3.6).

Bacteria possess an extraordinary diversity of lifestyles. They now inhabit almost all of the earth's environments and probably adapted to numerous subaqueous and subaerial conditions early in their history. Consequently, they have a variety of effects on biogeological processes. We can illustrate this diversity by describing the different methods by which bacteria obtain and use energy.

Organisms obtain the energy needed for life in five basic ways, all of which are used by different types of bacteria. One method is *fermentation*, by which complex organic molecules are broken down into simpler compounds accompanied by release of energy (see Section 3.6 for further discussion of sugars and other carbohydrates). A simple example is the production of methyl alcohol and formic acid from sugars by the general reaction

$$2CH_2O = CH_3OH + HCOO^- + H^+.$$

This process produces far less energy than complete oxidation of the sugars to CO_2 and H_2O. Because of its inefficiency, fermentation probably could have been a major process by which bacteria obtain energy only during the very earliest stages of life on earth, and possibly not even then.

A second method of bacterial energy production is *photosynthesis*. Photosynthesis refers to any reaction of the type

$$CO_2 + 2H_2X (+ light) = (CH_2O) + H_2O + 2X,$$

where (CH_2O) refers to carbohydrate cell material and X is the oxidation product. Light provides the necessary energy. Use of light energy is possible because some compound or compounds in the organism act as catalysts. In all algae and higher plants, those compounds are chlorophyll-a plus other varieties of chlorophyll that expand the range of light wavelengths that can be used (in green algae and plants, the second compound is the green-pigmented chlorophyll-b). The only prokaryotic organisms that contain chlorophyll are blue-green algae, otherwise referred to as cyanobacteria, which contain chlorophyll-a plus non-chlorophyll compounds that enable the organism to use a wide range of sunlight energies. Organisms that contain chlorophyll are 'oxygenic'. They produce oxygen by using H_2O as the oxidant in the above reaction and, therefore, produce free O_2; that is, X = O in the reaction shown above.

True bacteria (eubacteria) do not possess chlorophyll and cannot use oxygen for the X in the photosynthetic reaction. Thus, photosynthetic bacteria are anaerobic, and many cannot live in the presence of oxygen. Instead of oxygen, some bacteria contain materials similar to chlorophyll that enable them to photosynthesize compounds such as H_2S, producing free sulfur as the oxidation product.

Three other methods of producing energy are commonly grouped under the term 'chemosynthesis' and are the most prevalent among all bacteria. Chemosynthetic organisms obtain energy from spontaneous oxidation–reduction reactions that are not dependent on light. Reactions of this type form the basis of the ecology at submarine hydrothermal vents, which presumably have operated throughout most of geologic time. One type of chemosynthesis is *lithotrophy*, in which inorganic carbon is processed by a reaction such as

$$2CO_2 + 6H_2 = (CH_2O) + CH_4 + 3H_2O.$$

Another type of chemosynthesis is *aerobic respiration*, which is the reverse of the photosynthetic reaction. The third type of chemosynthesis is *anaerobic respiration*, typified by a denitrification reaction such as

$$5CH_2O + 5H_2O + 4NO_3^- + 4H^+ = 5CO_2 + 2N_2 + 12H_2O.$$

Bacteria in the general category of 'chemosynthetic' use a wide variety of both reductants and oxidants. The oxidants include oxygen, sulfate (sulfate-reducing bacteria), nitrate (denitrifying bacteria) and CO_2. Reductants include hydrogen, ammonia, organic compounds, Fe^{2+}, and S^{2-}, and methane from the reduction of CO_2. These various reactions enable bacteria to produce ferric oxides and hydroxides, native sulfur and sulfide ions, nitrate ions, etc. This diversity of methods of obtaining energy was obviously useful in the early stages of earth evolution.

Classification and evolution of bacteria

The topics of classification and evolution must be considered together because the taxonomy used by an investigator determines the evolutionary scheme proposed. First we will review ideas that were dominant until the past few years and then consider recent major challenges to these concepts.

The prevailing view has been that prokaryotes, including certain bacteria and cyanobacteria (blue-green algae), were the first forms of life on earth, with some strains evolving into the eukaryotes (with nuclei and organelles). Diversification within the eukaryotes did not take place until after a primitive form was established. At that time, the eukaryotes began to branch into their present varieties, including protists (protoctists), fungi, plants, and animals (Sections 6.3 to 6.5).

The principal question resulting from the preceding classification was the mechanism of evolution of the

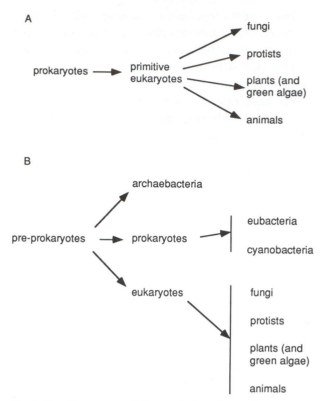

Fig. 5.17. Two proposals for evolution of organisms.

(A) Evolution from primitive prokaryotes to eukaryotes. This evolutionary path implies some mechanism for the development of cells that contain nuclei and other specialized organelles (eukaryotes) from cells that lack nuclei (prokaryotes). One possibility is incorporation of small prokaryotes (which developed into the organelles) into larger prokaryotes (which formed the major part of the cell). After development of eukaryotes, evolutionary diversification led to the major groups of organisms that now characterize the earth's biota.

(B) Direct evolution from an ancestral organism to archaebacteria, other prokaryotes, and eukaryotes. This scheme places the differentiation of a common ancestor into archaebacteria, prokaryotes, and eukaryotes at a very early time in earth history. This evolutionary path does not include sequential development of eukaryotes from the types of prokaryotes now living.   □

eukaryote cell. We can mention two hypotheses (Fig. 5.17). One is that some prokaryotes learned to infold parts of the cell walls, which gradually pulled apart and engulfed separate organs that had mingled freely within the cell. These isolated organs became the nucleus and differentiated organelles of the eukaryote cell. A second hypothesis is that large prokaryotes such as cyanobacteria, with diameters of several tens of microns, engulfed other types of prokaryotic bacteria, all of which have diameters of a few microns.

These incorporated small bacteria became internally symbiotic and gradually evolved into special organelles (mitochondria, etc.) and the nucleus.

The view that eukaryotes evolved from prokaryotes of the type living today has been challenged in the past decade. The impetus has come largely from the ability to obtain detailed information on the molecular nature of cells. A particularly useful technique is based on the sequences of nucleotides along RNA molecules in ribosomes, which are present in all cells (Section 3.6). These sequences can be compared quantitatively to obtain degrees of similarity between various organisms.

Based on molecular studies, all organisms can be subdivided into three groups more fundamental than the five kingdoms (prokaryotes, protoctists, fungi, plants, animals). The three proposed subdivisions are: 1) archaebacteria (Archaea), which contain several varieties of bacteria with special properties (see below); 2) all other prokaryotes (Bacteria), including blue-green algae (cyanobacteria); and 3) all eukaryotes (Eucarya), including protoctists, fungi, plants, and animals.

The construction of an archaebacterial group is based on similarities in molecular structure in addition to the RNA sequencing mentioned above. For example, the cells of all living organisms are partly constructed of fatty molecules referred to as lipids, but only the archaebacteria contain lipids in which carbon components are linked as ethers (across oxygen atoms). Three different types of organisms constitute the archaebacteria. The most common varieties generate methane in swamps. The other two varieties include those that flourish either in extremely saline water and/or in extremely hot water. The association of archaebacteria with methane, high temperature (near boiling), and high salinity appears to be compatible with their early evolution in a very primitive earth.

The differences between the Archaea, other prokaryotes, and eukaryotes are so fundamental that the three groups appear to have differentiated in the very earliest stages of earth history, probably well before 3500 Ma. Presumably they all arose from a common ancestor, but the nature of that ancestor is highly speculative. Even the primitive archaebacteria contain such highly complex molecules (e.g. DNA) plus complex interrelationships among cellular processes that it seems impossible for them to have acquired all of these intricate functions at the same time and become the original 'life' (Section 3.6). Presumably, more primitive organisms preceded the archaebacteria. One possibility is that these earliest organisms managed to survive with the operations of sets of fairly simple nucleic acids that controlled error-prone reproductive and metabolic processes. (Lifespan of individuals was probably short.) More complex genetic patterns evolved within these early cells, quickly diversifying into the archaebacteria and eubacteria. Eukaryotes developed either by the engulfment or infolding of primitive organisms similar to the processes described above or by some undetermined process.

Preservation of Proterozoic organisms

Indications of life in the Precambrian are commonly quite different from the well-preserved shells and bones of more recent ages. The only Precambrian organisms directly associated with the production of minerals are the blue-green algae (cyanobacteria), which were responsible for the widespread development of stromatolite mounds ('algal mats'). Cyanobacteria evolved by the middle Archean, and in the Proterozoic they occupied virtually all of the niches (reefs, platforms and ramps) now occupied in carbonate environments by a variety of metazoan organisms. The cyanobacteria form stromatolites not by secreting $CaCO_3$ but by two other methods: 1) by withdrawing CO_2 from the environment, thus causing 'inorganic' precipitation of $CaCO_3$; and 2) by forming a sticky mat that traps carbonate particles formed elsewhere. The abundance of stromatolites in the stratigraphic record appears to have decreased toward the later Proterozoic, over the same time range in which the taxonomic diversity increased.

With the exception of stromatolites, we can recognize three methods by which Precambrian organic activity is preserved. The first is imprints and/or pseudomorphs of organisms. Late-Proterozoic impressions, such as the Ediacaran and correlative biota, are clearly organic, although their classification is controversial (Section 6.2). Other imprints and pseudomorphs can be equivocal. For example, structures in cherts have rodlike and spherical shapes and sizes similar to those of modern eubacteria, but some 'dubiofossils' could be either organic or inorganic.

A second method of detecting organic activity is by examining the isotopic effect of metabolic processes, particularly on the fractionation of C and S isotopes. Preferential assimilation of ^{12}C by organisms causes organic C to have much lower $^{13}C/^{12}C$ ratios than inorganic C. Consequently, withdrawal of ^{12}C into organic material causes an increase in the $^{13}C/^{12}C$ of carbonates, which are precipitated from water from which the metabolic C has been removed. Metabolic enrichment in ^{12}C is presumably responsible for the observation that non-carbonate C in sedimentary rocks has had approximately the same $^{13}C/^{12}C$ ratio from the early Archean to the present. The inference is that metabolic fractionation of C has occurred at approximately the same rate throughout nearly all of geologic time. Thus, evolution to new forms of organisms has not increased the total level of organic activity but merely changed it from a solely bacterial biota to a diversified biota.

The effects of biological activity on sulfur isotopes are more difficult to interpret than on carbon. The problem is that sulfur in sediments may be both the result of biological reduction and of the later (e.g. hydrothermal) reduction of sulfate. Biological reduction yields S that has lower $^{34}S/^{32}S$ than the average of magmatic rocks, whereas high-temperature reduction of sulfates yields higher ratios. Variations in S isotope ratios through time show the oldest appearance of low $^{34}S/^{32}S$ ratios slightly earlier than about 2000 Ma. This time may correspond to the earliest development of sulfate-reducing bacteria (Section 5.6).

The third, and most controversial, indication of Precambrian organic activity is the 'molecular fossil'. The term signifies the preservation of organic molecules that have resulted from biologic activity. The controversy results partly from the difficulty of proving that samples have not been contaminated by younger (including modern) molecules and partly from the changes that can occur in complex organic molecules over time. Among the compounds most useful for these studies are 'terpenes', which constitute a diverse suite of molecules characterized by a complex multi-ring structure and extractable from cells into organic solvents. Their usefulness stems from the fact that some varieties of terpenes are uniquely synthesized by archaebacteria, other varieties by eubacteria, and others by eukaryotes; different groups of eukaryotes also have unique signatures.

The study of molecular fossils is still preliminary. At present, however, they show the expected dominance

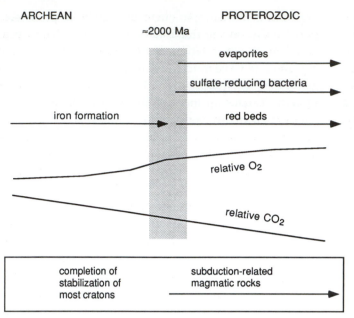

ARCHEAN   PROTEROZOIC

≈2000 Ma

evaporites

sulfate-reducing bacteria

iron formation   red beds

relative O₂

relative CO₂

completion of   subduction-related
stabilization of   magmatic rocks
most cratons

Fig. 5.18. Diagrammatic representation of major changes in the earth's surface environment during the Proterozoic. A broad time range near 2000 Ma is shown as a time of major transition to a more 'modern' earth with sufficient oxygen in the atmosphere and oceans to oxidize iron during weathering and transportation. This transition was accompanied by the development of sulfate-reducing bacteria. The change in the surface environment may have occurred at the same time that most continental cratons achieved tectonic stability and also as subduction-related (calcalkaline) batholiths became widespread.   □

of archaebacteria and other prokaryotes in most of the Precambrian. Eukaryotic derivatives may occur in the middle Proterozoic and earlier, and derivatives from heterotrophic protists (organisms that live on organic C from other organisms) have been reported from the latter part of the Proterozoic.

Evolution of Proterozoic biota

The major evolutionary patterns in Proterozoic fossils can be summarized as follows.

Archaebacteria and eubacteria evolved early in the Archean and were present throughout the Proterozoic.

Blue-green algae (prokaryote cyanobacteria) also were present throughout the Proterozoic after evolution in the middle Archean (or earlier).

The presence of eukaryotes is indicated by preserved organic molecules that they presumably synthesized and by the presence of films and imprints of cells that are probably too large to be bacteria. No cells containing nuclei have been preserved, however, and the evidence is inferential.

Metazoan (multi-celled) organisms had evolved by the end of the Proterozoic (Ediacaran, or Vendian, biota; Section 6.3), but their exact classification is controversial.

[**References** – Brock and Madigan (1988) provide a general survey of microorganisms, and Tappan (1980) discusses plant protists. The volumes edited by Schopf (1983) and Schopf and Klein (1992) contain integrated surveys by many investigators on the development of the biosphere; Schopf (1992) provides a recent review of Archean organisms. Review papers on the biosphere and its relationship to the early earth are also in the volume edited by Nagy *et al.* (1983). Margulis and Olendzenski (1992) discuss the general aspects of organic evolution, and the Proterozoic record is summarized by Knoll and Ostrom (1990) and Knoll and Walter (1992). Walter, Buick and Dunlop (1980) and Buick (1991) discuss early organic(?) forms in Western Australia. Interpretation of organic evolution by the study of sulfur isotopes is provided by Cameron (1982) and Ueda, Cameron and Krouse (1990). Archaebacteria and the general evolution of bacteria is discussed by Woese (1987) and Woese, Kandler and Wheelis (1990). Discussion of vents is provided by Jannasch and Mottl (1985) and of stromatolites by Walter and Heys (1985).]

5.8 Evolution of atmosphere and oceans

THE RECORD of Proterozoic weathering and sedimentation provides us with a (very imperfect) set of data that can be used to reconstruct the condition of the earth's surface environment at various times. In this section, we collect the data from observations mostly described in Sections 5.6 and 5.7 and propose general properties of both atmosphere and oceans. The atmosphere and oceans are considered together because interchange of volatile components between them is instantaneous on a geologic time scale, signifying chemical equilibration at all times. A diagrammatic summary of the proposed history is given in Fig. 5.18.

Early atmospheres on the earth probably contained a potent mixture of chemicals that almost all modern organisms would view with alarm (Section 3.2). Nevertheless, by the start of the Proterozoic, such species as methane, ammonia, hydrogen peroxide, etc. were presumably reduced to low or zero concentrations. Thus, for the purpose of our discussion, the only significant components of the Proterozoic atmosphere were N_2, CO_2, O_2, and H_2O. Furthermore, the only methods of monitoring N_2 variations are probably by study of bacteria involved in the N_2-to-NO_3 system, for which historical data are not adequate. Thus, we concentrate on O_2 and CO_2.

We can place approximate upper limits on atmospheric CO_2 partial pressures. Concentrations greater than present atmospheric levels (PAL; 0.03%) cause heating by a greenhouse effect. The earth's surface, however, probably was never hotter than about 40 °C (average) during the Proterozoic. If solar radiation was the same in the Proterozoic as it is now, then atmospheric CO_2 could not have been more than 100 times PAL, and probably not that high, in order to remain below this average temperature. Furthermore, if glaciation occurred in the early Proterozoic, then presumably the atmospheric temperature and CO_2 levels were not significantly higher than present ones. Some astronomers have argued that solar radiation has increased during the history of the earth, starting at about 25% less than present in the Archean. This reduction in radiation would permit somewhat higher levels of atmospheric CO_2 without excessive greenhouse heating, but exact figures are impossible to determine.

An additional constraint on atmospheric and dissolved oceanic CO_2 can be based on the observation that some limestone has been precipitated throughout the Archean and Proterozoic. Although many factors influence the solubility of $CaCO_3$, there must be some upper limit on dissolved CO_2 to allow any kind of limestone accumulation. In a highly simplified system containing only $CaCO_3$, CO_2, and H_2O, we can calculate the equilibrium values of pH and Ca^{2+} from the two ionization constants of H_2CO_3 (first constant = 4.2×10^{-7}, second constant = 5×10^{-11}) and the solubility product of $CaCO_3$ (5×10^{-9}). For a CO_2 pressure of 3000 PAL (~1 atm; yielding a total atmospheric pressure considerably greater than 1 atm), the result is a pH of 5

to 6 and a Ca^{2+} of 10^{-1} to 10^{-2} (compared with pH values of 8.3 and Ca^{2+} near 10^{-2} in modern oceans). The actual concentration of Ca^{2+} would be much higher than is calculated for this simple system because other components of the water complex with the calcium ion and modify the activity coefficients of all species present.

A pH of 6 seems much too low to permit any carbonate accumulation, and we are tempted to conclude that atmospheric CO_2 could never have been two orders of magnitude greater than PAL. Some confirmation of low pH values, however, has been provided by the observation of co-precipitation of amorphous SiO_2 and illites (smectites) with low K/H ratios in late-Archean rocks. H-rich illite cannot co-exist with amorphous SiO_2 unless the pH of the water in which they are deposited is low, perhaps near 6 or even lower. Thus, an acidic ocean is possible in the Proterozoic, presumably requiring very high Ca^{2+} concentrations for carbonate precipitation. Conversely, a low-pH ocean was also possible if alkalinity was maintained by high Na^+ values (see below).

Placing exact limits on O_2 is as difficult as for CO_2. The initial terrestrial atmosphere was certainly reducing. It may have been completely anoxic, although photolytic decomposition of H_2O and CO_2 could have generated some O_2. The first organisms that emitted O_2 as a result of photosynthesis were blue-green algae (cyanobacteria; Section 5.7), which had evolved by the middle of the Archean and possibly earlier. Thus, a biologic mechanism for filling the atmosphere with oxygen had been established long before the start of the Proterozoic. Before the evolution of sulfate-reducing bacteria (possibly at ~2000 Ma), however, atmospheric oxygen could have been irretrievably withdrawn into insoluble sulfates, and thus the net effect of O_2 production is unclear. One proposal for the sudden appearance of skeletal parts in organisms at the start of the Cambrian is that atmospheric O_2 values reached a few percent of PAL at that time (Section 6.2). If that proposal is correct, then the net effect of oxygen emission into the atmosphere from the middle Archean until about 600 Ma was only a very minimal increase.

Another restraint on atmospheric O_2 is provided by the time distribution of BIF and red beds, by the observations of removal of acid-titratable cations from soils, and by the presence of oxidizable minerals in mature clastic sediments. As we discussed in Section 5.6, iron was apparently mobile in surface waters before ~2000 Ma, implying inability of water and atmosphere to oxidize it from the Fe^{2+} state except under conditions of extreme evaporation, by photolysis, or by biological mediation. Most BIF formed before 2000 Ma. After that time, iron appears to have been converted rapidly into Fe^{3+} by surface exposure, permitting Fe retention in soils and the formation of ferruginous clastic sediments. The O_2 level needed for this transition is not certain but is probably in the range of 1% to 10% PAL.

Estimates of the dissolved solutes of Proterozoic sea water are even more inferential than those of the volatiles. The observation that evaporites have a similar mineralogy to that produced by modern water pertains only to deposits within the past 1000 million years, and the early and middle Proterozoic yield little information. Extreme acid leaching before ~2000 Ma probably delivered large quantities of Ca^{2+}, Mg^{2+}, Fe^{2+} and Na^+ to the oceans. The Ca was removed by limited carbonate precipitation before ~2000 Ma and by a combination of carbonate and sulfate precipitation after that time. These precipitations, if confined to continental platforms, seem incapable of maintaining calcium balance against the supply, and the mechanism of distributing Ca between various reservoirs in the Proterozoic remains obscure.

Balancing of ions other than Ca is simpler to discuss (but not to understand). Removal of Fe from a mobile state was achieved either by precipitation in iron formations or red beds. The Mg probably was not completely balanced by precipitation as dolomite, but cycling of sea water through oceanic crust could have caused considerable Mg loss. Mass balance of Na is difficult in present oceans and no easier in the Proterozoic. High concentrations of Na^+, combined with high dissolved CO_2 contents, could have given the oceans high 'salinity' in the form of dissolved sodium carbonate and thus a high pH.

[**References** – See references for Section 5.6 plus books on evolution of the atmosphere by Walker (1977) and on the atmosphere and oceans by Holland (1984). Argast and Donnelly (1983) provide information on the composition of early oceans, and Kempe and Degens (1985) discuss the possibility that early oceans were sodic.]

5.9 Summary – and importance of events at ~2000 Ma and the end of the Proterozoic

IN REACHING the end of the Proterozoic, we have surveyed some 2000 m.y., roughly half of earth history. The time period may actually be longer if we adopt a tectonic definition of the start of the Proterozoic as ~3000 Ma in some cratons (Section 3.0). During the Proterozoic, the earth changed from the possibly very mobile tectonic regimes, reducing atmosphere, and acidic oceans of the Archean (Chapter 3) to the unquestioned condition of lithospheric rigidity and near-modern air and water that characterized the Phanerozoic (Chapter 6). Two episodes in the Proterozoic appear to have been particularly important: 1) ~2000 Ma, when atmospheric/oceanic and lithospheric changes brought the earth to a more modern state; and 2) at the end of the era, ~550 Ma, represented by a major extinction event and followed by the explosive radiation of skeletal organisms. These episodes are discussed after a brief review of the major characteristics of Proterozoic history.

Despite the crustal rigidity that characterized most, if not all, of the Proterozoic, an enduring controversy is the nature of orogenies and the extent of crustal growth. Most activity prior to ~2000 Ma is recorded only by dates in materials incorporated into older crust and overprinted by younger events. This early Proterozoic apparently was a time of separation of new crust, possibly with major increase in continental volume, but few details are available. Orogenic belts formed since ~1900 Ma seem to be mostly the result of collision, but some 'intracontinental' belts may represent largely the reworking of older continental crust rather than the creation of new crust; these belts do not contain the suture zones characteristic of collisional orogens.

Sedimentary basins in the Proterozoic are similar to those of the Phanerozoic. We have classified them into four types: 1) clastic wedges deposited along comparatively undeformed continental margins probably similar to modern passive margins; 2) combined eugeoclinal/miogeoclinal suites associated with compressive belts; 3) successor basins on former sutures; and 4) basins, possibly rift-floored, within cratons. A generalized stratigraphy in the basins shows high abundance of volcanic rocks in older suites, restriction of banded iron formation largely to rocks older than ~2000 Ma, and increase in the abundance of limestones in younger suites.

Two very characteristic, and possibly related, features of the brittle lithosphere of the Proterozoic are crustal rifting and anorogenic magmatism. Sets of intracontinental rift valleys formed in the early to middle Proterozoic may be more closely spaced than those formed more recently, but the relationship is not clear. Many of the rifts occur as aulacogens extending into the continental margins, and they are locally responsible for the development of some broad intracontinental basins. The establishment of a brittle lithosphere at the start of the Proterozoic is particularly well shown by emplacement of sets of basalt dikes, some of which can be traced for several hundred kilometers; comparable suites do not occur in the Archean. Another important anorogenic magmatic suite in the Proterozoic contains anorthosites and rapakivi granites (in variable proportions) and is particularly characteristic of the fused North America/Baltica in the middle Proterozoic.

The dominant life form in the Proterozoic was bacteria (prokaryotes; without nuclei in their cells). Eukaryotes (with nuclei) had certainly evolved by the end of the era, but evidence for their earlier existence is inferential. The most noticeable prokaryote was cyanobacteria (blue-green algae) that formed stromatolites. Other bacteria include: the primitive archaebacteria, capable of living in hot, saline, and/or reducing conditions; and eubacteria, which constitute most modern varieties. By the last few hundred million years of the Proterozoic, the earth was occupied by a widespread biota of soft-bodied organisms (generally referred to as Ediacaran; further discussion in Sections 6.1 and 6.2). The nature of the organisms in this biota is highly controversial, but regardless of the classification, few similar varieties survived into the Paleozoic.

Major changes in atmosphere/ocean compositions during the Proterozoic include: increase in O_2/CO_2 ratio; absolute increase in O_2 and absolute decrease in CO_2; and evolution of sulfate-reducing bacteria and mobilization of sulfur into the environment. Variations in the O_2 and CO_2 concentrations appear to have been progressive throughout the era, although a critical value of O_2 concentration probably was exceeded at ~2000 Ma, causing the transition from deposition of banded iron formations to deposition of redbeds. The evolution of sulfate-reducing bacteria apparently occurred at the same time. The decrease in CO_2 concentrations must have corresponded to a decrease in acidity of ocean water.

In all discussions of the Proterozoic, a period of time at ~2000 Ma seems to have been a critical transition on both the earth's surface and internally (Fig. 5.18). In addition to the increase in atmospheric O_2, decrease in CO_2, and development of sulfate reduction, this period was also a time when still-mobile Archean shields seem to have completed the incorporation of early-Proterozoic orogenic belts into stable cratons. Possibly, it was also a time when calcalkaline batholithic suites were first developed (Section 3.5). In many ways, ~2000 Ma may be the time at which the earth passed from an 'ancient' state into a 'modern' one.

The end of the Proterozoic is a period of biotic change from the soft-bodied organisms of the Precambrian to the skeletal organisms of the Paleozoic (this change is discussed in detail in Sections 6.1 and 6.2). At this time, or shortly before: the earth underwent a major period of glaciation; the continents had assembled into a supercontinent of controversial configuration (Section 7.0); and virtually all varieties of Proterozoic biota disappeared, making this time period one of the great extinction events in earth history. Unlike the Permo-Triassic (Section 6.9) and Cretaceous/Tertiary (Section 8.7) extinction events, however, the Precambrian/Cambrian boundary occurs in sections of continuous sedimentation and is not marked by evidence of volcanic, meteoritic, or other catastrophic activity.

References

Aleinikov, A. L., Ballavin, O. V., Bulasevich, Yu. P., Tavrin, I. F., Maksimov, E. M., Rudkevich, M. Ya., Nalivkin, V. D., Shablinskaya, N. V. & Surkov, V. S. (1980). Dynamics of the Russian and West Siberian platforms. In *Dynamics of Plate Interiors:*, vol. 1, ed. A. W. Balley, P. L. Bender, T. R. McGetchin & R. I. Walcott, pp. 53–71. American Geophysical Union Geodynamics Series.

Argast, S. & Donnelly, T. W. (1983). Javanahalli quartzites: evidence for sedimentary mica and implications for the chemistry of Archean ocean water. In *Precambrian of South India*, ed. S. M. Naqvi & J. J. W. Rogers, pp. 158–68. Geological Society of India Memoir 4.

Armstrong, N. V., Wilson, A. H. & Hunter, D. R. (1986). The Nsuze Group, Pongola sequence, South Africa: geochemical evidence for Archaean volcanism in a continental setting. *Precambrian Research*, **34**, 175–203.

Bertrand-Sarfati, J. & Moussine-Pouchkin, A. (1983). Platform-to-basin facies evolution: the carbonates of late Proterozoic (Vendian) Gourma (West Africa). *Journal of Sedimentary Petrology*, **53**, 275–93.

Bickford, M. E. (1988). The formation of continental crust: Part 1. A review of some principles; Part 2. An application to the Proterozoic evolution of southern North America. *Geological Society of America Bulletin*, **100**, 1375–91.

Bonhomme, M. G., ed. (1982). Geochronological correlation of Precambrian sediments and volcanics in stable zones. Special issue of *Precambrian Research*, **18**, 194 pp.

Bowden, P. & Kinnaird, J., eds. (1987). African Geology Reviews: Thematic issue of *Geological Journal*, **22**, 578 pp.

Bowring, S. A. & Karlstrom, K. E. (1990). Growth, stabilization, and reactivation of Proterozoic lithosphere in the southwestern United States. *Geology*, **18**, 1203–6.

Brock, T. D. & Madigan, M. T. (1988). *Biology of Microorganisms, 5th ed*. Englewood Cliffs, New Jersey: Prentice Hall, 635 pp.

Bronner, G., Roussel, J., Trompette, R. & Clauer, N. (1980). Genesis and geodynamic evolution of the Taoudeni cratonic basin (Upper Precambrian and Paleozoic), Western Africa, in *Dynamics of Plate Interiors*, vol. 1, ed. A. W. Balley, P. L. Bender, T. R. McGetchin & R. I. Walcott, pp 81–92, American Geophysical Union Geodynamics Series.

Buick, R. (1991). Microfossil recognition in Archean rocks: an appraisal of spheroids and filaments from a 3500 M.Y. old chert-barite unit at North Pole, Western Australia. *Palaios*, **5**, 441–59.

Butov, Yu. P., Zanvilevich, A. N. & Litvinovskiy, B. A. (1974). Problem of the Baikalides in the light of new data on stratigraphy and magmatism in the central part of the Baikal Mountain region. *Geotectonics*, no. 2, pp. 81–6.

Cahen, L., Snelling, N. J., Delhal, J. & Vail, J. R. (1984). *The Geochronology and Evolution of Africa*. Oxford: Oxford University Press, 512 pp.

Cameron, E. M. (1982). Sulphate and sulphate reduction in early Precambrian oceans. *Nature*, **296**, 145–8.

Cameron, E. M., ed. (1983). *Uranium exploration in Athabasca basin, Saskatchewan, Canada*: Geological Survey of Canada Paper 82-11, 310 pp.

Cameron, E. M. & Garrels R. M. (1980). Geochemical comparisons of some Precambrian shales from the Canadian shield. *Chemical Geology*, **28**, 181–97.

Campbell, F. H. A., ed. (1981). *Proterozoic Basins of Canada*. Canada Geological Survey Paper 81-10 (and Supplement), 444 pp.

Chadwick, B., Ramakrishnan, M., Vasudev, V. N. & Viswanatha, M. N. (1989). Facies distributions and structure of a Dharwar volcanosedimentary basin: evidence for late Archaean transpression in southern India? *Geological Society of London Journal*, **146**, 825–34.

Chen, J., Zhang, H., Xing, Y. & Ma, G. (1981). On the Upper Precambrian (Sinian suberathem) in China. *Precambrian Research*, **15**, 207–28.

Chumakov, N. M. & Semikhatov, M. A. (1981). Riphean and Vendian of the USSR. *Precambrian Research*, **18**, 229–53.

Condie, K. C. & Sun, D., eds. (1990). Geochemistry and mineralization of Proterozoic mobile belts. Special issue of *Precambrian Research*, **47**, 155–320.

Coward, M. P. & Ries, A. C., eds. (1986). *Collision Tectonics*: Geological Society of London Special Publication 19, 415 pp.

Crow, C., Condie, K. C., Hunter, D. R. & Wilson, A. H. (1989). Geochemistry of volcanic rocks from the Nsuze Group, South Africa: arc-like volcanics in a 3.0-Ga intracratonic rift. *Journal of African Earth Sciences*, **9**, 589–97.

de Brito Neves, B. B. & Cordani, U. G. (1991). Tectonic evolution of South America during the Late Proterozoic. *Precambrian Research*, **53**, 23–40.

De Duve, C. (1991). *Blueprint for a Cell*. Burlington, North Carolina: Neil Patterson Publishers.

Deynoux, M., ed. (1985). Glacial Record. Special issue of *Palaeogeography, Palaeoclimatology, and Palaeoecology*, **51**, 461 pp.

Emslie, R. F. (1978). Anorthosite massifs, rapakivi granites, and late Proterozoic rifting of North America. *Precambrian Research*, **7**, 61–98.

Emslie, R. F. & Hunt, P. A. (1990). Ages and petrogenetic significance of igneous mangerite–charnockite suites associated with massif anorthosites, Grenville province. *Journal of Geology*, **98**, 213–31.

Eriksson, P. G., Callaghan, C. C. & Zawada, P. K., eds. (1991). Precambrian sedimentary basins of Southern Africa. Special issue of *Journal of African Earth Sciences*, **13**, 156 pp.

Fabre, J., ed. (1983). *Afrique de l'Ouest – West Africa*. Oxford, Pergamon Press, 396 pp.

Fahrig, W. F. (1987). The tectonic setting of continental mafic dyke swarms: failed arm and early passive margin, in *Mafic Dyke Swarms*, ed. H. C. Halls & W. F. Fahrig, pp. 331–48. Geological Association of Canada Special Paper 34.

G-Farrow, C. E, & Mossman, D. J. (1988). Geology of Precambrian paleosols at the base of the Huronian Supergroup, Elliot Lake, Ontario, Canada. *Precambrian Research*, **42**, 107–39.

Gaal, G. & Gorbatschev, R., eds. (1987). Precambrian geology and evolution of the central Baltic shield. Special Issue of *Precambrian Research*, **35**, 382 pp.

Garrels, R. M. (1987a). A model for the deposition of the microbanded Precambrian iron formations. *American Journal of Science*, **287**, 81–106.

Garrels, R. M. (1987b). Some factors influencing biomineralization in earth history, in *Origin, Evolution, and Modern Aspects of Biomineralization in Plants and Animals*, ed. R. E. Crick. New York: Plenum Press, 536 pp.

Gibbs, A. K. & Barron, C. N. (1983). The Guiana shield reviewed. *Episodes*, 1983, no. 2, 7–14.

Gower, C. F., Rivers, T. & Ryan, B., eds. (1991). *Mid-Proterozoic Laurentia-Baltica*. Geological Association of Canada Special Paper 38.

Greenberg, J. K. & Brown, B. A., (1984). Cratonic sedimentation during the Proterozoic: an anorogenic connection in Wisconsin and the Upper Midwest. *Journal of Geology*, **92**, 159–71.

Grew, E. S. & Manton, W. I. (1986). A new correlation of sapphirine granulites in the Indo-Antarctic metamorphic terrain: late Proterozoic dates from the Eastern Ghats province of India. *Precambrian Research*, **33**, 123–37.

Grobler, H. J., van der Westhuizen, W. A. & Tordiffe, E. A. W. (1989). The Sodium Group, South Africa: Reference section for Late Archaean–Early Proterozoic cratonic cover sequences. *Australian Journal of Earth Sciences*, **36**, 41–64.

Grotzinger, J. P. (1990). Geochemical model for Proterozoic stromatolite decline. In *Proterozoic Evolution and Environments*, ed. A. H. Knoll & J. H. Ostrom. American Journal of Science vol. **290**A, pp. 80-103.

Grotzinger, J. P. & McCormick, D. S. (1988). Flexure of the early Proterozoic lithosphere and the evolution of Kilohigok basin (1.9 Ga), northwest Canadian shield. In *New Perspectives in Basin Analysis*, ed. K. L. Kleinspehn & C. Paola, pp. 405–30. New York: Springer Verlag.

Halls, H. C. & Fahrig, W. F. (1987). *Mafic Dyke Swarms*. Geological Association of Canada Special Paper 34, 508 pp.

Hambrey, M. J. & Harland, W. B., eds. (1981). *Earth's pre-Pleistocene Glacial Record*. Cambridge: Cambridge University Press, 1004 pp.

Hegner, E., Kroner, A. & Hofmann, A. W. (1984). Age and isotope geochemistry of the Archaean Pongola and Usushwana suites in Swaziland, southern Africa: a case for crustal contamination of mantle-derived magma. *Earth and Planetary Science Letters*, **70**, 267–79.

Hinze, W. J. & Braile, L. W. (1988). Geophysical aspects of the craton: U.S. In *Sedimentary Cover – North American Craton: U.S.: The Geology of North America Volume D-2*, ed. L. L. Sloss, pp. 5–24. Boulder, Colorado: Geological Society of America.

Hoffman, P. F. (1988). United Plates of America, the birth of a craton. *Annual Reviews of Earth and Planetary Sciences*, **16**, 563–603.

Holland, H. D. (1984). *The Chemical Evolution of the Atmosphere and Oceans:*. Princeton, New Jersey: Princeton University Press, 582 pp.

Holland, H. D., Feakes, C. R, & Zbinden, E. A. (1989). The Flin Flon paleosol and the composition of the atmosphere 1.8 BYBP. *American Journal of Science*, **289**, 362–89.

Hoppe, A., Schobbenhaus, C. & Walde, D. H. G. (1987). Precambrian iron formation in Brazil. In *Precambrian Iron Formations*. ed. P. W. Uitterdijk Appel & G. L. LaBerge, pp. 347–90. Athens: Theophrastus Publications, Athens.

Hunter, D. R., ed. (1981). *Precambrian of the Southern Hemisphere: Developments in Precambrian Geology 2*. Amsterdam: Elsevier, 882 pp.

Jahn, B. M., Zhou, X. H. & Li, J. L. (1990). Formation and tectonic evolution of Southeastern China and Taiwan: isotopic and geochemical constraints. *Tectonophysics*, **183**, 145–60.

James, H. L. & Sims, P. K., eds. (1973). Precambrian iron-formations of the world: *Economic Geology*, **68**, 913–1220.

Jannasch, H. G. & Mottl, M. J. (1985). Geomicrobiology of deep-sea hydrothermal vents. *Science*, **229**, 717–25.

Johnson, P. R., Scheibner, E. & Smith, E. A. (1987). Basement fragments, accreted tectonostratigraphic terranes, and overlap sequences: Elements in the tectonic evolution of the Arabian shield. In *Terrane Accretion and Orogenic Belts, Geodynamics Series, vol. 19*, ed. E. C.Leitch & E. Scheibner, pp. 323–43. American Geophysical Union .

Kale, V. S. & Phansalkar, V. G. (1991). Purana basins of peninsular India; a review. *Basin Research*, **3**, 1–36.

Kalsbeek, F., Pidgeon, R. T. & Taylor, P. N. (1987). Nagssugtoqidian mobile belt of West Greenland: a cryptic 1850 Ma suture between two Archaean continents - chemical and isotopic evidence. *Earth and Planetary Science Letters*, **85**, 365–85.

Karlstrom, K. E. & Bowring, S. A. (1988). Early Proterozoic assembly of tectonostratigraphic terranes in southwestern North America. *Journal of Geology*, **96**, 561–76.

Keller, G. R., Lidiak, E. G., Hinze, W. J. & Braile, L. W. (1983). The role of rifting in the tectonic development of the midcontinent, U.S.A. *Tectonophysics*, **94**, 391–412.

Kempe, S. & Degens, E. T. (1985). An early soda ocean? *Chemical Geology*, **53**, 95–108.

Key, R. M., Charsley, T. J., Hackman, B. D., Wilkinson, A. F. & Rundle, C. C. (1989). Superimposed Upper Proterozoic collision-controlled orogenies in the Mozambique orogenic belt of Kenya. *Precambrian Research*, **44**, 197–225.

Khain, V. E. (1985). *Geology of the USSR; First Part – Old Cratons and Paleozoic Fold Belts – Beitrage zur Regionalen Geologie der Erde*. Berlin: Gebruder Borntraeger, 272 pp.

Klevtsova, A. A. (1979). Late Riphean stage of development of the Russian plate. *International Geology Review*, **21**, 167–80.

Knoll, A. H. & Ostrom, J. H., eds. (1990). Proterozoic Evolution and Environments. Special Volume of *American Journal of Science*, **290A**, 332 pp.

Knoll, A. H. & Walter, M. R. (1992). Latest Proterozoic stratigraphy and Earth history. *Nature*, **356**, 673–8.

Korkutis, V. (1981). Late Precambrian and Early Cambrian in the East European platform. *Precambrian Research*, **15**, 75–94.

Kukla, P. A. & Stanistreet, I. G. (1991). Record of the Damaran Khomas Hochland accretionary prism in central Namibia: refutation of an 'ensialic' origin of a Late Proterozoic orogenic belt. *Geology*, **19**, 473–6.

Laayoki, K. (1986). The Precambrian supracrustal rocks of Finland and their tectono-exogenic evolution. *Precambrian Research*, **33**, 67–85.

Laayoki, K. (1990). Early Proterozoic tectofacies in eastern and northern Finland. In *Precambrian Continental Crust and its Economic Resources: Developments in Precambrian Geology, vol. 8*, ed. S. M. Naqvi, pp. 437–52. Amsterdam: Elsevier.

Ledru, P., N'Dong, J. E., Johan, V., Prian, J. P., Coste, B. & Haccard, D. (1989). Structural and metamorphic evolution of the Gabon orogenic belt: collision tectonics in the Lower Proterozoic? *Precambrian Research*, **44**, 227–41.

Lewry, J. F. & Stauffer, M. R., eds. (1990). *The Early Proterozoic Trans-Hudson Orogen of North America*. Geological Association of Canada Special Paper 37, 505 pp.

Lindh, A. (1987). Westward growth of the Baltic shield. *Precambrian Research*, **35**, 53–70.

Lindsley, D. H. & Simmons, E. C., eds. (1990). Anorthosites and associated rocks. Special Section of *American Mineralogist*, **75**, 1–58.

Mabesoone, J. M., de Brito Neves, B. B. & Sial, A. N., eds. (1981). The geology of Brazil. Special Issue of *Earth-Science Reviews*, **17**, 219 pp.

Machado, N., Lindenmayer, Z., Krogh, T.E. & Lindenmayer, D. (1991). U–Pb geochronology of Archean magmatism and basement reactivation in the Carajas area, Amazon shield, Brazil. *Precambrian Research*, **49**, 329–54.

Margulis, L. & Olendzenski, L. (1992). *Environmental Evolution – Effects of the Origin and Evolution of Life on Planet Earth*. Cambridge, Masachusetts: The MIT Press, 400 pp.

Maynard, J. B., Ritger, S. D. & Sutton, S. J. (1991). Chemistry of sands from the modern Indus River and the Archean Witwatersrand basin: Implications for the composition of the Archean atmosphere. *Geology*, **19**, 265–8.

Medaris, Jr., L. G., ed. (1983). *Early Proterozoic Geology of the Great Lakes Region*. Geological Society of America Memoir 160, 141 pp.

Miller, R. McG., ed. (1983). *Evolution of the Damara Orogen of South West Africa/Namibia*. Geological Society of South Africa Special Publication 11, 515 pp.

Moore, J. M., Davidson, A. & Baer, A. J., eds. (1986). *The Grenville Province*. Geological Association of Canada Special Paper 31, 358 pp.

Mukhopadhyay, D., Baral, M. C. & Ghosh, D. (1981). A tectono-stratigraphic model of the Chitradurga schist belt, Karnataka, India. *Geological Society of India Journal*, **22**, 22–31.

Nagy, B., Weber, R., Guerrero, J. C. & Schidlowski, M., eds. (1983). Development and interactions of the Precambrian atmosphere, lithosphere, and biosphere: results and challenges. Special Issue of *Precambrian Research*, **20**, 105–588.

Nalivkin, D. V. (translated by S.I. Tomkeieff) (1960). *The Geology of the U.S.S.R. – A Short Outline: International Series of Monographs on Earth Sciences*. New York: Pergamon Press, 170 pp. plus geologic map of U.S.S.R.

Nalivkin, D. V. (translated by N. Rast) (1973). *Geology of the USSR* . Edingurgh: Oliver & Boyd, 855 pp.

Naqvi, S. M. & Rogers, J. J. W. (1987). *Precambrian Geology of India*. New York: Oxford University Press, 223 pp.

Nicolaysen, L. O. & Reimold, W.U., eds. (1990). Cryptoexplosions and catastrophes in the geological record, with a special focus on the Vredefort structure. Special Issue of *Tectonophysics*, **171**, 422 pp.

Ojakangas, R. W. (1988). Glaciation: an uncommon 'mega-event' as a key to intracontinental and intercontinental correlation of early Proterozoic basin fill, North American and Baltic cratons. In *New*

Perspectives in Basin Analysis, ed. K. L. Kleinspehn & C. Paola, pp. 431–44. New York: Springer Verlag.

Oliver, R. L., James, P. R. & Jago, J. B., eds. (1983). *Antarctic Earth Science.* Cambridge: Cambridge University Press, 697 pp.

Olszewski, W. J., Wirth, K. R., Gibbs, A. K. & Gaudette, H. E. (1989). The age, origin, and tectonics of the Grao Para Group and associated rocks, Serra dos Carajas, Brazil: Archean continental volcanism and rifting. *Precambrian Research,* **42**, 229–54.

Onstott, T. C., Hall, C. M. & York, D. (1989). $^{40}Ar/^{39}Ar$ thermochronometry of the Imataca complex, Venezuela. *Precambrian Research,* **42**, 255–91.

Palmer, J. A., Phillips, G. N. & McCarthy, T. S. (1989). Paleosols and their relevance to Precambrian atmospheric composition. *Journal of Geology,* **97**, 77–92.

Park, R. G. & Tarney, J., eds. (1987). *Evolution of the Lewisian and Comparable Precambrian High Grade Terrains.* Geological Society of London Special Publication 27, 315 pp.

Parker, A. J., Rickwood, P. C. & Tucker, D. H., eds. (1990). *Mafic Dykes and Emplacement Mechanisms.* Rotterdam: A.A. Balkema, 541 pp.

Pin, C. & Poidevin, J. L. (1987). U–Pb zircon evidence for a Pan-African granulite facies metamorphism in the Central African Republic: a new interpretation of the high-grade series of the northern border of the Congo craton. *Precambrian Research,* **36**, 303–12.

Plumb, K. A. (1985). Subdivision and correlation of late Precambrian sequences in Australia. *Precambrian Research,* **29**, 303–29.

Porada, H. (1989). Pan-African rifting and orogenesis in southern to equatorial Africa and eastern Brazil. *Precambrian Research,* **44**, 103–36.

Rainbird, R. H., Nesbitt, H. W. & Donaldson, J. A. (1990). Formation and diagenesis of a sub-Huronian saprolith: comparison with a modern weathering profile. *Journal of Geology,* **98**, 801–22.

Ramo, O. T. (1991). Petrogenesis of the Proterozoic rapakivi granites and related basic rocks of sFennoscandia: Nd and Pb isotopic and general geochemical restraints. *Geological Survey of Finland Bulletin* **335**, 161 pp.

Retallack, G. J., ed. (1986). Precambrian paleopedology. Special Issue of *Precambrian Research,* **32**, 95–259.

Rogers, J. J. W. (1991). Comparison of the Indian and Nubian–Arabian shields. In *Precambrian Continental Crust and Its Economic Resources: Developments in Precambrian Geology 8*, ed. S. M. Naqvi, pp. 223–43. Amsterdam: Elsevier.

Rogers, J. J. W., Dabbagh, M. E., Olszewski, W. J., Jr., Gaudette, H. E., Greenberg, J. K. & Brown, B.A. (1984). Early poststabilization sedimentation and later growth of shields. *Geology,* **12**, 607–9.

Rogers, J. J. W., Ghuma, M. A., Nagy, R. M., Greenberg, J. K. & Fullagar, P. D. (1978). Plutonism in Pan-African belts and the geologic evolution of northeastern Africa. *Earth and Planetary Science Letters,* **39**, 109–17.

Ross, C. P. (1963). *The Belt Series in Montana.* U.S. Geological Survey Professional Paper 346, 122 pp.

Roy, S. (1981). *Manganese Deposits.* London: Academic Press, 408 pp.

Roy, S. (1988). Manganese metallogenesis: a review. *Ore Geology Reviews,* **4**, 155–70.

Schau, M. & Henderson, J. B. (1983). Archaean chemical weathering at three localities on the Canadian shield. *Precambrian Research,* **20**, 189–224.

Schopf, J. W., ed. (1983). *Earth's Earliest Biosphere – Its Origin and Evolution.* Princeton: Princeton University Press, 543 pp.

Schopf, J. W. (1992). Paleobiology of the Archean. In *The Proterozoic Biosphere. A Multidisciplinary Study,* ed. J. W. Schopf and C. Klein, pp. 25–39. New York: Cambridge University Press.

Schopf, J. W. & Klein, C., eds. (1992). *The Proterozoic Biosphere. A Multidisciplinary Study.* New York: Cambridge University Press.

Southgate, P. N. (1989). Relationships between cyclicity and stromatolite form in the Late Proterozoic Bitter Springs Formation, Australia. *Sedimentology,* **36**, 323–39.

Stern, R. J & Dawoud, A. S. (1991). Late Precambrian (740 MA) charnockite, enderbite, and granite from Jebel Moya, Sudan: a link between the Mozambique belt and the Arabian–Nubian shield. *Journal of Geology,* **99**, 648–59.

Stewart, A. J. (1979). A barred-basin marine evaporite in the Upper Proterozoic of the Amadeus Basin, central Australia. *Sedimentology,* **26**, 33–62.

Stewart, J. H. (1972). Initial deposits in the Cordilleran geosyncline: Evidence of a late Precambrian (<850 m.y.) continental separation. *Geological Society of America Bulletin,* **83**, 1345–60.

Stoeser, D. B. & Stacey, J. S. (1988). Evolution, U–Pb geochronology, and isotope geology of the Pan-African Nabitah orogenic belt of the Saudi Arabian shield. In *The Pan-African Belt of Northeast Africa and Adjacent Areas,* ed. S. El-Gaby & R. O. Greiling, pp. 227–88. Braunschweig/Wiesbaden: Fred. Vieweg & Sohn.

Tankard, A. J., Eriksson, K. A., Hunter, D. R., Jackson, M. P. A., Hobday, D. K. & Minter, W. E. L., eds. (1982). *Crustal Evolution of Southern Africa.* New York: Springer Verlag, 523 pp.

Tappan, H. (1980). *The Paleontology of Plant Protists.* San Francisco: W.H. Freeman and Co., 1028 pp.

Texeira, W., Tassinari, C. C. G., Cordani, U. G. & Kawashita, K. (1989). A review of the geochronology of the Amazonian craton: tectonic implications. *Precambrian Research,* **42**, 213–27.

Thomson, M. R. A., Crame, J. A. & Thomson, J. W. (1991). *Geological Evolution of Antarctica.* Cambridge: Cambridge University Press, 722 pp.

Tobi, A. C. & Touret, J. L. R., eds. (1985). *The Deep Proterozoic Crust in the North Atlantic Provinces.* Dordrecht: D. Reidel, 603 pp.

Trendall, A. F. & Morris, R. C., eds. (1983). *Iron-Formation Facts and Problems.* Amsterdam: Elsevier, 558 pp.

Ueda, A., Cameron, E. M. & Krouse, H. R. (1990). ^{34}S-enriched sulphate in the Belcher Group, N.W.T., Canada: evidence for dissimilatory sulphate reduction in the early Proterozoic ocean. *Precambrian Research,* **49**, 229–33.

Van Biljon, W. J. & Legg, J. H., eds. (1983). *The Limpopo Belt.* Geological Society of South Africa Special Publication 8, 203 pp.

Van Reenen, D. D., Roering, C., Ashwal, L. D. and de Wit, M. J. (1992). The Archaean Limpopo granulite belt: tectonics and deep crustal processes. Special Issue of *Precambrian Research,* **55**, 1–587.

Van Schmus, W. R. & Hinze, W. J. (1985). The midcontinent rift system. *Annual Reviews of Earth and Planetary Sciences,* **13**, 345–83.

von der Borch, C. C. (1980). Evolution of late Proterozoic to early Paleozoic Adelaide foldbelt, Australia: comparisons with post-Permian rifts and passive margins. *Tectonophysics,* **70**, 115–34.

Walker, J. C. G. (1977). *Evolution of the Atmosphere.* New York: Macmillan Publishing Co., 318 pp.

Walter, M. R., Buick, R. & Dunlop, J. S. R. (1980). Stromatolites 3,400–3,500 Myr old from the North Pole area, Western Australia. *Nature,* **284**, 443–5.

Walter, M. R. & Heys, G. R. (1985). Links between the rise of the metazoa and the decline of stromatolites. *Precambrian Research,* **29**, 149–74.

Wang, H. & Qiao, X. (1984). Proterozoic stratigraphy and tectonic framework of China. *Geological Magazine,* **121**, 599–614.

Wang, Z., Cheng, Y. & Hongzhen, W. (1986). *The Geology of China.* Oxford: Oxford University Press, 303 pp.

Wiggering, H. & Beukes, N. J. (1990). Petrography and geochemistry of a 2000–2200-Ma-old hematitic paleo-alteration profile on Ongeluk basalt of the Transvaal Supergroup, Griqualand West, South Africa. *Precambrian Research*, **46**, 241–58.

Woese, C. R. (1987). Bacterial evolution. *Microbiological Reviews*, **51**, 221–71.

Woese, C. R., Kandler, O. & Wheelis, M. L. (1990). Towards a natural system of organisms: Proposal for the domains Archaea, Bacteria, and Eucarya. *Proceedings of the National Academy of Sciences*, **87**, 4576–9.

Wyborn, L .A. I. & Etheridge, M .A., eds. (1988). The early to middle Proterozoic of Australia. Special Issue of *Precambrian Research*, vols. **40/41**, 588 pp.

Young, G. M. (1981). Upper Proterozoic supracrustal rocks of North America: a brief review. *Precambrian Research*, **15**, 305–30.

Young, G. M. (1988). Proterozoic plate tectonics, glaciation and iron-formations. *Sedimentary Geology*, **58**, 127–44.

Zhang, M., Liou, J. G. & Coleman, R. G. (1984). An outline of the plate tectonics of China. *Geological Society of America Bulletin*, **95**, 295–312.

transition

Precambrian/Cambrian sedimentary sequence on Burin Peninsula, Newfoundland, Canada. The zone of first appearance of complex burrowing organisms is shown by the arrow (labeled 'transition') as the lowest Cambrian. (Courtesy of E. Landing, New York State Geological Survey.)

6

THE PALEOZOIC — PART I.
LIFE, CLIMATES AND OCEANS

6.0 Introduction

LIFE as we know it, and an earth's surface that would be familiar to us, both began in the Paleozoic. The era started with a purely soft-bodied fauna and a land surface barren of life (except for bacteria). It ended with a group of marine and terrestrial animals and plants that are at least broadly similar to modern forms; indeed, some varieties are still in existence. All of the animal phyla and plant divisions had evolved by the end of the Paleozoic, most of them much earlier in the era, and many of the subphyla and classes had also appeared.

The Paleozoic also witnessed a reorganization of continental blocks from some conjectural end-Proterozoic starting point to an end-Paleozoic Pangea whose broad outlines are reasonably well known. The effort to decipher this tectonic activity is both tantalizing and frustrating. Because of the comparative 'modernness' of the Paleozoic, and the fossil content of its rocks, we have the feeling that we should be able to construct a detailed history of Paleozoic plate movements, sealevel variations, and orogeny. Because of its age, however, the Paleozoic record commonly is buried beneath younger sediments, or uplifted and eroded, or overprinted by a later orogeny. Paleozoic

ocean crust can be found only in ophiolites, and none remains in ocean basins. Thus, we cannot make statements about the Paleozoic with the precision that is possible for the Mesozoic and Cenozoic.

This chapter discusses the evolution of Paleozoic organisms, their provincial distribution into realms, and their relationships to climate and ocean history. The next chapter (Chapter 7) discusses the tectonic evolution of the earth in the Paleozoic, including plate movements, orogenic belts and the development of cratonic platforms. Some information from Chapter 7, particularly regarding plate movements, is needed for this chapter; we will borrow it without further ado.

We start this chapter with the transition from the Proterozoic to the Cambrian. This boundary is one of the most remarkable in earth history because it shows a profound change in the nature of organisms without any significant change in the sedimentary section. All of a sudden, animals learned to secrete skeletal parts! Why did they do so? Why can we not detect some environmental change that accompanied, or perhaps caused, this evolution? Why did nearly all of the major types of animals appear in the geologic record almost immediately after this transition? We discuss these issues in Sections 6.1 and 6.2.

The next three sections are somewhat detailed descriptions of the evolution of invertebrates (6.3), vertebrates (6.4), and plants (6.5). This information is mostly biologic and is intended to set the stage for the summary of evolutionary changes in Section 6.6.

After our discussion of the evolution of separate groups of organisms, we combine this information into a summary of biotic changes through the Paleozoic (Section 6.6). During the Paleozoic, the evolutionary process developed: 1) a fish–amphibian–reptile lineage; 2) an enormous diversity of trilobites, all of which became extinct at the end of the era; 3) true plants that could thrive on land; and 4) an invertebrate fauna largely dominated by brachiopods. No calcareous plankton are known during the entire Paleozoic. Land plants did not evolve until the end of the Silurian or beginning of the Devonian, and the first animals followed them onto the land almost immediately afterward; insects may have evolved with the plants.

Section 6.6 summarizes the nature of life at various times in the Paleozoic that can be separated from each other by 'mass extinction' events, a term that may be inappropriate for most biotic transitions. The problem is whether many organisms became extinct at one 'instant' or whether the evolutionary changes proceeded at a more gradual pace. This problem is contentious even for the better-preserved record of the event between the Cretaceous and Tertiary (Section 8.7), and the Paleozoic data are far less clear.

The evolutionary changes occurred as continents and ocean basins were being reorganized by seafloor spreading. These plate movements caused the establishment of various biotic 'realms' characterized by different types of organisms. (Some of the plate movements are, in actuality, inferred from the faunal and floral distributions.) Most of the earth's continental fragments spent most of the Paleozoic in roughly equatorial positions except for the continuously South Polar, and glaciated, position of parts of Gondwana. Section 6.7 shows the relationships between plate positions and movements, provincialism among animals and plants (biotic realms), and climate.

Section 6.8 discusses ocean and atmosphere compositions. We have very little evidence for inferring compositions of either air or sea water that are greatly different from those of the present. Most of the climatic change indicated by animals or plants is probably the result of plate movements across latitudes rather than global changes in the atmosphere. Furthermore, Paleozoic organisms are sufficiently similar to modern ones that neither oceans nor atmosphere could have been enormously different from their present states.

The end of the Paleozoic is almost as enigmatic as its beginning. The largest extinction in earth history occurred over a span of time whose duration is controversial but could not have been greater than a few million years. It occurred immediately following a major glaciation, and the emergence of most continental areas above sealevel prevented preservation of continuous sections across in the boundary except in rare locations. We discuss this event in Section 6.9.

6.1 The beginning of the Cambrian

IN the early days of geological investigations, most geologists proposed (or assumed) that there was a major break between the old, 'pre-Cambrian', lifeless rocks and the fossil-bearing rocks of Cambrian and younger ages. This hiatus was commonly referred to as the 'Lipalian Era' (or 'Lipalian Interval') and was regarded as a time of major erosion following an end-of-Proterozoic orogeny and preceding the encroachment of life-bearing ocean waters back onto the continents. Indeed, many continental platforms exhibit a major unconformity below Cambrian rocks, and an interval of geologic time that was not recorded in the stratigraphic record provided an explanation for the sudden appearance of a diversified fauna in the earliest Cambrian.

These early concepts of universal unconformity and sudden appearance of life had to be discarded as more geologic work was done. Now, numerous areas are known in which sedimentation appears to have occurred continuously across the Proterozoic/Cambrian boundary (see below). Furthermore, the finding of abundant soft-bodied fauna in the late Precambrian (Ediacaran) demonstrates a much longer period of organic evolution that probably resulted in the formation of all five of the kingdoms of organic life (Section 5.7). The time of transition from Precambrian to Cambrian is variously referred to as 'Eocambrian' or 'Infracambrian'.

Fig. 6.1 diagrammatically shows relationships in two sections (southeastern China and eastern Newfoundland) that have been proposed as strato-

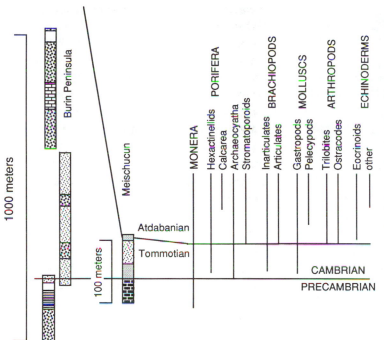

Sandstone, siltstone, shale

variegated red and green

gray/black, with pyrite and phosphate

undifferentiated

Gray/green sandstone

Dolomite (mostly cherty and phosphatic)

Dolomite and limestone

Fig. 6.1. Stratigraphic sections across the Precambrian/Cambrian boundary at the Burin Peninsula, Newfoundland, and Meishucun, China. Faunal ranges are shown diagrammatically.

The Precambrian/Cambrian boundary is commonly within a conformable sequence of sediments. It is defined by a change from trace fossils and purely soft-bodied forms upward through a zone of enigmatic 'small shelly' fossils to skeletal remains of recognizable Paleozoic genera. This diagram shows no lithologic change across the boundary at the Burin Peninsula and a minor upward decrease in the abundance of phosphatic, cherty, dolomite at Meishucun.   ☐

Fig. 6.2. *Phycodes pedum* burrow at Burin Peninsula, Newfoundland. The ability of organisms to burrow into sediments, rather than only graze on top of them, is the earliest indication of the transition from Precambrian to Cambrian. (Courtesy of E. Landing, New York State Geological Survey.)   ☐

Fig. 6.3. 'Small shelly' fossils of the Placentian Series at the Burin Peninsula from horizons slightly above the first development of burrowing organisms and far below the first appearance of trilobite skeletons. Most of the small shelly fossils are a few millimeters in diameter. C, E, G, J and N are conchs, including internal casts, of primitive hyolithids. All other specimens are of uncertain affinity: A, B and D are conchs and operculum of '*Ladatheca*'; F, I, K and O are calcareous conical tubes of *Conotheca*; H, M and P–R are apatite-cemented tubes of *Plinthoconion*; and L is an internal cast of the calcareous *Anabarites*. (Courtesy of E. Landing, New York State Geological Survey.) □

African fragment that was shed from Gondwana and collided with North America in the Paleozoic by closure of the Iapetus Ocean (Section 7.3). Other stratotype sections under consideration at this time (1992) are in Siberia and Morocco and are not depicted here.

The major changes upward (toward younger rocks) associated with the Precambrian/Cambrian boundary can be summarized briefly (Fig. 6.1). One is that trace fossils (ichnofossils), become more diverse and more complex upward, particularly in clastic facies (Fig. 6.2). These traces include burrows, trails, scratch marks, and resting places, all of various origins. A number of the ichnogenera in these sequences occur only in rocks below the presumed Cambrian boundary, and genera that occur above the boundary commonly have long ranges into the Phanerozoic. A notable feature is that vertical tubes, used by efficient burrowers, occur only above the base of the Cambrian.

A second change at the base of the Cambrian is that 'small shelly fossils' appear in the section slightly above the horizon of increase in diversity of the ichnofossils (Fig. 6.3). These shelly fossils consist of a diverse fauna, many of which cannot be related to any major living or fossil groups. Apparent molluscs are common, some of them related to gastropods; some forms appear to have been unrelated to major molluscan classes and to have become extinct early in the Cambrian. Tube-like organisms of uncertain classification include organic-walled forms (sabelliditiids) and phosphatic-walled forms (tommotiids, named for the Tommotian, the basal Cambrian section of Siberia). Archaeocyathids were sponge-like organisms (but probably a separate phylum) that formed calcareous mats and became extinct by the Middle Cambrian. Organic- (and phosphatic?) walled microorganisms and siliceous sponge spicules were also present among the earliest skeletal fossils. The fossils that traditionally mark the start of the Cambrian are trilobites, but they appear much higher in the section than the small shelly fossils (1000 m higher in the clastic rocks of Newfoundland).

As implied by the non-specific term, the 'small shelly fossils' are of problematic origin. Although originally regarded mostly as the secretions of organisms of about the same size as the shells (e.g. similar to a modern clam), paleontologists have always considered the possibility that the individual 'shells' might be platelets or other partial coverings of larger organ-

types for the Precambrian/Cambrian transition. The section in China is a comparatively thin suite of phosphatic carbonate (mostly dolomite) and black shale in the South China block (Section 7.2), which was probably attached to Gondwana at the start of the Cambrian. The section in Newfoundland is a thick suite of shallow-marine siliciclastic rocks (sandstones and siltstones) in the Avalon terrane, an apparently Panisms. Such hard coverings may have been unarticulated or poorly articulated and simply dropped off and dispersed when the organism died. Evidence that at least some of the 'shelly' material comprised platelets on larger organisms has recently been found in the impressions of organisms preserved in shales.

Most geologists now place the Precambrian/Cambrian boundary in the sedimentary interval of increasing diversity of the ichnofossils and the first appearance of shelly fossils. Small amounts of iridium have suggested a meteorite event, but the anomalies are too small and scattered to be definitive. Carbonate rocks at the boundary record a very high $^{87}Sr/^{86}Sr$ ratio, indicative of great exposure of continental surface and erosion of radiogenic Sr into the oceans; this observation is consistent with sealevel lowering during a late Proterozoic glaciation (Varanganian glaciation; Section 5.6) and subsequent sealevel rise. The best estimate of the age of the transition is currently ~570 Ma, but very recent work suggests a younger age of about 550 to 545 Ma.

[**References** – The late Precambrian and the transition to the Cambrian are discussed in volumes edited by Trompette and Young (1981) and Cowie and Brasier (1989) and a paper by Crimes (1987). Specific studies include Narbonne *et al.* (1987) and Landing *et al.* (1989) on the Precambrian/Cambrian transition zone in Newfoundland, Brasier *et al.* (1990) on the transition in China, Latham and Riding (1990) on a transition in Morocco, and Conway Morris and Peel (1990) on the discovery of articulated halkieriid fossils.]

6.2 Organic extinction and radiation across the Precambrian/Cambrian boundary

REGARDLESS of its exact stratigraphic definition, the beginning of the Cambrian was a time of major change in the types of organisms and the ways in which they lived. We can discuss these changes by comparing: 1) late-Proterozoic fossil suites (Ediacaran; Section 5.5); 2) organisms that evolved and began to secrete skeletal material during the earliest Cambrian (Figs. 6.2 and 6.3); and 3) the remarkably well-preserved fossils of the Middle-Cambrian Burgess Shale. This comparison provides an age range of some 100 to 150 million years, from about 600 to 650 Ma in the Ediacaran suite to about 500 to 550 Ma in the Burgess Shale. This profound evolutionary change in the

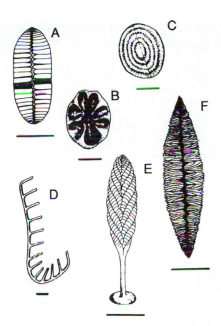

Fig. 6.4. Ediacaran-age fossils at Mistaken Point, Newfoundland. The fossils are described simply as spindle- or disc-shaped plus reference to the distribution of lobes and depressions. Specimen E is *Charnia Masoni* Ford. The bars show the length of three centimeters on each fossil. (From King, 1988.) ☐

nature of the soft-bodied organisms occurred within the same time range as organisms began to secrete skeletons.

The soft-bodied fossils preserved in the fine-grained quartz sandstones of the Ediacara Hills of South Australia, and now correlated with numerous other locations, provide an exceptional view of the diversity of late-Proterozoic life (Fig. 6.4). The Ediacaran biota are controversial organisms that range up to fairly large sizes (several tens of centimeters). Many of the forms have been regarded as medusoids (jellyfish) or other cnidarians, or as worms. Some paleontologists, however, have proposed that the Ediacaran organisms contain few, if any, metazoans and are so different from modern forms that they must be regarded as wholly separate phyla that did not survive past the end of the Proterozoic. Furthermore, it is possible to interpret the frond-like shape of some fossils as an attempt to increase their surface area so that they could feed autotrophically (i.e. by absorbing light or dissolved chemicals). An absence of predators is indicated by the large size of these exclusively soft-bodied organisms, which should have provided a tasty meal for anything capable of eating them.

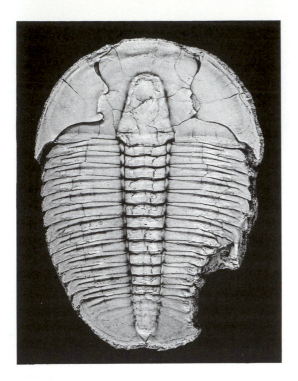

Fig. 6.5. A trilobite, *Elrathia kingii*, from the Middle Cambrian of western Utah, showing a healed bite mark on the right side. The injury is inferred to have been inflicted by the large, enigmatic, predator *Anomalocaris*. Preponderance of healed injuries on the right side of trilobites suggests that behavioral lateralization (handedness), which is related to right–left lateralization of the nervous system, was well established in organisms by the Early Cambrian. Magnification X2. (Courtesy of L. E. Babcock, Ohio State University.)   □

No representatives of Ediacaran fossils have ever been found above the base of the Cambrian, although microfossils in the general age range of 800 to 600 Ma show similarities to forms from Middle-Cambrian and younger suites. These observations lead to the important conclusion that the Precambrian/Cambrian boundary was a time of major organism extinction as well as a time of the beginning of radiation of new forms.

Whatever organisms were responsible for the 'small shelly fossils' at the base of the Cambrian apparently did not survive for more than a few million years. Within a period of about 10 million years in the Early Cambrian, they were replaced by organisms that secreted recognizable hard parts and included most of the animal phyla alive today (most of the remainder of the phyla appeared by the start of the Ordovician). This 'explosive radiation' could signify either that all of these phyla were established in the Precambrian and merely learned to secrete skeletal material in the Early Cambrian or that rapid evolution produced these diverse phyla within a few million years from a more primitive ancestry. All of the major minerals now found in skeletal material, including phosphate, silica, and calcite, were used in one or another of the Early-Cambrian organisms, although phosphatic material was more abundant than later in the Phanerozoic.

The Middle-Cambrian Burgess Shale is a dark-gray, basinal rock deposited against the flanks of reef structures in the southern Rocky Mountains of Canada. It was deposited rapidly by turbidity flows into the basinal deeps, which permitted both preservation of hard parts and also detailed impressions of soft parts on the shaly surfaces. Many of the preserved impressions are of animals that had only soft parts and do not fit into any modern classification (one of these has been enticingly named *Hallucigenia*.). Most of the Burgess Shale fauna are arthropods, with other varieties including sponges (very abundant), brachiopods and annelid worms. As in other Cambrian rocks, molluscs tend to have simple (primitive?) shells. Similar fauna have not been found elsewhere except in a newly discovered Middle-Cambrian suite in China.

Regardless of the classification of organisms in the Burgess Shale and in other Cambrian rocks, one important characteristic of the fauna is clear. Cambrian faunas were dominantly heterotrophic (dependent on other organisms for food). This heterotrophy was manifested in various ways, including grazing on algal mats, suspension- and deposit-feeding on organic remains in the sea water or the substrate, and predation of living animals. The first predators were sufficiently complex that they exhibited 'behavioral lateralization', particularly demonstrated by the preponderance of 'healed injuries' on the right sides of trilobites rather than on their left sides (Fig. 6.5). Whether heterotrophy existed in Precambrian faunas is not certain, but if it did not exist, then this mode of nutrition developed as rapidly as the new phyla and their ability to secrete hard parts.

Summary and discussion of changes across the Precambrian/Cambrian boundary

Let us list the observations that must be accounted for in any explanation of the changes that occurred at the beginning of the Cambrian.

1. At the start of the Cambrian, organisms began to secrete articulated skeletal material that could be preserved in the geologic record. This ability to form an (exo- or endo-) skeleton represented a significant evolutionary advance from the previous ability of organisms merely to induce an environment in which precipitation could occur (biologically mediated precipitation). Examples of the latter are calcareous stromatolites formed by blue-green algae and ironstones, possibly mediated by bacteria (Section 5.6).

2. Virtually all of the minerals that are now secreted by organisms were also used by Early-Cambrian forms of life, including chitinophosphatic materials, calcium carbonate, and silica. Phosphate was a much higher percentage of secreted material than it has been after the Cambrian.

3. The Cambrian period contained representatives of all of the animal phyla, and nearly all of the classes, that now exist on the earth (this statement depends somewhat on the classification used). Most of the plant divisions and many of the plant-like forms of Protoctista, however, did not evolve until later in the Paleozoic or the Mesozoic.

4. The base of the Cambrian may have been a change from autotrophic to heterotrophic animal nutrition. If so, then the multitude of Cambrian phyla must have evolved from unknown precursors within a few millions, or tens of millions, of years. Predation on other animals and grazing on algae, however, may have developed as an evolutionary adaptation within existing phyla (without introduction of new phyla).

5. The start of the Cambrian was a time of rapid sealevel rise following a long period of continental emergence and late-Proterozoic glaciation. Also at this time, virtually all continental blocks were in equatorial or temperate latitudes (Section 6.7), indicating that the seas were advancing onto the cratons in areas of abundant heat and light.

6. Many of the Precambrian life forms appear to have become extinct, indicating that the Precambrian/Cambrian transition was both an extinction and a radiation event.

Why did all of these events occur within a period of time no longer than a few tens of millions of years and possible shorter? The explanations basically fall into two categories. One assumes that all biologic changes were caused by environmental changes, such as sealevel rise, changes in composition of oceans and/or atmosphere, etc. The other set of arguments proposes that organisms were capable of rapid evolution without a specific external stress. The evolutionary, non-environmental, argument is difficult to address, and we merely discuss possible effects of environmental changes below.

One possibility is that rapid sealevel rise permitted the establishment of many new ecologic niches on the newly flooded continents. These niches caused faunal diversification (an 'explosive evolution' of new forms) and the development of predators that permitted survival only of organisms that managed to protect themselves by the secretion of a skeleton. This rise in sealevel is firmly established in platform sedimentary sequences (Section 7.1), but whether it was responsible for the development of organism skeletons is problematic.

A second explanation is based on the warmth of Early-Cambrian epicontinental seas. There is no firm evidence, however, that simple warming could cause the organic changes that accompanied the transition to the Paleozoic. Furthermore, the sealevel rise may not have been caused by glacial melting but could have been associated with rifting of a late-Proterozoic supercontinent in much the same way as the Middle-Cretaceous sealevel rise occurred during the enhanced seafloor spreading that accompanied the last rifting of Gondwana (Section 8.1); in that situation, no climatic inference can be drawn.

A third explanation is that the oxygen content of the atmosphere, and consequently of the oceans, increased at the Proterozoic/Cambrian boundary. This proposal is based on the observation that modern metazoan animals do not secrete skeletons in aquatic environments in which the oxygen content of the water is less than an equivalent pressure of about 0.004 atm., in comparison with a normal present atmosphere containing O_2 at 0.2 atm. A further consideration is that animals with exoskeletons would probably require higher O_2 pressure than animals without skeletons in order to pass O_2 into their systems. Thus, an increase in atmospheric and oceanic oxygen may be required for the evolution of skeletons but might not cause that evolution.

[References – General books on invertebrate paleontology are by House (1979) and Clarkson (1986). Books on biomineralization are by Lowenstam and Weiner (1989) and Simkiss and Wilbur (1989). A general discussion of the evolution of the atmosphere and oceans is presented by Holland (1984). Specific discussions of the paleontology are from: a paper by Brasier (1982) and a book by McMenamin and McMenamin (1989) on the paleontology of the transition zone; a book on the evolution of animal life, including the Ediacara biota, by Glaessner (1984); a study of very old trace fossils in China by Junyuan et al. (1991); a book on the Burgess Shale by Whittington (1985), and a paper on behavioral lateralization by Babcock (1993). Drawings of Ediacaran-age biota are from King (1988).]

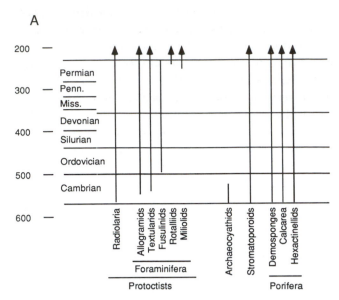

Fig. 6.6. Ranges of invertebrates and protoctists in the Paleozoic. Virtually all of the varieties of invertebrates that have occupied the earth are shown here, but a few groups that existed only in post-Paleozoic time are in Section 8.6. Most of the criteria for distinguishing higher taxonomic levels (e.g. phyla) are given in the text, but additional information and a more detailed classification is presented here. To the right of the column that defines the period boundaries, the horizontal lines represent boundaries between biotic intervals discussed in the text ('extinction events'?). These boundaries are: Precambrian to Cambrian; Cambrian to Ordovician; Ordovician to Silurian; Frasnian to Famennian (Late Devonian); and Permian to Triassic; the boundary between the primitive fauna of the earliest Cambrian and later Cambrian organisms is discussed in Section 6.1.

(A) Protoctists and porifera. Protoctists include: 1) Foraminifera, which are heterotrophic and either benthic or planktic ; 2) Radiolaria, which are heterotrophic, planktic, and composed of a meshwork of opaline silica; 3) Diatomacea, which are siliceous, planktic, autotrophic, and consist of two opaline shells that fit into each other like a pill box; and 4) Coccolithophoridae, which are calcareous, planktic, autotrophic and secrete tests composed of highly ornamented, shield-like plates, referred to as coccoliths. Foraminifera are classified as follows: Allogromids have a purely organic test. Textularids have a test formed by agglutinated sand and silt (arenaceous test). Fusulinids formed a test of small calcite grains, perhaps partly by agglutination. Two foraminiferal groups have tests of chemically precipitated calcium carbonate, the rotaliids and the miliolids. Rotaliids have their calcite tests perforated by pores, whereas the miliolids are imperforate.   □

6.3 Organic evolution in the Paleozoic – invertebrates

AFTER the extraordinary events that marked the transition from the Precambrian to the Cambrian, the fossil record begins to show continuity of organic evolution and extinction. Although there is no complete proof that much of the framework for this evolution took place in the Precambrian, the establishment of the five kingdoms of organic life was discussed in Section 5.7. In this chapter, we discuss Phanerozoic diversification, both with respect to the origins of the animal (metazoan) phyla and plant (metaphyte) divisions and to the ranges of the major groups; we will not discuss the Monera (prokaryotes which do not seem to have changed very much) or the Fungi (which have virtually no fossil record). We start our discussion with the Protoctista, the most primitive organisms except the Monera, and then turn to major trends in the evolution of the metazoans, which presumably had a eukaryotic protoctist ancestor. The ranges of the various groups are shown in Fig. 6.6A–F. (The taxonomy used here is only one of many that have been proposed.)

Protoctista

The simplest organisms are the protoctists (also referred to as protists), which had developed as eukaryotes by the end of the Proterozoic (Section 5.7). They diversified during the Paleozoic and can be grouped into four phyla (subkingdoms?): foraminifera, radiolaria, diatoms, and coccolithophores (Fig. 6.6A). The foraminifera and radiolaria are heterotrophic (depending on other organisms for food), and the diatoms and coccolithophores are photosynthetic (autotrophic). Both the diatoms and the coccolithophores developed in the Mesozoic and are discussed more completely in Section 8.6.

The protoctists provide one indication of the incompleteness of our knowledge of evolution – the absence of any record of autotrophic (including plant) microorganisms in the Paleozoic. The abundant animal life in the Paleozoic oceans, including filter-feeding benthos, required a primary, planktic, base of the food supply that presumably included photosynthetic organisms such as modern diatoms and coccolithophorids. No such organisms, however, are preserved in Paleozoic sedimentary rocks with the possible exception of uncertain chitinous forms such as acritarchs and chitinozoans.

Foraminifera have been so abundant and evolved so rapidly that they are ideal index fossils. Caution must be used, however, because the foraminifera reproduce both sexually and asexually, leading to 'dimorphic' tests of the same species that are slightly different. A first-order classification is based on the material used for tests. One variety uses sand and silt to produce 'agglutinated' forms; the second variety precipitates calcareous tests. All planktic forms are calcareous, but benthic forms are of both types. Benthic foraminifera have been present throughout the Phanerozoic record, but planktic foraminifera did not evolve until the Mesozoic.

Parazoa

Some of the simplest multicellular organisms are an association of cells without a regular shape, such as the Parazoa, whose major fossil representative is the Porifera (sponges; Fig. 6.6A). Because these animals commonly lack regular skeletons, most of their fossil record consists of the siliceous or, less commonly, calcareous spicules that they secrete as stiffening agents. Sponges with calcareous spicules belong to the Calcarea. The Demospongea secreted isolated siliceous spicules, but the more advanced Hexactinellida formed a network of six-rayed siliceous spicules.

The sponges are similar to two other groups of marine organisms that possibly should be regarded as separate phyla. One was the Archaeocyatha, which flourished in the Early and Middle Cambrian and became extinct before the end of the Cambrian. The archaeocyathids formed cup-shaped calcareous skeletons that have been, at various times, classed with the sponges or primitive coelenterates (cnidaria). The second group is the Stromatoporoidea, which formed layered and structured calcareous masses, mostly in the middle Paleozoic.

Cnidaria

A more complex multicellular organization than the Parazoa is the Cnidaria (mostly 'coelenterates'), a phylum that includes modern sea anemones, jellyfish, and corals (Fig. 6.6B). A principal feature of the cnidaria is the presence of stinging cells that assist the animal in feeding and defense. The 'primitive' character, which distinguishes the phylum from all 'higher' animals, is a 'diploblastic' body structure. Diploblastic animals have only an outer layer (ectoderm) separated by an undifferentiated mesogloea from an inner layer (endoderm) which surrounds their body cavity (enteron); thus the body cavity approximates the shape of the outer part of the animal. The Hydrozoa have a smooth enteron, and the Anthozoa (corals) have an enteron

divided by nearly radial wedge-shaped spaces between which most forms secrete calcareous septa. Some of the Hydrozoa form branching or encrusting colonies that secrete calcite (millepores or hydrocorals), but the major fossil record of the Cnidaria is found in the corals. The major Paleozoic corals, tabulate and rugose varieties, became extinct at the end of the era, but one living order (Scleractinia, or hexacorals) evolved in the Triassic and now approximately occupies the niches of the Paleozoic forms.

Lophophorata (Brachiopoda and Bryozoa)

Two groups of organisms are distinguished by the presence of lophophores. The lophophore is constructed somewhat differently in different phyla but basically is an organ with cilia that propel water for feeding and respiratory purposes. The phyla that contain this organ include the brachiopods (known from the Early Cambrian to the present), the bryozoa (Ordovician to present), and a group of soft-bodied animals known as phoronids, which have virtually no fossil record. Because the phoronids seem to be the most 'primitive', they may be the ancestors of other lophophorates.

The lophophorates, and all multicellular animals except the sponges and cnidaria, are 'triploblastic'. The term indicates the presence of a mesoderm lining an area between the ectoderm and endoderm, thus forming a body cavity known as the 'coelom' that characterizes animals referred to as 'coelomates'. (In humans, the coelom is the part of the body between the skin and the alimentary canal and intestines.)

The evolution of triploblastic anatomy is conjectural. The most primitive known organisms that can be described as triploblastic are flatworms (Turbellaria, Platyhelminthes), which are bilaterally symmetrical and have a mass of undifferentiated tissue between their endoderm and ectoderm. The flatworms need little more than a shell and gills to become molluscs, but their relationship to lophophorates and other animals is less clear.

The brachiopods can be divided easily into two classes, Inarticulata and Articulata, depending on whether or not they have teeth and sockets that 'articulate' their two valves (Fig. 6.6B). All varieties are solitary (non-colonial), and most attach themselves to the substrate by means of a single foot 'pedicle'. Their two shells are not identical to each other, but both are generally bilaterally symmetrical (in contrast to the pelecy-

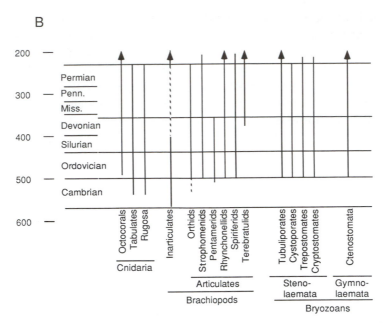

Fig. 6.6B. Cnidaria, brachiopods and bryozoa. Cnidaria include the Hydrozoa (e.g. anemones), Scyphozoa (e.g. jellyfish), Anthozoa (corals), and possibly Conulariida. The term 'coral' should technically be restricted to the Zoantharia (rugose, tabulate and scleractinian corals) but also is commonly applied to the Octocorallia (sea fans, organ-pipe corals, etc.), which are difficult to preserve and have a poor fossil record. The tabulate corals are colonial and calcitic, show very minor septa, and are characterized by numerous calcite partitions (tabulae) perpendicular to the length axis of the coral; they pack together in colonies so that individual animals adjoin each other; a related order (Heliolitidae) that lived in the Ordovician to Devonian is similar except that the individual organisms were separated by additional calcite. The rugose corals have prominent septa that ideally show a bilateral symmetry resulting from growth of four initial septa; they are both colonial and individual. The scleractinians have a six-fold symmetry of septa and are major constituents of modern reefs.

The lingulid brachiopods secrete a chitinophosphatic shell and have a slightly different form of pedicle attachment than the acrotretids, which can be either phosphatic or calcareous. The orthid brachiopods are characterized by primitive features such as a straight hinge line and simple internal structures. The five other brachiopod orders include: Strophomenida, which have a straight hinge line, simple or reduced teeth, and commonly one valve that is concave (in distinction with the biconvex nature of most other brachiopods); Pentamerida, which mostly have non-straight hingelines and an incurved beak; Rhynchonellida, which have a non-straight hinge line and a spoon-shaped internal muscle support; Spiriferida, which have a spiral internal support for the lophophore; and Terebratulida, which have a very short hinge area and a loop support for the lophophore.

The classification of Bryozoa is difficult because no single characteristic is diagnostic of the three classes. The two classes with fossil records are the Stenolaemata and the Gymnolaemata. The Stenolaemata contain several orders, all of which appeared at virtually the same time in the Ordovician and, thus, must have evolved from some common (but unknown) ancestor. The orders of Stenolaemata are: Tubuliporata (Cyclostomata), with long, closely packed zooidal structures; Cystoporata, which are similar to the Tubuliporata but have more material between the zooids and form encrusting or massive structures; Trepostomata, which form 'stony', mostly dendroidal colonies that show little pattern of the zooids; and Cryptostomata (including Fenestrata), which form delicate, dendroidal colonies. The only order of the Gymnolaemata that occurs in the Paleozoic is the Ctenostomata, whose members do not secrete calcite and are preserved almost entirely as borings. An order that secretes calcite, the Cheilostomata, formed encrusting or erect colonies of short, 'box-like' zooids and appeared in the Jurassic, presumably by evolution from other older varieties of Gymnolaemata. ☐

pods, whose shells are not symmetrical but are commonly mirror images of each other). Inarticulate and primitive articulate brachiopods suggest a common ancestry by their similar anatomies and presence in the Early Cambrian. Most inarticulates died out in the Cambrian, but two orders (Lingulida and Acrotretida) have persisted relatively unchanged to the present. The articulate brachiopods all have calcareous shells, and

the caption for Fig. 6.6 shows a division into six major orders. The orthids appeared in the Early Cambrian and appear to have been ancestral to the other five orders, which evolved at later times in the Paleozoic.

The bryozoa (Ectoprocta) are all colonial aggregations of individual 'zooids' and form encrusting, branching, and some massive growths. Some bryozoa are soft-bodied, but many secrete calcite, and they have been an important part of the fossil record throughout much of the Phanerozoic (Fig. 6.6B).

Annelida and Arthropoda

At first glance, the annelids (a type of 'worm') and the arthropods seem to have little in common except that they are both triploblastic. The relationship between them is shown by body segmentation. In the simplest case of the annelids, this segmentation appears as a replication of body parts in segments perpendicular to the length of the organism. The segmentation becomes more complex in the arthropods, which are also characterized by chitinous body coverings and specialized, jointed, appendages. The annelids do not secrete hard skeletons but have a trace-fossil record from the late Proterozoic to the present; they now include the common earthworm. The arthropods are an exceptionally diverse group of organisms that has successfully colonized most ecological niches. Among the 'achievements' of the arthropods are adaptation to land in the Devonian, adaptation to flight in the Pennsylvanian, and development of the largest number of species (mostly insects) of any modern phylum. Some workers regard the arthropods as several unrelated phyla.

Because of the numbers and complexity of the arthropods and the limited fossil record of many groups, we cannot devote much space to their classification. A first division may be made between arthropods that: 1) have exclusively non-branching (uniramous) appendages, which includes the Myriapoda (modern centipedes and millipedes) and Hexapoda (modern insects); and 2) have some branching (biramous) appendages, including Chelicerata, Crustacea (including ostracods) and Trilobita.

The marine groups with the largest fossil record are the ostracods, which have been stratigraphically useful throughout the Phanerozoic, and the trilobites (Fig. 6.6C). The trilobites, characterized by three-lobed exoskeletons, were dominant in the Cambrian, less important in the Ordovician, and became extinct at the end of the Paleozoic. The most primitive trilobites appear to be the olenellids (Fig. 6.7A), related to the order (?) Redlichiida. Because so many trilobite genera died out at the end of the Cambrian, it is not clear whether these early trilobites were ancestral to younger ones or whether the younger trilobites developed from varieties that are not preserved.

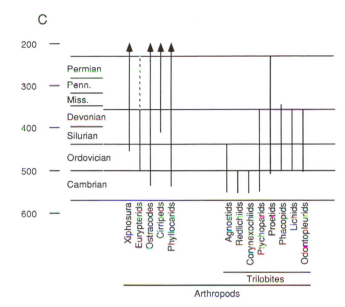

Fig. 6.6C. Arthropods. The chelicerates are characterized by a pair of jointed pincers at the head. They include the Arachnida (modern spiders), with eight appendages and a limited fossil record, and the Merostomata, which have a fused head and thorax (prosoma). One group of merostomes is the Xiphosura, including the modern horseshoe 'crab', with a comparatively large prosoma; the other group is the Eurypterida, with a smaller prosoma, which was important only from the Ordovician to the Devonian.

The most diversified arthropods are the Crustacea, characterized by their varied appendages and carapaces. They can be divided into three groups (classes?) with geologic significance. One is the Ostracoda, which are small crustacea with their bodies laterally compressed and enclosed in a calcified, bivalved, skeleton. A second is the Cirripedia (barnacles), which have enclosed their bodies within calcitic plates. The third, and largest, crustacean group is the Malacostraca, which have a carapace over the head and thorax. Among the Malacostraca, the Phyllocarida are characterized by having a bivalved carapace and have a limited fossil record throughout most of the Phanerozoic. The other major malacostracan group is the Eumalacostraca, which includes modern shrimps, crabs and lobsters, but has a meager fossil record.

A simple two-fold classification of the trilobites is into the Agnostida, which were eyeless, and the Polymerida, which had eyes. A further classification of the Polymerida is based on a variety of characteristics into the orders (?) Redlichiida, Corynexochiida, Ptychoparida, Proetida, Phacopida, Lichida and Odontopleurida.

Mollusca

The Mollusca are a highly diverse group of triploblastic (coelomate) animals whose modern representatives range from oysters to octopi (Fig. 6.6D and E). They are second to the arthropods in terms of species diversity.

D

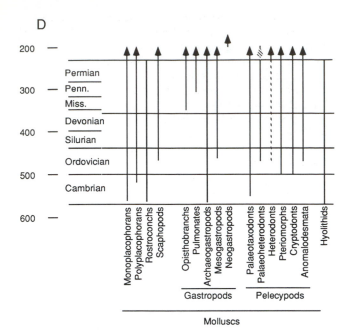

E

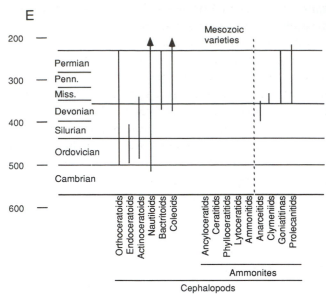

Fig. 6.6D. Molluscs (except cephalopods). The molluscs can be subdivided into seven groups. The Monoplacophora are simple, cap-shaped, organisms. The Polyplacophora are slightly more complex, with seven to eight calcareous plates on their backs (modern chitons). The Rostroconchia are similar to the Pelecypoda but with their shells fused across the back; rostroconchs occur only in the Paleozoic. The Scaphopoda are cone-shaped shells open at both ends, with the mouth at one end and the anus at the other. Three other groups have undergone a more-complex evolution (gastropods, pelecypods, and cephalopods).

Two of the three gastropod subclasses have limited fossil records, namely: the Opisthobranchia, which are marine organisms that have undergone 'detorsion' and mostly lost their shells; and the Pulmonata, which are terrestrial forms. The major subclass is the Prosobranchia, containing an enormous variety of shell forms in the orders Archaeogastropoda, Mesogastropoda and Neogastropoda.

Among the pelecypods, the Palaeotaxodonta have numerous small teeth as well as primitive features of their soft-bodied anatomy. Most modern pelecypods (the Heteroconchia) have a heterodont dentition, with two or three large teeth near the beak (apex) of the shell and lateral teeth elongated along the dorsal edge of the shell. These heterodonts appear to have evolved from palaeoheterodonts, which first appeared in the Ordovician. The Pteriomorphia include a wide range of hinge dentitions and other characteristics; they include such modern forms as oysters and scallops. Two other groups include: Isofilibranchia (mussels) and Anomalodesmata, which are deep-burrowing or rock-boring forms with thin shells.

A Paleozoic group that may be related to the molluscs is the hyolithids They have small, tapering, conical, bilaterally symmetrical, calcareous shells that are open at the large end and covered by an operculum. In addition, they have narrow calcareous 'arms' ('helens') that extend out from either side of the large end. The hyoliths appeared in the earliest Cambrian, and paleontologists have regarded them either as ancestral to the molluscs or as an early offshoot of the mollusc evolutionary line.□

Fig. 6.6E. Cephalopods. The cephalopods first appeared in the Cambrian as the subclass Orthoceratoidea, which was probably ancestral to all other cephalopods. Primitive features include straight or slightly curving shells with simple septa. Other non-ammonite orders include: the Endoceratoidea, with straight shells up to 9 m long, a large siphuncle, and relatively simple sutures; the Actinoceratoidea, with mostly straight shells, very large siphuncles, simple sutures, and partly distinguished from the endoceratoids by the type of deposits in the siphuncle; and the Nautiloidea, with straight to curving shells, simple sutures and a small siphuncle. Three subclasses of more complex cephalopods appeared in the middle Paleozoic and largely replaced the earlier forms. They include: the Bactritoidea, which have straight to slightly curved shells and more complex sutures than the earlier groups – they are probably ancestral to the ammonites; the Ammonoidea, which became the most diversified cephalopod subclass, with sutures of moderate to extreme complexity; and the Coleoidea, which have taken their shells inside the body or lost them completely (squids, octopi, etc.). The complex and variable sutures of the ammonites have made them extremely valuable in paleo-stratigraphic zonation. The principal coleoid fossil is the cone (rostrum) of the belemnite, which was important in the late Mesozoic.   □

The common characteristics of these organisms are a body partly or wholly surrounded by a mantle that secretes a calcareous shell; in some varieties, the shell has been secondarily lost or placed inside the animal (e.g. octopi and squids). Most of the major groups range from the Cambrian to the present. Marine molluscs reproduce via a 'trochophore' larva, which is important in determining relationships with other phyla.

Evolution of the molluscs centers around several

controversies, of which we will discuss only two. One problem is the pre-mollusc ancestor, with both flatworms and annelid worms serving as candidates. A flatworm ancestor can be transformed into a primitive mollusc largely by developing a mantle over the top of the body, evolving gills, and secreting a (presumably cap-shaped) shell from the mantle. This modification produces an animal similar to Monoplacophora, which were presumed to have become extinct in the Triassic until a living variety (*Neopilina*) was recently dredged from the deep sea. An annelid worm has been regarded as an ancestor by some paleontologists for two reasons: 1) primitive molluscs show apparent segmentation similar to that of the annelids; 2) annelids share with molluscs (and arthropods) the fact that they propagate through a trochophore larva stage. The trochophore larva, however, may merely indicate that the three phyla were derived from a common ancestor rather than serially via annelids.

A second major issue regarding the evolution of the molluscs is the concept of a primitive ancestral mollusc from which all other classes evolved. This mollusc, which may have lived in the Early Cambrian, might have characteristics that could be inferred by extrapolating younger forms 'backward' to their starting point. Some of these projections reconstruct an animal very similar to that proposed as the natural evolutionary development of a worm (flatworm or annelid), with a cap-shaped shell overlying a straight organism with mouth at one end and anus at the other end. Other theories propose a shell-less, largely unsegmented, ancestor similar to modern aplacophoran molluscs. Regardless of the nature of the original mollusc, all classes had established themselves by the start of the Ordovician.

Four of the molluscan classes appear to be relatively unmodified from their primitive ancestor (monoplacophorans, polyplacophorans, rostroconchs, and scaphopods). Three classes (gastropods, cephalopods and pelecypods) have undergone extensive differentiation.

- The gastropods (snails) underwent a twist of the straight, ancestral, body by 180° and then a folding of the twisted body over itself in order to place the anus above the head.
- The pelecypods (Lamellibranchia, Bivalvia) are untwisted molluscs surrounded by two separate shells that commonly consist of aragonite or both aragonite and calcite. In most varieties, the shells are identical, with a mirror symmetry plan between them. The principal

difference of the pelecypods from the Rostroconchia is in the suturing of the two pelecypod shells by an organic ligament that permits the shells to open and shut. The pelecypods may have a rostroconch or laterally compressed monoplacophoran ancestor. Pelecypods are sparse in Cambrian rocks, but primitive forms (subclass Palaeotaxodonta) from the Early Cambrian may be ancestral to all other forms. With only minor exceptions, pelecypods are marine, benthic, deposit- and suspension feeders.

- The cephalopods are similar to the gastropods in having a body that is folded over so that the anus is near the mouth. They are not twisted, however, and secrete a shell that is generally bilaterally symmetrical, with septa closing the shell at various intervals and with the animal living in an outermost chamber 'backed' against the last septum. The septa are joined by a tubular 'siphuncle' that permits the mantle to communicate with the chambers behind the living animal. The septa intersect the wall of the organism to form sutures of variable complexity. The septa, the siphuncle, and particularly the form of the sutures are diagnostic in the classification of many cephalopods. The cephalopods have a mass of tissue that constitutes a primitive 'brain' and have been regarded as the 'intellectuals' of the invertebrate world. They are all marine, and almost all are predatory.

Echinodermata

The echinoderms are marine organisms that have a basically bilateral symmetry but show a superficially five-fold symmetry. Most secrete calcite in the mesoderm, leaving an external organic layer outside of the protected skeleton. The general-purpose organs of the echinoderms are the 'tube feet', which are responsible for feeding, movement, and respiration. The tube feet protrude through holes in 'ambulacral areas', which alternate around the animal with 'interambulacral areas'. The ambulacra and interambulacra intersect at the top of the organism in an apical disk. In many echinoids in which the anal hole is within the apical disk, the organism has an apparent pentameral symmetry and is called 'regular'. In most modern varieties, however, the anal hole has migrated out of the disk, giving the organism a striking bilateral symmetry ('irregular').

Primitive varieties of several types of echinoderms were present in the Cambrian, and their ancestry is unclear. Some paleontologists group echinoderms with hemichordates (and thus with vertebrates) as a 'superphylum' that evolved from a sipunculid worm. The evidence is largely from the larval stage of the echinoderms, which bears some resemblance to the sipunculids and which appears to be capable of modi-

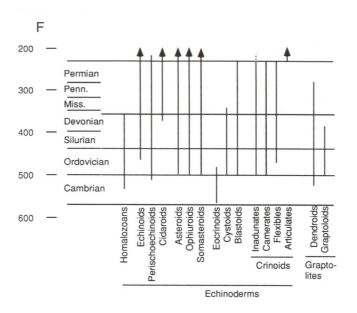

Fig. 6.6F. Echinoderms and Graptolites. The echinoderms can be placed into three broad groups: Echinozoa, which are mostly globular; Asterozoa, which have long arms radiating from a central body; and Crinozoa and related varieties, which commonly are attached to stalks. The Echinozoa include Echinoidea (sea urchins), Holothuroidea (sea cucumbers, which have a poor fossil record because they have calcareous plates on their body but no articulated skeleton), and several extinct groups. Primitive (Paleozoic) echinoids may be grouped as Perischoechinoidea (including the Edrioasteroidea), characterized by regular form and flexible skeletons in which the plates can slide past each other. The Asterozoa include the Asteroidea (starfish) and the Ophiuroidea (brittle stars). The asteroids have a relatively large body with five radiating arms; the skeleton is flexible, possibly a late development as the asteroids evolved from deposit feeders to active predators. The ophiuroids have a small body and five long, sinuous, arms. Both groups may have evolved from an ancestor (the Somasteroidea) that had characteristics intermediate between those of asteroids and those of crinoids. The somasteroids occurred only in the Early and Middle Ordovician. In addition, a group of organisms (Homalozoa; 'carpoids') that lived in the early and middle Paleozoic are probably echinoderms but also have been related to the chordates.

Stalked echinoderms include eocrinoids, cystoids, blastoids and crinoids. Classifications are based on such features as the arrangement of plates in the calyx (the body attached to the stalk) and the arrangements of ambulacra, which become long arms in more evolved forms. The most primitive are the Eocrinoidea, with highly irregular calyx and limited food-gathering ability. A slightly more advanced form is the cystoids (Diploporita and Rhombifera), with pentameral symmetry but an irregular arrangement of perforated plates in the calyx and short arms. The Blastoidea have a more regular arrangement of plates and ambulacra in various patterns along the calyx. The crinoids can be divided into four subclasses (orders?). The most primitive are the inadunates, which have a rigid calyx, which is not invariably pentameral, and completely free arms. The camerates have a large, rigid, pentameral calyx that includes some of the arm structure. The flexibles were a minor group characterized by a calyx of plates that were only loosely sutured. Post-Paleozoic crinoids are articulates, with a small calyx and highly flexible arms; some articulates (comatulids) no longer form stalks.

Graptolites occurred in three different forms. Some minor orders were largely encrusting and ranged only from the Cambrian to the Silurian. One order (Dendroidea) was benthic and formed fans with numerous attached branches. The major graptolites, however, are in the order Graptoloidea, which was planktic.   □

fication to a hemichordate and chordate anatomy (further discussion in Section 6.4). The sipunculids have also been proposed to be related to hyolithids and to be ancestral to the lophophore-bearing brachiopods and bryozoans.

One order (subclass?) of Paleozoic echinoids, the Cidaroidea, appears to have been ancestral to all modern forms (Fig. 6.6F). The cidaroids are all regular and characterized by primitive features such as small, sinu-

ous, ambulacra. During the Paleozoic, the cidaroids evolved from flexible non-articulated skeletons to articulated ones, and a few varieties survived the end of the Paleozoic, giving rise to the modern regular and irregular echinoids.

The principal contribution of echinoderms to the stratigraphic record was made by crinoids in the Paleozoic (Fig. 6.7B). Crinoids are the most advanced type of stalked echinoderms, which include such prim-

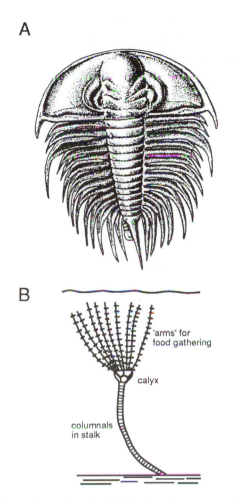

A

B

'arms' for
food gathering

calyx

columnals
in stalk

Fig. 6.7 Invertebrate fossils. (A) *Olenellus thompsoni*, a trilobite
~5 cm long, characteristic of the Early Cambrian (described by
Walcott, 1910). (B) Diagrammatic sketch of typical crinoid,
averaging ~20 cm high, showing 'arms' that gathered food and a
stalk that attached the animal to the seafloor and provided calcite
columnals as major components of Paleozoic limestones (see
Lane and Burke, 1976).

Hemichordata

The hemichordates are similar to the chordates (vertebrates) in the possession of some type of cord, possibly
with a nervous system, along the length of the organism. The only living examples are acorn worms and
pterobranchs. The major fossil group placed in the
phylum is the graptolites, of uncertain origin and possibly not related to modern hemichordates (Fig. 6.6F).
The graptolites are colonial organisms with individual
zooids arranged along a stem (sicula). Because all
graptolites died out by the Pennsylvanian and are
largely preserved as imprints in organic-rich shales,
details of their soft parts are unknown. Graptolites are
extremely widespread both as benthic and planktic
forms during the Ordovician and Silurian, and the
variety of morphologies has permitted extensive use as
index fossils.

Conodonta

Conodonts are small (millimeter-sized), jagged pieces
of carbonate–apatite that were clearly parts of some
organism. Generally, these mineralized 'elements' are
found individually, but a few preservations indicate
that the conodont was a bilaterally symmetrical and
coelomate metazoan. The diversity of shapes of conodonts has made them extremely useful for dating
during the Paleozoic.

[**References** – General books that provide information on invertebrate paleontology include those of House (1979), Clarkson (1986),
Kuhn-Schnyder and Rieber (1986), Boardman, Cheetham and
Rowell (1987), and Stearn and Carroll (1989). Brasier (1980) discusses
microfossils, and McKerrow (1978) discusses the ecology of fossils.
Fig. 6.7A is based on Walcott (1910) and Fig. 6.7B on Lane and Burke
(1976).]

itive groups as eocrinoids, cystoids and blastoids. The
crinoids all have a regular arrangement of plates in a
calyx of pentameral symmetry and long arms that
direct food toward the feeding area. The stems of
Paleozoic varieties were well developed and provided
abundant crinoid columnals into the shallow carbonate environments of that time. The most primitive
crinoids were the inadunates (caption to Fig. 6.6F),
which appear to have been ancestral to the Paleozoic
camerate and flexible crinoid orders. One inadunate
genus survived the end of the Paleozoic and was probably ancestral to the modern articulate crinoids.

6.4 Organic evolution in the Paleozoic – vertebrates

WE are vertebrates. Specifically, however, what does
the term 'vertebrate' mean? How did this condition
evolve? What types of vertebrates preceded us?
Because the vertebrate lineage was established in the
Paleozoic, we address these questions in this section.

Vertebrates are part of a group of organisms classified as 'chordates' (Fig. 6.8). The term implies the presence of a dorsal (backside) neural tube and, at least

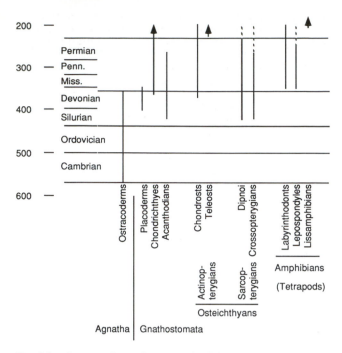

Fig. 6.8. Ranges of vertebrates in the Paleozoic. The most primitive vertebrates are jawless fish (Agnatha). Fish with jaws (Gnathostomata) appeared in the middle Paleozoic; they include the Acanthodii, Placodermi, Chondrichthyes (Elasmobranchii), and Osteichthyes. The Acanthodii were characterized by a poorly developed jaw and tooth structure, paired lateral fins supported by large spines, and bone-like scales. The Placodermi also had a primitive jaw structure and had a head and body sufficiently separated that the animals could lift their heads in order to catch food. A major group of placoderms, the arthrodires, had heavy armor covering both the head and body. The Chondrichthyes (mostly sharks) have a cartilaginous skeleton instead of bone and do not have lungs or swim bladder. The Osteichthyes are true bony fish that include the lobe-finned Sarcopterygii (Choanichthyes) and the ray-finned Actinopterygii. The Sarcopterygii include the Dipnoi (modern lungfish) and the Crossopterygii (modern coelacanth).

Amphibians developed after colonization of the land by plants. The amphibian lineage presumably led to the development of reptiles in the late Paleozoic (Section 8.6). □

during some stage of its development, a flexible dorsal rod (the 'notochord') and gill slits. True vertebrates surround the dorsal neural tube with a 'vertebral' column of cartilage or bone, and its development was one of the most significant events in earth history. In addition to vertebrates, the chordates include several groups of 'protochordates' (pre-vertebrates) that exhibit one or more chordate features in their juvenile or adult forms. A modern protochordate is *Amphioxus*, which has a primitive neural tube, notochord and gills and lies on the sea bottom to filter feed.

The origin of the chordates is not entirely clear but can be inferred largely from studies of embryonic development in various phyla. Suggested ancestors include arthropods, molluscs, annelids and echinoderms. Both echinoderms and chordates share several developmental features, including radial cleavage of the egg, similarity of larval stages, and gastrulation of the early embryo by invagination; creation of a 'gastrula' by infolding of an early, more spherical, embryo forms the initial gut and places the precursors of various organs in their appropriate positions. Because organisms similar to *Amphioxus* occur in the Middle-Cambrian Burgess Shale, a divergence of chordate and echinoderm lineages must have occurred before (probably long before) that time. Definite vertebrates did not appear until the Late Silurian.

Vertebrates (also referred to as 'Craniata') have a variety of characteristics in addition to a vertebral column. They include: bilateral symmetry; two pairs of locomotor appendages; internal segmentation of the skeletal, muscular, and nervous systems; an internal axial skeleton of bone or cartilage; a well-developed brain enclosed in a skull; well-developed sense organs on the head; gills or lungs connected to the pharynx (throat); and a closed circulatory system with a ventral (front) heart and median dorsal artery.

Vertebrates can be classified in various ways (Fig. 6.8). One is into the Agnatha, represented by jawless fish, and the Gnathostomata, which have jaws and include all other vertebrates. Another classification is into the Anamniota, which do not have amnions and are represented by amphibia and fish, and Amniota, which have amnions in their developmental stages and are represented by all other vertebrates. We discuss amnions more completely below.

The earliest, and least-evolved, vertebrates are a variety of jawless fish (Agnatha) that are grouped under the general term 'ostracoderms' because of their complete covering by a bony skeleton (Fig. 6.9A). Because of the lack of jaws, they presumably lived by scooping decaying material from lake or sea bottoms, similar to the feeding habits of the few modern agnathic representatives. The Agnatha were particularly important in the Devonian, after which they were mostly replaced by fish with jaws (the Gnathostomata); the only modern representatives of the jawless fish are the lamprey and the hagfish.

The jaw-bearing fish diversified into a variety of

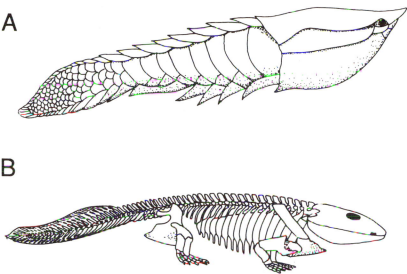

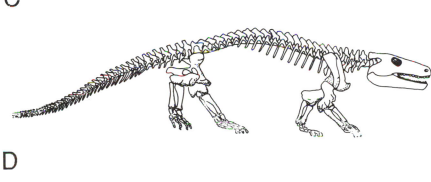

Fig. 6.9. Vertebrates. (A) *Anglaspis*, an agnathic (jawless) fish that was one of the first vertebrates to appear on the earth; ~15 cm long (from Carter, 1993, based on original work by Kiaer, 1928). (B) *Ichthyostega*, a crossopterygian with legs for walking and a fish-like tail, that presumably represents the transition from fish to amphibians; ~1 m long (from Carter, 1993, based on original work by Jarvik, 1952). (C) *Seymouria*, a cotylosaur, or 'stem reptile', that was among the first vertebrates to hatch an egg not laid in water and thus represents the transition from amphibians to reptiles; ~0.5 m long (from Carter, 1993, based on original work by White, 1939). (D) *Eozostrodon* (*Morganucodon*), a primitive Triassic mammal recognized by having a jaw consisting of one bone on each side; ~15 cm long (from Carter, 1993, based on original work by Jenkins and Farrington, 1976). □

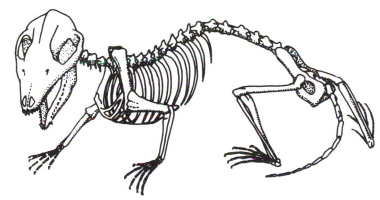

groups in the Devonian (Fig. 6.8). All of the primitive bony fish had both lungs and gills, thus enabling them to breathe air as well as extract oxygen from water. It seems likely that these fish evolved in shallow, brackish or fresh, water and migrated into the oceans (modern fish) or developed into amphibians. The Acanthodii and Placodermi did not survive the Paleozoic, and the two living groups of jawed fish that evolved in the Paleozoic are the Chondrichthyes and the Osteichthyes (Fig. 6.8). The Chondrichthyes (sharks, rays) are poorly preserved except for an abundance of shark teeth in some rocks. The shark lineage

the earth's surface was the occupation of the land by living organisms. The earliest record of this event is in fossils from the Silurian (or perhaps Early Devonian), but the possibility of older (not preserved) land plants must be considered. One highly disputed line of evidence for older land plants is the finding of trilete spores in Ordovician sediments, which are proposed to have been produced by the earliest forms of bryophytes.

Primitive plants are always surrounded by water, and the colonization of the land required the development of special adaptations to enable the plants to retain moisture, stand erect, and distribute nutrients. One is in the simple Bryophyta (mostly mosses), in which the plant spreads along the ground or some other material (e.g. a tree), absorbs water and nutrients across its entire surface, and generally requires frequent wetness in order to survive. The more advanced adaptation in the Tracheophyta, or vascular plants, is the development of cells with lignin stiffening (e.g. wood), a moisture-retardant outer surface, and an internal distribution system for water (xylem cells) and nutrients (phloem cells). The development of vascular structures provided an opportunity for preservation of plants in the stratigraphic record. Non-vascular plants, such as mosses, probably would have left no trace of their existence, and the pre-Silurian earth may have been covered in some now-unknown, mossy, green fuzz. Plants more advanced than the bryophytes also developed fertilization methods that did not require the gametes to travel through water.

Although this section is about plants, we must realize that an understanding of their evolution begins with marine organisms variously regarded as plants or plant-like protoctists (Section 5.7). Consequently, we start with a discussion of algae and then proceed to the true plants.

Algae

Eukaryotic organisms whose reproductive structures consist of single cells (or primitive multicells) can be divided into algae and fungi. Two distinctions are: 1) the cells of algae contain 'plastids' and those of fungi do not; and 2) most fungi have chitin in their cell walls, and algae do not. The major plastid in algae is the chloroplast, a chlorophyll-containing body that provides the plant with an opportunity to conduct photo-

synthesis. The absence of these bodies in fungi makes it necessary for fungi to nourish themselves by decomposing surrounding (attached) organic matter and incorporating the nutrients produced. The symbiotic (?) relationship between fungi and algae in lichens uses the ability of algae to make food and the fungi to protect the algae while they do so. Fungi have a very limited fossil record except for some unusual occurrences of other fossils with fungal infestations.

The earth presently contains an extraordinary variety of algae, all of which are aquatic, but few of them have a fossil record. Much of the classification is based on the variety of pigments that the algae contain in addition to chlorophyll. These pigments contribute to the photosynthetic process by absorbing radiant energy at different wavelengths and sending that energy to the chlorophyll-a, the variety of chlorophyll that primarily initiates the photosynthetic reactions. Different varieties of algae are either unicellular or multicellular, and the important division Chlorophyta (green algae) can be both. (The blue-green 'algae', or cyanobacteria, are prokaryotic and were discussed in Section 5.7.)

Despite the virtual absence of a fossil record, the green algae are extremely important in the evolutionary history of plants. Green algae share three important characteristics with true (land) plants: 1) they contain both chlorophyll-a and chlorophyll-b (the pigment that gives them their green color); 2) they have cell walls constructed of cellulose; and 3) they use starch for storage of food energy. Furthermore, the green algae occur both in marine and fresh water and undergo the same sexual reproduction cycle as land plants, although with a much more significant haploid phase. The major difference between green algae and land plants is the development of specialized structures that allowed the more advanced plants to survive out of water. For these reasons, one group of green algae is commonly regarded as ancestral to land plants.

Several groups of algae secrete calcium carbonate. Among the red algae (Rhodophyta), the Corallinaceae secrete calcite around the margins of their cells, thus forming a calcareous encrustation that is an important component of the wave-resistant parts of modern reefs. The corallinaceae are known to have existed in the Jurassic and may have appeared much earlier in the Paleozoic. Green algae commonly do not form articulated calcareous masses, but some varieties generate calcium carbonate in semi-articulated or individ-

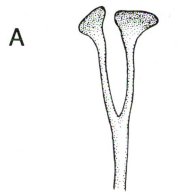

Fig. 6.10. Paleozoic plants. (A) *Cooksonia*; one of the oldest land plants, showing sporangia at the tips of a simple branching system; average height of a few centimeters (from P. Gensel, based on original work by Edwards, 1970). □

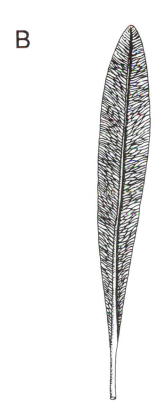

Fig. 6.10B. *Glossopteris* leaf, whose distribution on various now-separate continents indicated the existence of Gondwana during the late Paleozoic; typical length ~20 cm (from Stewart, 1983; based on Gould and Delevoryas, 1977). □

ual grains around the algal cells. Thus, the green algae are major contributors of calcite and/or aragonite to a carbonate environment, but entire preserved fossils are sparse.

Bryophyta

Modern bryophytes include mosses, liverworts and hornworts. Because of their lack of a well-developed (lignified) vascular system, they are generally restricted to wet environments, and their gametophyte stage is dominant. The bryophytes first appear in the geologic record in the Early Devonian, at about the same time as the vascular plants, and it is possible that both groups of land plants evolved separately and synchronously. Some Devonian bryophyte genera are strikingly similar to modern ones, as is common with many other primitive organisms.

Tracheophyta (primitive varieties)

Primitive vascular plants (tracheophytes) are characterized by an upright form, lack of true leaves, and prominent sporangia. The oldest proposed land plant is *Cooksonia* (Fig. 6.10A), which is dated as Late Silurian to Early Devonian. *Cooksonia* was a few centimeters high, had only a few branches and no lateral branches or leaves, and had kidney-shaped (reniform) sporangia at the tips of the branches. Shortly after the appearance of *Cooksonia*, two basically different architectures are found in the primitive plants; it is not clear whether *Cooksonia* was ancestral to either or both types.

One group of post-*Cooksonia* plants continued to show sporangia at the tips of branches but had more complex branches, clusters of sporangia at the ends of branches, and spiny 'enations', instead of leaves on the branches. This group evolved toward later forms, such as *Psilophyton* and *Trimerophyton*, that may have been ancestral to virtually all other vascular plants except the lycopods. The second major type of primitive plant is represented by *Zosterophyllum* , which was characterized by sporangia arranged along the sides of the outer parts of the branches, instead of on the tips. *Zosterophyllum* and related plants were covered with enations that are regarded as precursors of 'microphyll' leaves (see below) and apparently evolved in lycopods. The present Psilophyta (whisk ferns) have similar properties to these primitive plants and have been proposed to be relatively unmodified descendents.

Lycophyta (lycopods) and Sphenophyta

Lycopods represent a distinct advance over the primitive Zosterophyllum-type of plant, with all adaptations favoring a life on land. Lycopods have a microphyll leaf, which increases photosynthesis, instead of enations. Early lycopods bore sporangia on the upper surfaces of leaves and produced spores of roughly equal size, but the later lycopods produced definite small spores (male) and large spores (female). Lycopod root systems are present although not very well developed. The sporophyte phase is dominant.

Modern lycopods are minor elements of modern flora (club moss and quillwort), but the group was a dominant member of Carboniferous forests. After a limited beginning in the Devonian, lycopods developed into 'trees' with heights of more than 30 m and diameters of more than 0.5 m. These trunks are well preserved in Carboniferous coal deposits, characterized by impressions that show the scars of leaves attached in spiral patterns around the branches. A general term for the forest lycopods is 'lepidodendroid', with the major genera being *Lepidodendron* and *Sigillaria*. The fate of lycopods after the Carboniferous is unknown, either because they were replaced by more competitive organisms or because the onset of Gondwana glaciation in the Permian terminated the coal swamps that characterized much of North America and Eurasia in the Late Carboniferous.

The sphenophytes (horsetails) are similar to the lycopods in having a Devonian beginning, a widespread expansion in the Carboniferous, and a survival in only a few modern forms. The sphenophytes have narrow microphyll leaves that are not photosynthetic and separate branches for the sporangia. A dominant characteristic is horizontal jointing of the trunks and branches, with the joints marked by protuberant rings. Branches and leaves emanate only from the joints, leaving large bare areas in between. The major genus is *Calamites*, which was 20 to 30 m tall during the Carboniferous.

Pterophyta

The Pterophyta are the largest group of living plants, and we discuss only their major subdivision, the Filicopsida, or true ferns. Ferns have been an enormously diverse group since their development in the Devonian and are abundantly preserved in Carboniferous coals. Ferns reproduce by spores, most of which are not very well differentiated into male and female. The sporangia are on the underside of the

leaves, and the resultant gametophyte phase is a relatively inconspicuous pad of green cells. Ferns require moist areas for fertilization.

Ferns represent an advance over the plants that we have discussed above in their development of macrophyll (megaphyll) leaves. Macrophylls and the more primitive microphylls are distinguished primarily by two features. One is that microphylls have only one vein in the leaf, whereas macrophylls have several. The second is that microphylls are simply branches off of the stem and do not leave 'leaf gaps' at the point of branching. Macrophylls form by removing a strand of phloem and xylem from the stem, leaving a gap just above the point of departure that is no longer part of the circulatory system of the tree. The macrophyll leaves of ferns are photosynthetic and generally 'frond-like'. Most ferns are small, but some 'tree ferns' reach heights of 10 m.

Progymnosperms and pteridosperms

Beginning in the Late Devonian, the *Psilophyton* and *Trimerophyton* groups of plants apparently evolved into two now-extinct groups with characteristics intermediate between those of primitive plants and those of the more advanced seed-bearing varieties. One of these groups was the progymnosperms, which had thick secondary xylem (see below) with internal characteristics of true gymnosperms but also the pinnate leaves and sporangia of ferns. The progymnosperms apparently evolved into the present gymnosperms, such as pines (with unprotected, 'naked', seeds). A second group was the seed ferns (pteridosperms), which had the limited secondary growth and leafy structure of ferns but produced seeds. Pteridosperms may have been ancestral to modern angiosperms (flowering plants with protected seeds).

The evolutionary advance produced two important features. One is a distinction between seeds and pollen. When primitive plants undergo meiosis to develop spores, the resulting male and female gametes are commonly very similar, germinate in the ground, and are produced on the same plant. In advanced plants, the female gametophyte is much larger and is produced on a different plant from the male. The male gametophyte is encased in a protective coating while attached to the plant and then is released as pollen. At the same time, the female gametophyte is also protected on the plant and, when fertilized by the pollen, develops into a seed.

A second development is the increasing importance of 'secondary' growth in the plants. Advanced plants consist of a thin inner core formed by direct growth from the plant tip and a generally thicker outer part formed by growth from a cylindrical sheath near the outer edge of the stem. This secondary growth produces more xylem on the inside of the sheath and more phloem on the outside. The secondary xylem constitutes most of what is commonly referred to as 'wood' and may have annual rings.

One of the best-studied groups of extinct plants is the *Glossopteris* flora (Fig. 6.10B), which is controversially regarded either as a type of pteridosperm or an intermediate between a pteridosperm and an angiosperm. *Glossopteris* apparently were small trees or large bushes and were characterized by a tongue-shaped leaf with a prominent rib in the middle and a network of veins on either side; the leaves were arranged in whorls around the stems. *Glossopteris* and related forms dominated the Permian and Triassic flora of circum-glacial parts of Gondwana. *Glossopteris* was originally discovered in the early 1800s, and the gradual realization that these distinctive plants occur in strata of similar age in South Africa, India, South America, Australia and Antarctica was an important early line of evidence for the existence of Gondwana.

Gymnosperms

The true gymnosperms are characterized by seed-bearing cones, many with the different sexes on different plants, and by trunks with considerable thicknesses of secondary growth. Varieties of leaves vary enormously from one group to another. The major modern group of gymnosperms is the conifers, with predominantly needle-like leaves. Two other groups, cycads and ginkgoes, evolved in the late Paleozoic and were relatively prominent in the Mesozoic. Cycads have gymnosperm-like wood and male and female cones, but their leaves are pinnate and very similar to those of a palm. Ginkgoes have a more branching, tree-like appearance with fan-shaped leaves containing numerous forking veins.

A fourth group of gymnosperms, the gnetophytes, is a very strange mixture of minor plants that have no fossil record except for some presumed pollen and possibly a few leaves. They include such diverse varieties as *Ephedra* and *Welwitschia*. Ephedra-like plants are small and jointed; some of them produce ephedrine, commonly used in nose drops. *Welwitschia* occurs only in the Namib Desert of southwestern Africa. Its stem barely protrudes above the sand, and leathery leaves, commonly whipped into long strands by the wind, lie on the sand; even a healthy *Welwitschia* looks dead.

Gnetophytes are important in plant evolution because they are gymnosperms that share two characteristics with angiosperms. One is that *Ephedra* undergoes double fertilization without producing endosperm (see below for comparison with angiosperms). The other is the presence of 'vessels' (long conducting tubes) in the secondary xylem, which apparently increase the efficiency of the xylem.

Angiosperms

Angiosperms include varieties familiar to us – flowering plants, hardwood trees, and grasses. Their ancestry is uncertain, the two major possibilities being primitive gymnosperms and pteridosperms (see above). The oldest undoubted angiosperms are Cretaceous, and there is a controversial history in the earlier Mesozoic.

Angiosperms show their advanced character in several ways. The most significant is that their seeds are covered and protected while they mature within the fruit part of the plant. The reproductive process involves 'double fertilization', in which one sperm unites with a female haploid spore to form an egg while a second sperm unites with another part of the female spore to produce a tissue with three sets of chromosomes (triploid). This triploid body is designated as an 'endosperm' and serves as nourishment for the new plant within the seed. These adaptations make angiosperm seeds more durable and productive than those of gymnosperms. A second special characteristic of angiosperms is the presence of both vessels and 'fibers' in their wood. Fibers are layers of dead cells with tough coverings that transect the living tissue and give rise to the term 'hardwood'.

[**References** – Information for this section was obtained from books on plants and their evolution by Stewart (1983), Scagel *et al.* (1984) and Stern (1985). The oldest land plants are discussed by Gensel and Andrews (1984, 1987). Other references include books by Wilson (1975) on carbonate rocks, by McKerrow (1978) on fossil ecology, and by Stearn and Carroll (1989) on general paleontology. Fig. 6.10A is based on Edwards (1970) and Fig. 6.10B on Stewart (1983).]

6.6 Episodes of Paleozoic life

THE PRECEDING sections identified the basic types of animals and plants that have evolved on the earth and, where possible, suggested some lineages that were followed in their sequential development. In this section, we essentially 'invert' the discussion by describing the associations of organisms in five intervals of the Paleozoic. The earliest Cambrian interval is discussed in Section 6.1. Other intervals include the remainder of the Cambrian to the start of the Ordovician, the Ordovician, the Silurian to Late Devonian (Frasnian/Famennian boundary), and the latest Devonian to the end of the Permian. The five intervals are separated by six boundaries, many of which have been regarded as 'extinction events' because of the major faunal and floral changes that occurred approximately at the same time.

The term 'extinction event' poses several questions. For example, at what taxonomic level should we attempt to recognize an extinction? Few animal phyla or plant divisions have ever become extinct; thus, boundaries cannot be recognized at this high level of classification. Class taxonomic levels are also too high, but orders and families can be used to determine extinction rates. Genera can be used to determine extinction and evolution rates on the basis of the number of genera a particular group of organisms contains during some interval of time.

All of the studies of extinction events assume that the taxonomy is adequate for the recognition of major changes. We cannot discuss taxonomic difficulties in depth but should mention two major problems. One is that different investigators use different hierarchies in classifying organisms. For example, one paleontologist's genus is another paleontologist's subfamily. Thus, counting of genera alive during some time interval is a very subjective effort. A second problem is that the amount of taxonomic work done is highly variable among different groups of organisms. Groups such as pelecypods that have been well studied tend to have numerous high-order levels of classification (e.g. orders), whereas less-studied forms have a simpler hierarchy. All statistical work on extinction and evolution is affected by these taxonomic decisions.

Extinction events are difficult to recognize for two other reasons. One is that a true extinction must be worldwide, and a local disappearance of a particular fossil group may merely mean a shift in environment, perhaps caused by plate movement. This problem is complicated by the fact that a high percentage of paleontologic studies has been concentrated in North America and Europe, providing a bias toward low-latitude locations in the Paleozoic (Section 6.7) and possibly not recognizing the continuation of a fossil group elsewhere. A second reason is that a group of organisms may not really become extinct but simply evolve into another taxonomically recognizable group.

The times of major organism change, which separate intervals of different faunal and floral characteristics, are generally associated with major changes in sealevel and climate. A true 'eustatic' sealevel variation should affect all parts of the earth simultaneously but is extraordinarily difficult to recognize because apparent sealevel is locally controlled by tectonics and possibly by geoidal variations (Section 2.4). Association of organic change with climate results partly from the the fact that sealevel is so dependent on the volume of glacial ice supported by continents. In addition, climate and plate movements are strongly related. During the Paleozoic, glaciation and glacio-eustacy were largely determined by the movement of Gondwana over the South Pole (Section 6.7). Most of the continents outside of Gondwana occupied tropical to temperate regions during much of the Paleozoic. There is no evidence that any of them were ever glaciated, but they did undergo latitudinal movements, resulting in local faunal and floral changes. Movements of the continents also affected ocean currents, with resulting climatic effects.

With these general principles in mind, we now review the major faunal and floral intervals of the Paleozoic and the boundaries between them. The earliest Cambrian and the Precambrian/Cambrian boundary have been described in Section 6.1, leaving four organic intervals for further discussion. The ranges of the major fossil groups are in Figs 6.6 and 6.8.

Cambrian, from the earliest trilobites to the Ordovician

The first (tentative?) efforts of animals to secrete skeletal material produced the 'small shelly' fossils and other enigmatic material of the earliest Cambrian. At some time, perhaps several million years later, the earliest trilobite produced a coherent skeleton. At this time, the nearly unclassifiable shelly material disappeared and was replaced by organisms typical of the remainder of the Paleozoic (in some groups, the entire Phanerozoic). The reason for this transition to a more 'normal' fauna is completely unknown, and stratigraphic sections that record the transition show no sedimentary or environmental breaks.

During the remainder of the Cambrian, the dominant animals were trilobites. They flourished in the fine-grained clastic sediments that were deposited as marine waters transgressed over the continental surfaces exposed during the latest Proterozoic; where actual skeletons are absent, traces of their burrowing and grazing are common. In more carbonate-rich environments, trilobites were accompanied by early corals (mostly tabulate, some rugose) and a variety of rostroconchs and other primitive molluscs (both gastropods and pelecypods). Cambrian strata also are characterized by hyolithids, eocrinoids, and the dominantly benthic form of graptolite (dendroids). The only vertebrates are ostracoderms. The land was completely uncolonized, except possibly by lichen or other organisms based on algae.

Trilobites underwent a major change at the end of the Cambrian. Two primitive orders became extinct, along with families and genera from other orders, and three new ones evolved. This development was accompanied only locally by a significant hiatus in sedimentation or apparent climate change. One explanation for the evolution is a simple marine transgression, which apparently caused reduction in the variety of habitats on the continental shelves and interior platforms. The reduction in environmental niches, in turn, led to reduction in genetic diversity, with the elimination of some forms that had survived only because of their occupation of niches in which they had no competition. Regardless of the reason, this Late Cambrian shift in trilobite fauna accompanied the development of an Ordovician period in which the faunal content of the world's oceans was vastly different from that of the Cambrian.

Ordovician

Despite diversification of trilobites following the Late Cambrian extinction event, Ordovician fauna are dominated by organisms that had been absent from, or minor components of, Cambrian sediments. The Ordovician System worldwide is basically a carbonate suite. The beginning of the Ordovician witnessed a sudden development of octocorals, lophophorates (both articulate brachiopods and bryozoa), pelecypods and gastropods more developed than Cambrian forms, most forms of cephalopods, virtually all forms of echinoderms, and graptoloid (mostly planktic) varieties of graptolites. The abundance of brachiopods makes them the principal stratigraphic markers of this interval. Shells and skeletal fragments of these animals constitute a large part of the Ordovician stratigraphic section.

The end of the Ordovician was an extended time of glaciation in the northern African part of Gondwana (Section 6.7). Limited evidence suggests four pulses of glacial advance and retreat, similar to those of the Pleistocene, with corresponding alternation of marine transgression and regression. No major extinction occurred at this time, but faunal diversity measured in terms of numbers of genera was greatly reduced. This reduction in diversity is similar to that shown by modern faunas from different latitudes, with diversity decreasing toward colder environments. Consequently, the faunal evidence is consistent with paleomagnetic data and the Ordovician record of northern Africa in showing a profound episode of glacial activity. Most of the generic extinction occurred during a time of falling sealevel, in contrast with the trilobite extinction during marine transgression at the end of the Cambrian.

Silurian to Late Devonian

Marine animals of the Silurian and most of the Devonian were broadly similar to those of the Ordovician, with major differences attributed to normal evolution. The Silurian and Devonian were the principal periods of distribution of the planktic graptoloids, which are best preserved in organic-rich shales. Several varieties of both marine and freshwater fish evolved at approximately the same time, possibly indicating that fish originally evolved in brackish water such as coastal swamps.

The major evolutionary event of this time interval was the colonization of the land, both by plants and animals. Primitive plants probably were present by the latter part of the Silurian, although definitive evidence is absent until the Devonian. Development of land plants was accompanied by migration of modified forms of arthropods onto the land, including insects and spiders. Arthropods apparently were the first forms of animal life to leave the ocean.

The boundary between the Frasnian and Famennian (the last two stages of the Devonian; Fig. 1.8) is commonly designated as the 'Kellwasser event' in Europe because of the extensive development of black shales. The principal transition between types of organisms occurred at this time rather than at the commonly recognized end of the Devonian. The extinction was gradual, perhaps over several millions of years, during a Frasnian transgression and a Famennian regression. This event is marked by the disappearance of pentamerid and some varieties of spiriferid brachiopods, all trilobites except for members of the proetid order, all graptoloid forms of graptolites, and cystoids.

The cause of the Frasnian/Famennian extinction is controversial. Suggestions that it may have been a meteorite impact have not been substantiated by direct evidence and would not be supported by the gradual character of the faunal changes. A major possibility is based on the plate movements that caused gradual closing of the ocean(s) between North America and Europe (Laurussia) during the Devonian (Section 6.7). This closure, along generally equatorial latitudes, would have disrupted westward-directed ocean currents and prevented warm water from circulating freely around most shores of the two continents. A gradual spread of cold water, perhaps mostly generated on Gondwana, may have occurred, but there is no certainty that it was responsible for the extinctions.

Latest Devonian, Carboniferous and Permian

The transition in fauna and flora during the last several million years of the Devonian produced both oceanic and terrestrial life forms that were quite different from earlier ones. The most extensive change was on land, where the primitive plants of the Devonian were succeeded by widespread vegetative covering. The dominant plants were lycopods, which established forests of comparatively tall trees. Interspersed with the lycopods were other flora consisting of ferns, tree ferns, and progymnosperms. Many of these plants are well preserved in the extensive Carboniferous coal deposits. The *Glossopteris* flora characterized the periglacial regions of Gondwana.

The forests provided an opportunity for the diversification of animals beyond the primitive insects and other arthropods that had been the first land residents. Some of the fish that had originated in fresh and coastal waters developed the ability to remain on land for extended periods of time and became the first amphibians. Further development, mostly in the Permian, produced such primitive reptiles as cotylosaurs, pelycosaurs, and therapsids. The therapsids are apparently ancestral to the primitive mammals that developed in the Mesozoic.

Faunal changes in the oceans were less dramatic than on land. A new order of brachiopods, the terebratulids, flourished, and fusulinids became important index fossils. Perhaps the major development was the spread and diversification of primitive ammonites and related cephalopods, which replaced the straight-coned and simply curved earlier cephalopods. Many new orders of fish appeared and diversified, including both the strangely armored placoderms and osteichthyan varieties.

From the Carboniferous into the Permian, world climates gradually became less tropical. The change was caused by continued expansion of Gondwana glaciation and disruption of ocean currents by plate movements. Near the end of the Permian, Pangea began the long-continued episode of rifting that has led to the present configuration of continents. This rifting occurred perhaps 25 million years before the massive extinction event at the end of the Permian and is discussed more completely in Section 6.9.

[**References** – Studies of evolution and extinction are provided by volumes edited by Holland and Trendall (1984), Valentine (1985), Walliser (1986), and Kauffman and Walliser (1990). Fossil ecology and biogeography are discussed by McKerrow (1978) and papers in the volume edited by McKerrow and Scotese (1990). Information on extinction and evolution rates is in Raup and Boyajian (1988) and Donovan (1989). Copper (1986) discusses the Frasnian/Famennian extinction.]

6.7 Plate movements, biotic provinces and climates

BEFORE outlining specific faunal and floral provinces that characterized various times in the Paleozoic, we need to discuss the general principles on which they have been proposed. Four types of evidence can be used: 1) paleomagnetism; 2) lithologic indicators of paleoclimates; 3) fossil indicators of pale-oclimates; and 4) recognition of faunal and floral realms. We also must discuss a few aspects of the zonal climate pattern that has apparently characterized the earth throughout Phanerozoic time.

Paleomagnetic determination of plate locations is based on the assumption that the earth's magnetic axis (between the N and S magnetic poles) has remained close to the rotational axis throughout geologic time. This assumption is, in general, consistent with positions determined paleoclimatically and is probably (hopefully?) valid. Paleomagnetic measurements can be used to locate paleolatitudes but not paleolongitudes (except in favorable situations in which longitudinal differences can be proven between plates). Thus, the plate locations discussed below generally are not accurate with respect to longitude.

Common *lithologic indicators* of paleoclimates must be used with care because some are not solely sensitive to climatic variations. One indicator of low latitudes (up to 30°) is Bahamian-type and related carbonate rocks. Characteristics include abundance of lime muds, ooids (where evaporation rates are high), reef-building with algae as binding organisms, and hermatypic reefs (based on corals that contain photosynthetic algae). The necessity of high light intensity is probably the determining factor in the location of these rocks in low latitudes.

Evaporites could form at any latitude, but many of the world's major salt and anhydrite deposits are also associated with carbonates, indicating non-polar locations. Evaporation rates are not high at the equator, and evaporite deposits are commonly regarded as indicative of precipitation in the arid regions of a climatically zoned earth, roughly in the range from 10° to 40°. Arid conditions also may be indicated by sand dunes and caliche.

Coal is an indicator of humid climates. Although forests can extend to high latitudes, most coal accumulation is probably restricted to tropical to low-latitude temperate regions. This restriction may have been particularly true of the lycopods and fern-like organisms of the Paleozoic.

The best indicator of high-latitude deposition is glaciogenic features. Widespread (not valley-type) tills, glacial scours, and related features signify a near-polar origin. Some 'tillites' are controversial, however, as similar unsorted deposits can also be formed by mud flows, landslides, etc.

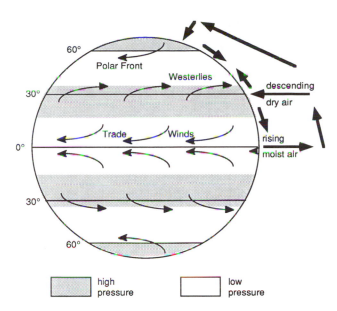

Fig. 6.11. Diagram of the earth's zonal climates. Hot moist air rising at the equator descends as dry air at ~30° north and south latitude, forming belts that contain most of the world's deserts. Further upper-atmospheric circulation causes descending air to form high-pressure zones in polar regions. The cold polar air moving toward lower latitudes collides with warmer air to form frontal systems. The trade winds result from 'slippage' of the rotating earth beneath the atmosphere. Imbalance between the angular velocities of air masses and the earth at different latitudes forms belts of westerlies (winds blowing from the west) in temperate regions.   □

The use of *fossils* as climatic indicators is very difficult. Similarity of an ancient organism to a modern one that has a specific environmental range is not sufficient to show that the ancient organism had the same environmental restrictions. The most general index of the temperature at which ancient faunal assemblages lived is the degree of diversity in the assemblage. Modern organisms commonly show larger numbers of species and genera in low latitudes than in high ones. For this reason, a faunal diversity gradient is probably a reasonable index of latitudinal variation in ancient rocks.

Recognition of distinct *faunal and/or floral realms* is an excellent method of determining relative movements of plates. A 'realm' is commonly defined as an assemblage of organisms that are 'endemic' to (occur only within) a particular area. Planktic organisms are commonly widespread, the most important controls being water temperature (mostly latitude) and salinity. Benthic organisms, conversely, may be unable to migrate long distances, and they develop assemblages in which the

A

EARLY PALEOZOIC

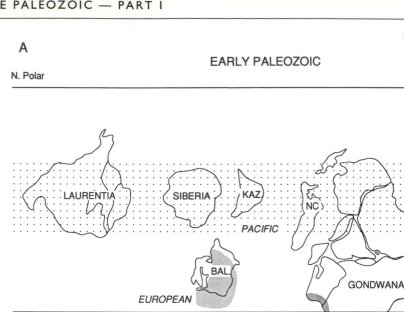

N. Polar

LAURENTIA   SIBERIA   KAZ

NC

Equatorial Region

PACIFIC

BAL.

EUROPEAN

GONDWANA

S. Polar

ATLANTIC

B

MIDDLE PALEOZOIC

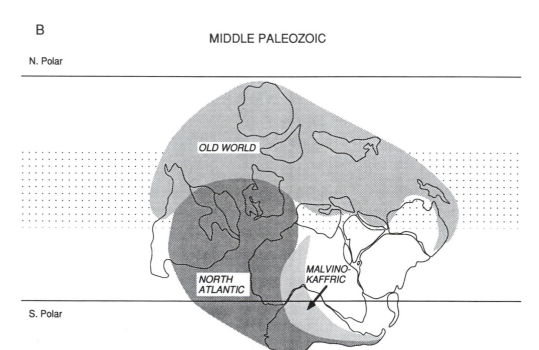

N. Polar

OLD WORLD

NORTH ATLANTIC

MALVINO-KAFFRIC

S. Polar

various organisms are associated with each other but not with organisms from another realm. Some migration of benthic organisms can occur, however, because many varieties, including brachiopods and pelecypods, have free-swimming larval stages. The distribution of a realm is determined by normal environmental factors such as temperature, light, water salinity, sediment supply, etc. The changing pattern of faunal and floral realms and its relationship to plate movements during the Paleozoic is discussed below.

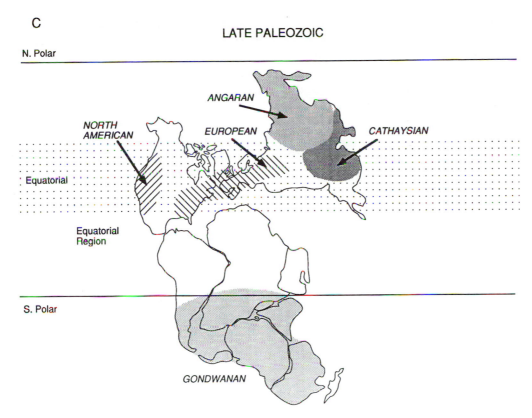

Fig. 6.12. Plate positions and biotic zonation in the Paleozoic. These diagrams show outlines of continents (and Gondwana) as they would appear if they were at low latitudes in a conventional Mercator projection. That is, the continental shapes do not show any distortion from being at high latitudes even in a projection centered at the equator. This type of view preserves the shapes of the continents reasonably well but distorts the relationships between continents at different latitudes. Abbreviations are: KAZ, Kazakhstan; BAL, Baltica; NC, North China.

The biotic realms (in italics) represent groups of 'endemic' organisms, which are associated with each other but not with organisms from other groups. Thus, organisms within a realm require similar environmental conditions, presumably related to latitude (climate) and configuration of ocean currents.

(A) Early Paleozoic. A temperate- to cold-water ('Atlantic') realm occurs on the margins of Gondwana, which underwent major glaciation beginning in the Ordovician. Most continents were in an equatorial ('Pacific') realm dominated by carbonate deposition. A separate ('European') biotic realm occurred in northern Europe (Baltica and terranes accreted to it) in the early Paleozoic. Europe moved northward during the early Paleozoic and changed from a cold-water biota to a warm-water one.

(B) Middle Paleozoic. Most of the world, including western North America, was dominated by a warm-water biota commonly referred to as 'Old World'. Continental closure during this time formed a 'North Atlantic' realm including 'eastern North America' and 'Europe' (not depicted). A cold-water 'Malvinokaffric' realm surrounded a position of the South Pole probably somewhere in southern Africa.

(C) Late Paleozoic. Five separate biotic realms are shown, indicating the high degree of provincialism that resulted from the fusion of the continents into Pangea and resultant interruption of oceanic circulation.   ☐

The efforts to use ancient climatic indicators to determine plate movements assumes that the broad zonal arrangements of climates on the earth has remained approximately the same throughout the Phanerozoic. The zonal climate of the present earth is established by the differential heating of equatorial regions relative to polar regions and the decrease in rotational velocity from the equator to the poles. These factors establish two atmospheric circulation cells in each hemisphere (Fig. 6.11). One cell is formed by heated, moist, air rising in a low-pressure belt near the equator. This air causes high tropical rainfall and moves northward and southward as drier air. It descends in a zone of high pressure at approximately

30° latitude. Because of differences of angular velocity between the air and the earth (Coriolis effect), air moving southward from 30° also moves westward, causing the trade winds (easterlies) of tropical areas. Conversely, air moving northward from 30° moves eastward as a belt of westerlies, generally in the temperate climate zone. The dry air descending at 30° and moving southward localizes most of the world's deserts in the 10° to 30° latitude range. The second atmospheric cell is caused by cold, high-pressure, air at the poles moving toward the equator. This air also moves westward and intersects the belt of westerlies in a rising (low-pressure) system at about 60° latitude.

This zonal arrangement of atmospheric circulation should have characterized the Paleozoic because the only major factors that would change it are changes in solar radiation and changes in angular rotation of the earth. Fluctuations in insolation are generally regarded as too small to have a significant effect on broad atmospheric patterns. Changes in rotational velocity, however, have occurred. A year in the middle Paleozoic is estimated to have contained about 400 days (Section 2.1). Because the actual length of the year has not changed, these measurements imply about a 10% higher velocity of rotation (22-hour day) in the middle Paleozoic than now. This increase in velocity should move climatic zones toward the equator by several degrees. A higher rotational velocity might establish a three-cell configuration of atmospheric circulation, but best estimates are that a velocity this high did not occur in the Phanerozoic.

Specific realms and plate positions

Fig. 6.12A–D shows approximate locations of continents and biotic realms for three time intervals during the Paleozoic. The time ranges for these intervals are long, and the maps do not indicate precise locations at any instant in time. Relative longitudinal separation between continental blocks is roughly inferred, but no absolute longitudinal positions are implied. The construction of these maps is different from that of customary projections, and the caption should be read carefully.

The early Paleozoic (approximately Cambrian/Ordovician) was a time of relative continental separation (Fig. 6.12A). We show Gondwana as a coherent block, which may not be correct depending on the con-

tinental configuration assumed for the end of the Proterozoic (Section 7.0). Other continental blocks are also shown separately, and their movement and aggregation during the Paleozoic is discussed throughout Chapter 7. Because of the generally equatorial position of most continental blocks, and a more polar position of much of Gondwana, the oceans contained normal westward ocean currents in tropical areas and circulation cells in temperate regions.

The oceanic circulation and continental distribution of the early Paleozoic established three areas of separate biota (Fig. 6.12A). One was in the temperate to polar waters of Gondwana (an 'Atlantic' realm), with its position varying as Gondwana moved across the South Pole. The pole is relatively well constrained as being in present North Africa during major glaciation in the Late Ordovician, but its movements before and after are controversial.

A second early Paleozoic realm was in carbonate-rich, generally equatorial, regions ('Pacific' realm). Most continental blocks were sufficiently separated from each other that they showed faunal provinciality, but the differences between the fauna are at relatively low taxonomic levels (including genus and species) rather than at a class or order level. In Fig. 6.12A, a third, 'European', realm is drawn beyond the limits of Baltica to show a progressive enlargement of Europe by accretion of island arcs, marginal volcano–sedimentary wedges, and exotic blocks during this time. Because of the northward movement of Baltica during the early Paleozoic, the biota changed from a cold-water variety to a warmer-water one by the end of the Ordovician.

By middle Paleozoic time, plate movement had altered continental positions and, thereby, the directions of oceanic currents (Fig. 6.12B). The 'Malvinokaffric' realm was a cool-water environment partly circumscribing the South Pole and dominated by siliciclastic sediments, which supported a low-diversity fauna rich in brachiopods and deficient in warm-water organisms such as corals, stromatoporoids, calcareous algae, bryozoa and conodonts. In the middle Paleozoic, the South Pole was in a highly disputed location, perhaps somewhere in southern Africa. In the Silurian and early Devonian, an 'Eastern North America' (Appalachian) realm was still distinct from Europe but may have extended into the northwestern parts of South America (not shown well on the projection used in Fig. 6.12B). During the Devonian, the progressive closure of North America and Europe, culminating in the Acadian/Variscan orogeny (Section 7.3), eliminated much of the faunal difference between Europe and North America and resulted in a 'North Atlantic' realm (Fig. 6.12B). Throughout the middle Paleozoic, an 'Old World' (Uralian–Cordilleran) realm included parts of Europe and much of the remainder of the world, including western North America.

By the end of the Devonian, closure of the Iapetus/Rheic Oceans had been caused by approach of North America, Europe and the African part of Gondwana (Section 7.2), thus uniting most fauna outside of the Malvinokaffric realm in one large group and terminating the westward flow of equatorial ocean currents. This redirection of ocean currents would have caused generally cooler oceans in the Late Devonian and may have contributed to the Frasnian/Famennian 'extinction event' (Section 6.6).

During Carboniferous and Permian time, continents and continental fragments continued closing on each other, ultimately forming the end-Paleozoic Pangea (Fig. 6.12C). By this time, the South Pole was probably located within the present Antarctic. Although the fusion of continents amalgamated some of the earlier biotic realms, it caused an increase in biotic provincialism because: 1) extensive land areas developed during the Permian, which isolated land fauna and flora and forced marine organisms to communicate around a long coastline; and 2) the supercontinent extended over more than 90° of latitude, yielding a complete range of climatic zones.

Five biotic realms can be recognized in the late Paleozoic (Fig. 6.12C). One was a 'Gondwanan' (Austral) province characterized by the *Glossopteris* flora and related organisms on land and by cold-climate fauna in siliciclastic sediments deposited around the high-latitude parts of Gondwana. A second biotic realm, the 'Angara' province in Siberia, was also characterized by cold-climate flora and by marine fauna of low diversity that lacked characteristic warm-water organisms.

Three realms can be recognized along the mostly tropical areas of North America, Europe and blocks moving toward Eurasia. Sediments in these areas were carbonate-dominated during the Carboniferous but became more siliciclastic as Permian glaciation caused progressive marine regression. Warm-climate fauna

included corals, fusulinids, algae, and crinoids in addition to large numbers of brachiopods and molluscs. A 'North American' realm was isolated from other areas by the emergence of a land barrier in central North America in the Permian. A 'European' realm developed where marine waters progressively retreated southward during the Permian. A third, 'Cathaysian', realm characterized the North China block and possibly other blocks that accreted farther south onto Eurasia. The Cathaysian province contained a typical warm-water, high-diversity, fauna during the Carboniferous, but the abundance of tropical forms such as corals and fusulinids decreased toward the Permian. This decrease can be attributed either to northward movement of the North China block or to progressive increase in siliciclastic sediment, which reduced the opportunity for reef-forming organisms to flourish.

[References – Reconstructions of the movements of continents during the Paleozoic are provided by numerous investigators. The discussion in this section was synthesized from: the volume edited by Hallam (1973), Rickard and Belbin (1980), Morel and Irving (1981), Smith, Hurley and Briden (1981), Irving and Irving (1982), Van der Voo, Scotese and Bonhommet (1984), Lawver and Scotese (1987), Piper (1987), the volume edited by Audley-Charles and Hallam (1988), Van der Voo (1988, 1990), and the volume edited by McKerrow and Scotese (1990). Additional information on specific intervals of time and specific places is from Bond, Nickelson and Kominz (1984) for the start of the Paleozoic, Copper (1986) for the end of the Devonian, and Mueller *et al.* (1991) for the end of the Paleozoic in North China. Parrish (1982) discusses controls on the earth's climate.]

6.8 Compositions of Paleozoic oceans and atmosphere

THE PROBLEMS of the compositions of oceans and atmosphere in the Paleozoic cannot be solved by direct measurements of the composition of ancient air and water. Fluid inclusions in sediments may have been affected by diagenetic and post-depositional processes; thus their present compositions may only dimly reflect original ones. Ancient air is virtually absent, although some efforts have been made to measure inclusions in amber. Consequently, the only evidence available for estimating oceanic and atmospheric compositions is indirect, and our conclusions must be regarded as inferential.

One type of evidence is simply the nature of

Phanerozoic organisms. During the Precambrian the compositions of oceans and atmosphere may have varied widely from those of the present. After the development of metazoan animals and land plants, however, atmospheric and oceanic chemistry can be constrained within the broad limits within which those organisms can survive. Modern organisms tolerate only small variations in O_2 contents in air and water, total dissolved solutes in water (salinity), and the ratio of ionic species in water. If (we emphasize 'if') ancient organisms had similar tolerance ranges, then atmospheric and oceanic compositions could not have been extremely different from present ones. We cannot prove, however, that organisms could not survive sudden changes in air and water composition or would not adjust their tolerances to very different oceanic and atmospheric chemistries over long periods of time.

Two broad constraints can be placed on the abundance of oxygen in the Paleozoic atmosphere. One is that O_2 concentrations must have been high enough throughout the era to permit adequately oxidated sea and fresh water for the development of marine life. Some oxidant is also necessary for plants to burn stored foods (e.g. starch). We have already discussed (Section 6.1) the possibility that increase in O_2 in the atmosphere permitted the development of calcareous skeletons at the beginning of the Cambrian. Thus, it is clear that O_2 was present in the atmosphere throughout the Phanerozoic, but the exact concentration needed for maintenance of organisms is unknown. A second constraint is the upper limit that can be placed on atmospheric O_2, perhaps at about 25% of total atmosphere, by the presence of land plants. Vascular plants can begin to burn spontaneously in atmospheres of greater than 25% to 50% O_2, and the continued presence of land plants from the end of the Silurian indicates that these concentrations were not attained.

Atmospheric CO_2 contents are more difficult to estimate than O_2 levels because of the tolerance of organisms for relatively large variations. Qualitatively, we can infer that the post-Silurian atmosphere contained lower CO_2 pressures, and possibly higher O_2 pressures, than older atmosphere because of the occupation of land by plants. A significant part of all photosynthesis is now conducted in the earth's land areas, and the advent of land plants presumably caused major reorganization of the CO_2–O_2 cycle. Inferences

can also be drawn from the fact that high CO_2 contents cause atmospheric temperature to rise because of a 'greenhouse effect'. How extreme might that range be? Clearly, organisms could not survive either freezing or boiling, and most modern organisms could not tolerate changes in average atmospheric or oceanic temperature of more than approximately 10 °C. Any restriction on atmospheric CO_2 contents from inferred temperatures, however, would not preclude variations in CO_2 pressure by one order of magnitude, particularly if the amount of solar radiation has changed over time.

Efforts to quantify atmospheric CO_2 content are complicated by the large number of influences on it. The carbon in the outer part of the earth can be considered to occupy three large reservoirs: 1) oceans and atmosphere; 2) carbonate rocks; and 3) organic carbon in living organisms and buried in rocks. Carbon dioxide equilibrates between ocean and atmosphere, is removed from the oceans by precipitation of calcite and dolomite, and is removed from the atmosphere by weathering of silicate rocks and by photosynthesis. Because the residence time of CO_2 in the atmosphere is only a few years, and equilibration between air and water is short on a geologic time scale, any variation in the rates at which organic carbon and carbonate rocks are precipitated or weathered are reflected immediately in oceanic and atmospheric CO_2 concentrations.

The preceding discussion essentially predicts that time intervals in which organic and carbonate carbon are removed from the environment should be times of low CO_2 in the atmosphere, and vice versa. If we assume that low-CO_2 periods allow the earth to lose heat (a reverse greenhouse effect), then carbon removal from the atmosphere and oceans should cause cold oceans, glacial conditions, and marine regression. Conversely, accelerated erosion and oxidation of buried organic carbon and dissolution of carbonate rocks should be times of relative global warming and marine transgression. (In Section 8.4, however, we discuss the possibility that the Cretaceous was a time of both high carbon precipitation and high temperature, thus shedding doubt on the assumptions discussed here.)

The early Paleozoic seems to have been a time of low preservation of carbon in rocks. If our preceding assumptions are valid, these data indicate high atmospheric CO_2 and globally high temperatures. Shortly after the development of land plants, this situation was reversed, with large-scale burial of organic matter, decrease in atmospheric CO_2, and cooling during the Permo-Carboniferous glacial intervals. These observations could be used – although we hesitate to do so – to imply that late-Paleozoic glaciation was partly the result of the development of land plants.

Oceanic compositions vary both with respect to total dissolved solute (salinity) and relative proportions of solutes. The possibility that salinity variations contribute to mass extinctions is discussed in Section 6.6. The amount of evaporite deposited during short intervals of time, such as the Permian, is sufficient to reduce ocean salinity by nearly 25%. Similarly, erosion of older evaporites, possibly during marine regression and continental exposure, might increase salinity by some 10% to 20%. Thus, salinity variations on the order of twofold from lowest to highest are possible, and many organisms would be unable to survive such changes, either because they could not tolerate brackish water or could not tolerate more saline water.

Despite apparent variations in total ocean salinity, ratios of dissolved components do not seem to have varied much. Some direct measurements of sea-water inclusions in sediments indicate similar composition to modern ocean water. Perhaps more conclusive is the observation that the sequence of minerals (sulfates, carbonates, halides, etc.) precipitated during evaporite deposition is relatively unchanged through time. This invariance would be impossible if the ratios of the involved ions had varied by more than a factor of two to three from present values.

[**References** – Information on the atmosphere is provided by the book of Walker (1977), and on both the oceans and atmosphere by the book of Holland (1984). Changes on the earth's surface are also discussed by papers in Holland and Trendall (1984). The history of CO_2 in the atmosphere is described by Sundquist and Broecker (1985) and Berner (1991).]

6.9 The end of the Paleozoic

THE THREE largest biotic changes in the history of the earth occurred at the end of the Proterozoic, the end of the Paleozoic, and the end of the Mesozoic. Indeed, these changes define the eras. Of the three, probably the most devastating to existing organisms

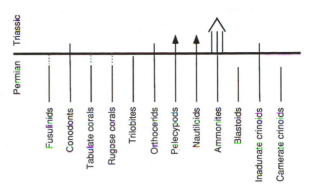

Fig. 6.13. Biotic changes at the end of the Paleozoic. Many groups of organisms did not survive the end of the Paleozoic. They did not become extinct, however, at one instant of time but over a range of several millions of years. Some groups, such as pelecypods and nautiloids, passed through the extinction event without significant change. Ammonites, which became a dominant group in the Mesozoic, had begun to diversify in the Permian. □

was the one that separated the Permian and the Triassic (Fig. 6.13). More than 50% of marine families, and perhaps more than 90% of marine species, became extinct near the Permian/Triassic boundary. Similar extinction rates also were characteristic of terrestrial organisms, although some fresh- and brackish-water varieties (fish, gastropods, etc.) may have suffered less than normal marine biota.

This crisis among the earth's organisms occurred after the assembly of continents into a Pangea supercontinent at some time in the Early to Middle Permian and the beginning of rifting in the Karoo area of Africa, India, and Antarctica shortly afterward (Section 8.1). Terrestrial deposits in Karoo rift basins contain both Permian and Triassic deposits, although nowhere in complete sequence. During the Late Permian, most of Pangea appears to have been land, with marine conditions confined to marginal areas. The South Pole at this time was probably in present Antarctica, and glacial sealevel lowering contributed to the wide exposure of continental surface.

Because of the limited availability of marine Permian and Triassic rocks, and the incompleteness of terrestrial clastic suites, study of the Permian/Triassic extinction has proven particularly difficult. Even the thickest stratigraphic sections, generally sequences of shale and limestone, lack the uppermost stage of Permian strata. The most complete, possibly continuous, sections are in southern China, where the transition from Permian to Triassic is marked by a clay bed (bentonite) only a few centimeters thick and containing magmatic zircons with an age of 251 Ma. Evidence of meteorite impact or other cataclysmic event has not been found at the Permo-Triassic boundary.

The paleontologic record also does not indicate an instantaneous devastation of life. Permian organisms died out over a period of several million years (possibly much of Permian time), and Triassic organisms took at least as long to establish themselves. Biota that became extinct include fusulinids, tabulate and rugose corals, and most families and genera of bryozoa and brachiopods (Fig. 6.6). Trilobites also became extinct but had declined to nearly zero abundance well before the Permian. Some organisms survived for a short part of the Triassic; they include orthid brachiopods, orthocerid cephalopods, hyolithids, and conodonts. Echinoderms declined in numbers but otherwise showed no particular change in evolutionary pattern at the Permian/Triassic boundary. Similarly, most molluscs were relatively unchanged except for a decline in numbers. Ammonites, which characterized the Mesozoic, had begun to diversify within the Permian rather than at the boundary with the Triassic. Vertebrates and plants also appear to have been comparatively unaffected. The major event in plant history near this time was the arrival of gymnosperms within the Late Permian, and not at its end.

Based on the preceding data, the end of the Paleozoic appears to have been a time without an instantaneous dramatic event but one in which there was a life-ending environmental change over a period of several million years. This change was particularly acute in the oceans and primarily threatened the more primitive organisms. These data are not consistent with a catastrophic event, such as meteorite impact, although eruption of the enormous Siberian basalts occurred at the end of the Permian and has been suggested as source of so much atmospheric dust that climatic modification caused widespread extinction (Section 8.2).

A variety of explanations for the end-Paleozoic biotic change have been proposed that are consistent with its gradation over time and its selectivity toward the more primitive marine organisms. One possibility is the rapid fluctuations of sealevel caused by the central-Gondwana position of the South Pole. Retreating and advancing ice sheets could have caused sudden marine transgressions and regressions, which would have put environmental stress on organisms and possibly caused rapid extinction of some groups followed by rapid evolution of successful species. The problem with this explanation is that the South Pole was within Gondwana through much of the Paleozoic, and at least some glaciation was continuous from the Ordovician through the Permian. There seems to be no reason for this pole position to have caused more rapid climatic fluctuations at the end of the era than earlier unless the aggregation of continents into Pangea accentuated the size and variability of ice sheets.

[**References** – Discussions of the end of the Paleozoic are in the books of Valentine (1985) and Donovan (1989) and the volumes edited by Walliser (1986) and Kauffman and Walliser (1990). The age of zircons at Meishan, China, is reported by Claoue-Long *et al.* (1991).]

References

Audley-Charles, M. G., & Hallam, A., eds. (1988). *Gondwana and Tethys*: Geological Society of London Special Publication 37, 317 pp.

Babcock, L. E., (1993). Trilobite malformations and the fossil record of behavioral asymmetry. *Journal of Paleontology*, **67**, 217–29.

Berner, R. A. (1991). A model for atmospheric CO_2 over Phanerozoic time. *American Journal of Science*, **291**, 339–76.

Boardman, R. S., Cheetham, A. H. & Rowell, A. J., eds. (1987). *Fossil Invertebrates*. Palo Alto, California: Blackwell Scientific Publications, 713 pp.

Bond, G. C., Nickelson, P. A. & Kominz, M. A. (1984). Breakup of a supercontinent between 625 Ma and 555 Ma: new evidence and implications for continental histories. *Earth and Planetary Science Letters*, **70**, 325–45.

Brasier, M. D. (1980). *Microfossils*. London: George Allen and Unwin, 193 pp.

Brasier, M. D. (1982). Sea-level changes, facies changes and the late Precambrian–early Cambrian evolutionary explosion. *Precambrian Research*, **17**, 105–23.

Brasier, M. D., Margaritz, M., Corfield, R., Luo, H., Wu, X., Lin, O., Jiang, Z., Hamdi, B., He, T. & Fraser, A. G. (1990). The carbon- and oxygen-isotope record of the Precambrian–Cambrian boundary interval in China and Iran and their correlation. *Geological Magazine*, **127**, 319–32.

Carroll, R. L. (1988). *Vertebrate Paleontology and Evolution*. New York: W.H. Freeman and Co., 698 pp.

Carter, J. G. (1993). *Evolution of Life – Lecture Notes for Geology 16*. Chapel Hill, North Carolina: Department of Geology, University of North Carolina, 120 pp.

Claoue-Long, J. C., Zhang, Z., Ma, G. & Du, S. (1991). The age of the Permian-Triassic boundary. *Earth and Planetary Science Letters*, **105**, 182–90.

Clarkson, E. N. K. (1986). *Invertebrate Paleontology and Evolution*, 2nd ed. Boston: Allen and Unwin, 382 pp.

Conway Morris, S. & Peel, J. S. (1990). Articulated halkieriids from the Lower Cambrian of north Greenland. *Nature*, **345**, 802–5.

Copper, P. (1986). Frasnian/Famennian mass extinction and cold-water oceans. *Geology*, **14**, 835–9.

Cowie, J. W. & Brasier, M. D., eds. (1989). *The Precambrian–Cambrian Boundary*. Oxford: Clarendon Press, 213 pp.

Crimes, T. P. (1987). Trace fossils and correlation of late Precambrian and early Cambrian strata. *Geological Magazine*, **124**, 97–119.

Donovan, S. K. (1989). *Mass Extinctions*. New York: Columbia University Press, 266 pp.

Edwards, D. (1970). Fertile Rhyniophytina from the Lower Devonian of Britain. *Palaeontology*, **13**, 451–61.

Gensel, P. G. & Andrews, H. N. (1984). *Plant Life in the Devonian*. New York: Praeger Publishers, 180 pp.

Gensel, P. G. & Andrews, H. N. (1987). The evolution of early land plants. *American Scientist*, **75**, 478–88.

Glaessner, M. F. (1984). *The Dawn of Animal Life*. Cambridge: Cambridge University Press, 244 pp.

Gould, R. E. & Delevoryas, T. (1977). The biology of *Glossopteris*; evidence from petrified seed-bearing and pollen-bearing organs. *Alcheringa*, **1**, 387–99.

Hallam, A., ed. (1973). *Atlas of Paleobiogeography*. Amsterdam: Elsevier, 531 pp.

Holland, H. D. (1984). *The Chemical Evolution of the Atmosphere and Oceans*. Princeton, New Jersey: Princeton University Press, 582 pp.

Holland, H. D. & Trendall, A. F., eds. (1984). *Patterns of Change in Earth Evolution*. Berlin: Springer Verlag, 431 pp.

House, M. R., ed. (1979). *The Origin of Major Invertebrate Groups*. Systematics Association Special Volume No. 12. London: Academic Press, 515 pp

Irving, E. & Irving, G. A. (1982). Apparent polar wander paths Carboniferous through Cenozoic and the assembly of Gondwana. *Geophysical Surveys*, **5**, 141–88.

Jarvik, E. (1952). On the fish-like tail in the ichthyostegid stegocephalians with descriptions of a new stegocephalian and a new crossopterygian from the Upper Devonian of East Greenland. *Meddelelser om Gronland*, **114**, no. 12, 1–90.

Jenkins, F. A., Jr. & Farrington, F. R. (1976). The post-cranial skeletons of the Triassic mammals Eozostrodon, Megazostrodon and Erythrotherium. *Philosophical Transactions of the Royal Society of London B*, **273**, 387–431.

Junyuan, C., Bergstrom, J., Lindstrom, M. & Xianquang, H. (1991). The Chengjiang fauna – oldest soft-bodied fauna on earth. *Research and Exploration* (National Geographic Society), **7**, 8–19.

Kauffman, E. G. & Walliser, O. H., eds. (1990). *Extinction Events in Earth History: Lecture Notes in Earth Sciences 30*. Berlin: Springer Verlag, 432 pp.

Kiaer, J. (1928). The structure of the mouth of the oldest known vertebrates, pteraspids and cephalaspids. *Palaeobiologica* **1**, 117–34.

King, A. F. (1988). *Trip A4 – Late Precambrian sedimentation and related orogenesis of the Avalon Peninsula, eastern Avalon zone*. St. John's, Newfoundland, Geological Association of Canada Newfoundland Section, 84 pp.

Kuhn-Schnyder, E. & Rieber, H. (translated by Kucera, F.) (1986). *Handbook of Paleozoology*. Baltimore, Maryland: Johns Hopkins University Press, 394 pp.

Landing, E., Myrow, P., Benus, A. & Narbonne, G. M. (1989). The Placentian Series: appearance of the oldest skeletalized fossils in southeastern Newfoundland. *Journal of Paleontology*, **63**, 739–69.

Lane, N. G. & Burke, J. J. (1976). Arm movement and feeding mode of inadunate crinoids with biserial muscular arm articulations. *Paleobiology*, **2**, 202–8.

Latham, A. & Riding, R. (1990). Fossil evidence for the location of the Precambrian/Cambrian boundary in Morocco. *Nature*, **344**, 752–4.

Lawver, L. A. & Scotese, C. R. (1987). A revised reconstruction for Gondwanaland. In *Gondwana Six: Structure, Tectonics and Geophysics*, ed. G. D. McKenzie, pp. 17–23. American Geophysical Union Geophysical Monograph 40.

Lowenstam, H. A. & Weiner, S. (1989). *On Biomineralization*. New York: Oxford University Press, 324 pp.

McKerrow, W. S., ed. (1978). *The Ecology of Fossils*. Cambridge, Massachusetts: MIT Press, 383 pp.

McKerrow, W. S. & Scotese, C. R., eds. (1990). *Paleozoic Palaeogeography and Biogeography*. Geological Society of London Memoir 12, 435 pp.

McMenamin, M. A. S. & McMenamin, D. L. S. (1989). *The Emergence of Animals – The Cambrian Breakthrough*. New York: Columbia University Press, 217 pp.

Morel, P. & Irving, E. (1981). Paleomagnetism and the evolution of Pangea. *Journal of Geophysical Research*, **86**, 1858–72.

Mueller, J. F., Jr., Rogers, J. J. W., Jin Yu-gan, Wang Huayu, Li Wenguo, Chronic, J. & Mueller, J. F. (1991). Late Carboniferous to Permian sedimentation in Inner Mongolia, China, and tectonic relationships between North China and Siberia. *Journal of Geology*, **99**, 251–63.

Narbonne, G. M., Myrow, P. M., Landing, E. & Anderson, M. M. (1987). A candidate stratotype for the Precambrian–Cambrian boundary, Fortune Head, Burin Peninsula, southeastern Newfoundland. *Canadian Journal of Earth Sciences*, **24**, 1277–93.

Parrish, J. T. (1982). Upwelling and petroleum source beds, with reference to Paleozoic. *American Association of Petroleum Geologists Bulletin*, **66**, 750–74.

Piper, J. D. A. (1987). *Paleomagnetism and the Continental Crust*. Liverpool: University of Liverpool, 434 pp.

Raup, D. M. & Boyajian, G. E. (1988). Patterns of generic extinction in the fossil record. *Paleobiology*, **14**, 109–25.

Rickard, M. J. & Belbin, L. (1980). A new continental assembly for Pangea. *Tectonophysics*, **63**, 1–12.

Romer, A. S. (1966). *Vertebrate Paleontology*. Chicago: University of Chicago Press, 468 pp.

Scagel, R. F., Bandoni, R. J., Maze, J. R., Rouse, G. E., Schofield, W. B. & Stein, J. R. (1984). *Plants – An Evolutionary Survey*. Belmont, California: Wadsworth Publishing Co., 757 pp.

Simkiss, K. & Wilbur, K. M. (1989). *Biomineralization – Cell Biology and Mineral Deposition*. San Diego, California: Academic Press.

Smith, A. G., Hurley, A. M. & Briden, J. C. (1981). *Phanerozoic Paleocontinental World Maps*. Cambridge: Cambridge University Press, 102 pp.

Stahl, B. J. (1974). *Vertebrate History: Problems in Evolution*. New York: McGraw-Hill Book Co., 594 pp.

Stearn, C. & Carroll, R. (1989). *Paleontology: The Record of Life*. New York: John Wiley and Sons, 453 pp.

Stern, K. R. (1985). *Introductory Plant Biology, 3rd edn*. Dubuque, Iowa: Wm. C. Brown Publishers, 517 pp.

Stewart, W. H. (1983). *Paleobotany and the Evolution of Plants*. Cambridge: Cambridge University Press, 405 pp.

Sundquist, E. T. & Broecker, W. S., eds. (1985). *The Carbon Cycle and Atmospheric CO_2: Natural Variations Archean to Present*. American Geophysical Union Monograph 32, 627 pp.

Trompette, R. & Young, G. M., eds. (1981). Upper Precambrian correlations. Special Issue of *Precambrian Research*, **15**, 187–422.

Valentine, J. W., ed. (1985). *Phanerozoic Diversity Patterns*. Princeton, New Jersey: Princeton University Press, 441 pp.

Van der Voo, R. (1988). Paleozoic paleogeography of North America, Gondwana, and intervening displaced terranes: comparison of paleomagnetism with paleoclimatology and biogeographical patterns. *Geological Society of America Bulletin*, **100**, 311–24.

Van der Voo, R. (1990). Phanerozoic paleomagnetic poles from

Europe and North America and comparisons with continental reconstructions. *Reviews of Geophysics*, **28**, 167–206.

Van der Voo, R., Scotese, C. R. & Bonhommet, N., eds. (1984). *Plate Reconstruction from Paleozoic Paleomagnetism*. American Geophysical Union Geodynamics Series, vol. 42, 136 pp.

Walcott, C. D. (1910). Olenellus and other genera of the Mesonacidae. *Smithsonian Miscellaneous Collection*, **53**, no. 6, 231–422.

Walker, J. C. G. (1977). *Evolution of the Atmosphere*. New York: Macmillan Publishing Co., 318 pp.

Walliser, O. H., ed. (1986). *Global Bio-Events: Lecture Notes in Earth Sciences 8*. Berlin: Springer Verlag, 442 pp.

White, T. E. (1939). Osteology of Seymouria baylorensis broili. *Harvard Museum of Comparative Zoology Bulletin*, **85**, 325–409.

Whittington, H. B. (1985). *The Burgess Shale*. Geological Survey of Canada and Yale University Press, 151 pp.

Wilson, J. L. (1975). *Carbonate Facies in Geologic History*. New York: Springer Verlag, 471 pp.

Paleozoic section in Grand Canyon of Arizona, showing stratigraphic sequences as defined by Sloss (1963). (Courtesy of M. Follo.) □

7

THE PALEOZOIC – PART II.
TECTONICS

7.0 Introduction

IN ORDER to trace Paleozoic lithologic and tectonic history, we need a beginning and an end. The end is not difficult to determine. During the Permian, the earth's continental crust was assembled into a supercontinent normally referred to as 'Pangea.' As discussed in Chapter 6, the maximum 'packing' of Pangea occurred within the Permian, and dispersal had begun before the paleontologic start of the Mesozoic. The configuration of Pangea can be determined with relative accuracy because of our ability to trace its fragmentation in the Mesozoic and Cenozoic (Chapter 8). Thus, all tectonic histories of the Paleozoic must arrive at this end-Paleozoic Pangea.

Our problem in this chapter is that we do not know where to begin. Most geologists presume that the earth's continental crust was aggregated into a supercontinent near the end of the Proterozoic. There is, however, little agreement on its configuration, and numerous possibilities have been proposed. The 'conventional' end-Proterozoic Pangea is similar to the one at the end of the Paleozoic (Fig. 7.1A). This configuration permits an opening of the Atlantic Ocean (Iapetus Ocean) between North America and Africa and retains Gondwana as an entity from some time in the Precambrian until the end of the Paleozoic.

In the middle of 1991, several geologists proposed alternative configurations for an end-Proterozoic Pangea (Fig. 7.1B and C). In these assemblies, North America was bordered by Antarctica on the west and South America, or parts of present South America, on the east (in present orientations). The configurations shown in Fig. 7.1B and C require the Iapetus Ocean to open between North America and the west coast of South America (or Brazilian shield), with creation of a new assembly of Gondwana at some time during the Paleozoic and a new configuration of a supercontinent (Pangea) near the end of the Paleozoic.

One consistent aspect of the proposed configurations in Fig. 7.1B and C is that the end-Proterozoic supercontinent was centered around Laurentia (North America, Greenland and the Baltic shield). Similarly, the end-Paleozoic supercontinent (Pangea) could be regarded as centered around the Congo craton. Thus, the basic plate reorganization of the Paleozoic may have been fragmentation from one nucleus (Laurentia) and coalescence around another one (Congo).

Evidence for or against the various configurations of an end-Proterozoic supercontinent is not overwhelming (to put the matter mildly). Resolution of the issue requires two types of data that are currently (in 1992) being eagerly sought. One is a possible correlation of

A

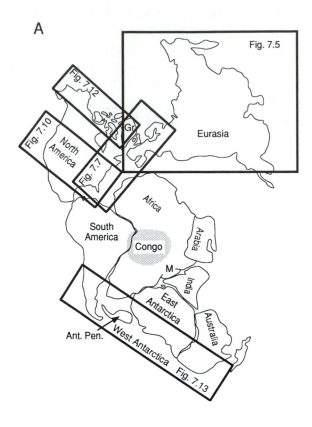

C

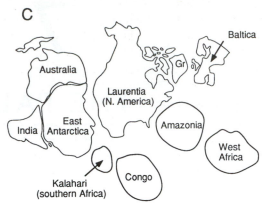

Fig. 7.1. Three configurations of supercontinental assemblies.

(A) Pangea at the end of the Paleozoic. This sketch is also an index map for other figures and areas discussed in the text. If configuration A also represents a late-Proterozoic supercontinent, then the Iapetus Ocean on the east coast of North America opened by rifting of North America and Europe/Africa in the late Proterozoic. These same two continental masses later reversed their separation and converged during the Paleozoic, ultimately redeveloping the same (Pangea) assembly at the end of the Paleozoic that it had at its beginning. Ant. Pen. is the Antarctic Peninsula; Congo is the Congo craton; Gr is Greenland; M is Madagascar.

(B) Simplified configuration of supercontinent at the end of the Proterozoic, showing North America in a central position. Configuration B implies formation of the Iapetus Ocean by rifting of North America and South America in the late Proterozoic and formation of the Pacific margin of North America by rift separation of a combined Antarctica/Australia/India. This configuration requires movement of Antarctica/Australia/India around the earth to converge with the east coast of Africa at some time in the latest Proterozoic or early Paleozoic. Most of central and eastern Asia is not shown in this diagram because of uncertainty about its relationship to the areas depicted here. Gr is Greenland.

(C) Configuration at the end of the Proterozoic assuming separate evolution of cratons that are now fused to form both South America and Africa. Configuration C implies a supercontinental configuration at the end of the Proterozoic that underwent widespread rifting, with later (Paleozoic) assembly of South America and Africa and suturing of Antarctica/Australia/India to Africa/South America to form Gondwana. Thus, the supercontinent (Pangea) formed at the end of the Paleozoic would have accreted around the Congo craton (see diagram A). □

B

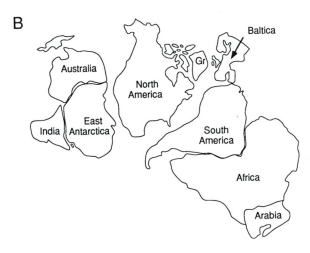

fragments rifted off of Laurentia with the North American nucleus. For example, the Grenville province has been proposed to extend into southern South America and the Antarctic. A second necessary type of data is ages of suturing in various orogenic belts in Gondwana. For example, if the west coast of India/Madagascar and the east coast of Africa have been joined since the middle Proterozoic, then the configurations of Fig. 7.1B and C are wrong; conversely, if

India/Madagascar and East Africa collided along a late Proterozoic/early Paleozoic (Pan-African) orogenic belt, then the configuration in Fig. 7.1A is impossible for the late Proterozoic.

Regardless of the configuration of a supercontinent at the beginning of the Paleozoic, a conventional assembly of Gondwana may have been achieved reasonably early in the Paleozoic, perhaps no later than the Early Ordovician. This age precedes the time of most Paleozoic orogenic activity. Consequently, we will describe the evolution of Paleozoic orogenic belts and the sources of far-travelled terranes in those belts as if Gondwana existed in its end-Paleozoic configuration through most of the Paleozoic.

Proceeding from this admittedly shaky beginning, this chapter discusses two aspects of the history of the Paleozoic. One is the evolution of generally unde-formed sedimentary cover sequences on cratonic plat-forms (Section 7.1). These platforms are the sites of quartzite/limestone/dolomite deposition and record a large part of the paleontologic history of the era. Stratigraphy in the platform suites is controlled by an interaction between eustatic sealevel changes and epeirogenic (purely vertical) movements of the conti-nental crust. The interaction between these two effects is difficult (some geologists would say impossible) to determine. A principal method of investigating the interaction between eustacy and epeirogeny is to attempt intercratonic correlation of the cover sequences, and we discuss this possibility in Section 7.1.

The second aspect of Paleozoic history is the evolu-tion of orogenic belts and the assembly of continental blocks. At the start of this orogenic history (but per-haps not the start of the Paleozoic), the earth contained numerous separate continental blocks: 1) Gondwana; 2) Laurentia, including North America and Greenland; 3) Baltica, including the Baltic shield and the Russian platform; 4) Siberia, or possibly two independent plates that fused early in the Paleozoic; and 5) the sepa-rate blocks of Kazakhstan, North China, South China and other blocks now accreted in Asia.

Starting from these independent continental blocks, the Paleozoic was dominated by the shedding of frag-ments from Gondwana and their accumulation into other landmasses. This process generated compressive orogenies throughout much of Eurasia and North America. In addition, the southern margins of Gondwana and the northern margins of North America and Eurasia (both referring to present orien-tations) underwent compressive deformation caused by spreading in the world-encircling ocean. Margins of all blocks facing Tethys, however, appear to have been passive, both along Gondwana and the retreating edges of blocks rifted from it.

The Paleozoic fragmentation of Gondwana appears almost everywhere to have been 'successful' (Section 4.2). Neither Gondwana, nor any other large continen-tal block that existed during the Paleozoic, was pene-trated by networks of 'unsuccessful' rifts of the type that affected Africa in the Mesozoic and Cenozoic and several shields in the Proterozoic (Section 5.3). With the exception of a few continental-margin aulacogens, par-ticularly in North America, all Paleozoic rifts older than the Permian opened into oceans that provided new continental fragments for the world's orogenic belts.

The discussion of Paleozoic orogeny is organized into four main topics (Fig. 7.1A). In terms of area affected, the major event was the accumulation of Eurasia (Section 7.2). During the fragmentation of Gondwana, many Pan-African (and some older) ter-ranes drifted away and fused with Laurentia and Baltica, which caused the formation of an entire conti-nent of Laurasia extending from North America to China by the end of the Paleozoic. Because of its remoteness, however, and the scarcity of modern geo-logic investigation except in a few areas, our discus-sion of Eurasia is restricted largely to general state-ments about the colliding blocks and their modes of suturing.

The most intensely deformed area of Paleozoic orogeny is the set of mountain belts around the present North Atlantic, referred to as the 'North Atlantic oro-genic system' (Section 7.3). This zone of repeated colli-sion between North America, Baltica and Africa is vir-tually a large-scale melange of disparate plates. Our principal effort here is to identify a sequence in which the various orogenic events occurred.

The northern, western and southern margins of North America underwent compressive orogeny in the late Paleozoic (Section 7.4). Although complexly deformed, these orogenic belts are more coherent than those around the North Atlantic, and the relationships between orogeny and sedimentation can be described with more certainty.

The southern margin of Gondwana (Australia, East Antarctica, southern Africa and southern South America), was generally the site of subduction at vari-ous times during the Paleozoic (Section 7.5). Extensive preservation of the Paleozoic orogens has occurred only in Australia, and we concentrate on that area.

The following discussion of Paleozoic tectonics shows two major episodes of apparent worldwide change. One event is at some time, probably early, in the Devonian. At this time: 1) a discontinuity occurred between Caledonian and Variscan (Hercynian) orogenies throughout much of Europe; 2) compression began on previously passive margins of North America; 3) most of the compressive suturing of plates began in Asia; 4) compression ended along the Antarctic portion of the Gondwana margin; and 5) orogenic activity underwent reorganization along the Australian and South American parts of the Gondwana margin. Assembly of all continents into a Devonian Pangea has been proposed from paleomagnetic evidence, and this suturing may be associated with the interruption of orogenic activity.

A second major worldwide event occurred during the Permian. Compressive tectonism stopped along many of the previously active belts at this time, including the North Atlantic orogen, much of northern Asia, and the margins of North America. To the extent that dates can be determined accurately, this (Middle?) Permian cessation coincided with the initiation of rifting in Africa and other areas in the Gondwana part of Pangea. This extraordinary tectonic reorganization of the earth seems to have occurred about 25 million years earlier than the paleontological event that terminated the Paleozoic (Section 6.9).

[**References** – Recent proposals for the configuration of a supercontinent at the end of the Proterozoic are discussed by Dalziel (1991, 1992), Hoffman (1991), Moores (1991), and Unrug (1993).]

7.1   Cratonic cover sequences

THE VARIOUS continental blocks that moved around the earth during the Paleozoic (Fig. 6.12) were the sites of deposition of platformal (cratonic-cover) sediments. This section addresses three important aspects of their evolution. One is the distinction between the lithology and source of the sediments of platforms and those of more tectonically active environments. The second is the relationship between events that control stratigraphy on cratons and events elsewhere in the world, including paleontologic transitions. The third aspect is the development of a terrestrial 'Gondwana' facies near the end of the Paleozoic.

Many of the cover sequences that formed during the Paleozoic are difficult to investigate because of later events. Most of them have not undergone major tectonism, but many have been broadly eroded, and several are almost completely blanketed by Mesozoic/Cenozoic suites deposited under the same platformal conditions. The two best-exposed and areally widespread suites are on the North American craton and the Russian platform (Baltica). Consequently, much of our information is derived from these two areas.

Lithology and source of platform sediments

Cratonic-cover sediments consist largely of quartz-rich sandstone, limestone, dolomite and/or evaporite, shale, and no volcanic rocks. This suite differs from the sediments of cratonic margins, where active tectonism produces foredeeps, continental slopes, and arc-trench environments. Sediments in these active environments: have higher clastic/carbonate ratios; contain abundant mixtures of sand, silt and shale (commonly in turbidites); rarely include dolomite or evaporites; and contain volcanic rocks and volcanogenic sediment.

The clastic components of cratonic cover sequences have two sources. One is the continental interior, which commonly is partly emergent throughout an entire marine transgressive episode or is covered only late in the sequence. This interior may supply sediment from exposed crystalline basement or from older platform sediments uplifted before marine encroachment. A second source is developing orogenic belts around craton margins, where the generally immature sands eroded from active uplifts can wash out of the foredeep basin (e.g. Section 7.4) and onto the craton.

Cratonic sedimentary suites develop in various depositional environments that we can classify broadly into 'basins' and 'shelves'. In many areas, networks of rifts active in the late Proterozoic in both North America and Russia (Fig. 7.1A, B and C; Section 5.3) localized the development of sedimentary basins that profoundly affected net depositional thickness in overlying Paleozoic rocks. Some cratonic basins, however, do not appear to overlie older rifts, and the specific reason for their development is unknown.

Typically, the shelf parts of platform-cover sequences are no more than a few thousand meters thick and commonly much thinner. Quartzites are generally interbedded with biogenic limestones. Organic-rich shales occur in areas where bottom anoxia developed, possibly because of density stratification caused by high influx of fresh water to the epeiric seas. Dolomites and/or evaporites form in bank areas, but many of the dolomites may be post-diagenetic. Aggregate sediment thickness in basins can be up to 10 000 m, and basins are more likely to contain shaly and silty clastic sediments. Evaporites and dolomites form in basins primarily during episodes of basin isolation or general marine regression from the craton, but dolomitization and deposition of thin evaporite suites can occur throughout the deposition of shelf sequences. Conversely, organic-rich sediments can form throughout basin development but are commonly restricted to a few, very widespread, chrono-stratigraphic intervals on shelves.

Depositional environments for cratonic sediments range from subaerial to shallow water. Many suites were deposited essentially at sealevel, and even in basinal environments water was probably never deeper than about 100 m. Small-scale sealevel fluctuations formed cyclical deposits (cyclothems). In part of North America, these cyclothems typically have thicknesses of a few tens of meters and pass repeatedly from shallow subaerial sands and silts through coal formed in marine swamps to shallow marine shales and carbonate rocks. The top of the sequence is covered by fluvial sediments formed during regression; they represent the base of the next cycle. The frequency of cyclothems in the Carboniferous may be closely related to the frequency of small-scale glacial advance and retreat in Gondwana.

In addition to small-scale, presumably glacio-eustatic, episodes, the cratonic cover sequences can be characterized in terms of major episodes of marine transgression and regression. These major episodes form chronostratigraphic units (sequences) separated by hiatuses. The breaks in the sedimentary record mostly represent periods of non-deposition within sequences of seemingly conformable strata, but they may include local angular unconformities, generally without evidence of major deformation. Some unconformities show evidence of soil formation and have a relief of several tens of meters caused by karst topography or other features of subaerial erosion. Basal sediments on top of the unconformities are commonly coarsely clas-

tic, although carbonate rocks may develop in craton-margin areas far from clastic sources. The basal sands are invariably time-transgressive toward the craton interior, commonly over a time scale of a few tens of millions of years.

Sealevel, depositional history and intercontinental correlation

The sedimentary sequences of cratons reflect a complex relationship among long- and short-term variations in sealevel, local tectonics, craton-wide tectonism, and worldwide tectonism that may have affected several cratons together. Understanding this mixture of controlling processes is complicated by the large number of variables and the small number of observations.

The time scales and possible controlling factors for eustatic sealevel variations are discussed more completely in Section 2.4. Glacio-eustatic variations can occur over a span of 10^4 to 10^6 years, with total possible sealevel rise or fall of about 300 to 400 m. Variations in ridge-crest volume occur over time spans of 10^6 to 10^8 years, with total vertical sealevel movements also about 300 to 400 m. The sum of these processes should yield differences in continental elevation relative to sealevel of somewhat less than one kilometer. The earth's surface is apparently about in the middle of this range at the present time.

This one kilometer of sealevel movement must be compared with the requirements for the accumulation of well-studied cratonic cover sequences. For this purpose, assume a thickness of two kilometers of sedimentary cover on a shelf area in a continental platform. (We omit basinal areas that clearly have been affected by higher rates of thermal subsidence.) Two kilometers of platform sediments would cause about 1200 m of load subsidence, depending on sediment density, leaving about 800 m of accumulation to be accounted for by sealevel rise or cratonic subsidence. This accumulation appears to have been permanent in most cratons, allowing retention of much of the sediment rather than complete erosion during marine regression.

Permanent continental subsidence that allows permanent accumulation implies some type of tectonic movement of the craton. The amount of movement required on typical cratons, however, is less than one

A

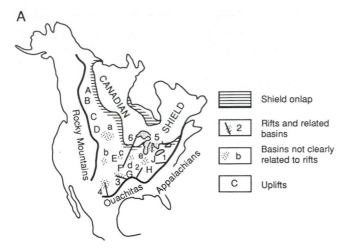

B

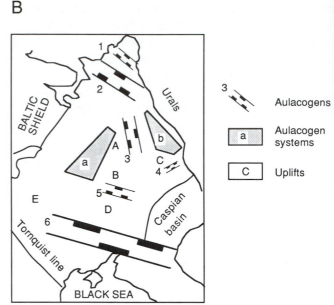

Fig. 7.2. Structures on cratons.

(A) North America. The cratonic interior of North America is bounded by orogenic belts (Rocky Mountains, Ouachitas, Appalachians) and consists of relatively undeformed sediments deposited on a stable Precambrian basement. Regardless of this stability, numerous basins (possibly rift-related) and uplifts are present, and the craton has been cut by several major late-Precambrian rifts. The stratigraphic sequence also shows several periods of craton-wide emergence that formed widespread unconformities. Basins designated as 'rifts and related basins' include: 1) Rome trough; 2) Reelfoot rift; 3) Anadarko (Arkoma, Ardmore) basin; 4) Tobosa basin; 5) Michigan basin; and 6) mid-continent gravity high. Basins designated as not related to rifts include: a) Williston basin; b) Denver basin (not active until the Kaskaskia sequence); c) Salina basin (not active until the Kaskaskia sequence); d) Forest City (Iowa) basin (not active until the Kaskaskia sequence); and e) Illinois basin. Uplifts include: A) Tathlina; B) Peace River arch; C) Lloydminster; D) Sweetgrass arch; E) Chadron arch; F) Nemaha uplift; G) Ozark dome; H) Nashville dome; and J) Cincinnati arch.

(B) Russia. Stratigraphic thickness in the Paleozoic sedimentary cover of the Russian platform is dominantly controlled by late-Proterozoic rift systems, which provided the locus for greatest subsidence. These rifts are divided into individual aulacogens and more-complex systems of aulacogens. Aulacogens include: 1) Pechora–Kolva; 2) Timan; 3) Vyatka; 4) Sernovodsk–Abulina; 5) Pachelma; and 6) Pripyat–Dnieper–Donets–Donbas–Karpinsky. Aulacogen systems include: a) central Russian; and b) Kama–Belaya. Uplifts include: A) Volga–Ural; B) Tokmovo; C) Zhiguli–Orenberg; D) Voronezh; and E) Mazurian–Belorussian. □

to two kilometers for the entire Paleozoic, in some places for the entire Phanerozoic. If the Paleozoic section contains several bounding unconformities, then tectonic movements for each one could have been only 100 to 200 m, a value comparable to glacial or ocean-ridge effects on sealevel. Thus, the potential controlling processes for stratigraphic boundaries all yield effects so small and so similar that separating them cannot be done with confidence.

Correlation of depositional histories between widely separated cratonic cover suites is complicated by numerous problems in addition to those discussed above. One is simply that investigators in different countries erect somewhat different paleontologic and stratigraphic nomenclatures for their various areas. A second issue is that separation and latitudinal difference between cratons place many of them in wholly different biotic realms.

The two areas in which detailed stratigraphic information is available for Paleozoic strata over entire cratons are North America and Russia (Fig. 7.2A and B). At various times during the Paleozoic, up to two thirds of each craton was covered by epicontinental seas. The North American sequences lap onto the Canadian shield and are bordered by orogenic belts (Section 7.4). The Russian platform laps onto both the Baltic and Ukrainian shields, which presumably are joined at depth and form the basement for the sedimentary cover (Section 3.0 and Fig. 5.3). The eastern edge of the Russian platform is the Ural orogen, and the western edge is the deformational belts formed along the Paleozoic Tornquist line (Section 7.3).

Structural features on the two cratons are shown in Fig. 7.2A and B. In North America, rift-related basins may have been largely inherited from Precambrian

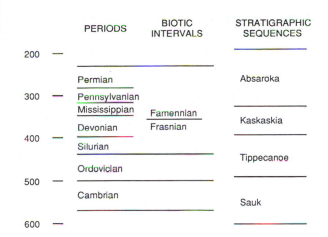

PERIODS	BIOTIC INTERVALS	STRATIGRAPHIC SEQUENCES
Permian		Absaroka
Pennsylvanian		
Mississippian	Famennian	Kaskaskia
Devonian	Frasnian	
Silurian		
Ordovician		Tippecanoe
Cambrian		Sauk

Fig. 7.3.   Relationships among absolute ages, stratigraphic periods, biotic intervals, and stratigraphic sequences on the North American craton. The sequences are bounded by unconformities that extend over virtually the entire craton. These unconformities do not correspond to times of abrupt paleontologic changes or to the conventional stratigraphic time scale.   □

structures and affected strata as old as Cambrian and Ordovician. Some non-rift structures appear to have developed during the later Paleozoic, probably largely as a result of the deformation that began in the Ouachita belt during the Mississippian. The generation of these later structures is one indication of the effect of small-scale, but broad-area, deformation within the craton. Structures in the Russian platform are shown mostly as aulacogen systems inherited from Riphean rift networks (Section 5.3).

In North America, the cover sequence spans the age range of late Proterozoic to Cenozoic and commonly is divided into six chronostratigraphic units ('sequences') named for Indian tribes; each of them is separated by a bounding unconformity (see above; Frontispiece for this chapter). Four of the stratigraphic sequences are wholly or partly in the Paleozoic, and their bounding unconformities are shown on Fig. 7.3, along with the normal geologic time scale and the paleontologic intervals used in Sections 6.3 to 6.6 to describe the evolution of organisms. The sequence boundaries probably are not restricted to the platformal areas of the mid-continent but can also be found in some of the marginal foredeep basins (miogeoclines; Section 7.3), although the greater thickness of sediment and more continuous deposition in the foredeeps make the boundaries more difficult to recognize.

The most striking relationship shown in Fig. 7.3 is that the major chronostratigraphic boundaries do not occur at the same time as the boundaries between biotic intervals. The boundaries also do not correlate with known Gondwana glacial periods in the Late Ordovician and Permo-Carboniferous. In fact, Gondwana glacio-eustatic sealevel fluctuations can be recognized only as secondary effects within the major stratigraphic intervals, and as many as sixty transgressive–regressive sequences in North America and elsewhere have been correlated with Permo-Carboniferous glacial fluctuations in Gondwana.

Assuming that Fig. 7.3 is correct, the major sequence-bounding unconformities on the North American craton are not solely the result of worldwide, eustatic, sealevel retreat. If they were, there should be a direct relationship with times of major biotic changes (extinctions and evolutionary radiations) and Gondwana glaciations. This conclusion requires that relative Paleozoic sealevel fluctuations on the craton were controlled at least partly by tectonic processes on the craton itself rather than by processes in surrounding oceanic areas. Such a process requires tectonic uplift and downwarp of the craton by a few hundred meters in addition to eustatic sealevel fluctuations.

One of the problems related to cratonic deformation is whether the entire craton was affected by a mild uplift or downwarp or whether the vertical movements were localized. The presence of basins and intervening arches (Fig. 7.2A) demonstrates some ability for differential movement, but much of this differential subsidence may have been the result of a thin crust inherited from Precambrian rifts rather than differential tectonic warping at different places in the craton. A firm answer to the question of the extent of tectonic movements would be a demonstration that periods of accelerated uplift or subsidence occurred at the same times in basin and shelf areas and were not correlated with known periods of eustatic sealevel variation, but this demonstration requires considerably more accuracy in the timing of vertical movements than is possible with present techniques.

If stratigraphic intervals on the North American craton are bounded by unconformities caused partly by vertical tectonic movements rather than by eustatic sealevel movements, then is there any possibility of correlating intercontinental stratigraphic events? Strong (very strong) statements have been made on both sides of this issue. Fig. 7.4 is an example of the data

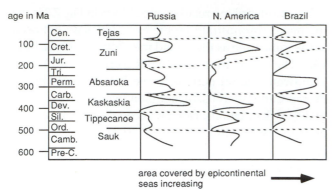

Fig. 7.4. Possible correlation of Paleozoic cratonic sequences in Russia, North America, and Brazil (adapted from Soares *et al.*, 1978). The dashed lines indicate possible correlation between stratigraphic events on different cratons, but the synchroneity has been questioned by some investigators. ☐

base that can be used in the argument. It shows times of major regression in North America, Russia, and several cratonic basins in Brazil where Paleozoic sediments can be studied around the basin margins. The dates of unconformities shown for the three areas are not identical, but they are very close. Geologists who feel that the dates are close enough (within the error of the measurements) propose intercontinental correlation of sequence boundaries; other geologists disagree.

In summary, we can make the following statement. If (emphasize if) major chronostratigraphic boundaries are caused partly by vertical tectonic movements of cratons, and if those boundaries correlate among cratons, then there must be some currently unexplained global process that causes the tectonic movements.

Gondwana sediments

The fusion of continental blocks to form Pangea appears to have occurred approximately in the Middle Permian, yielding an assembly similar to the one shown in Fig. 6.12C. At this time, Gondwana joined with North America and the western part of Eurasia to form a supercontinent that extended over more than 90° of latitude. At the same time, global sealevel began to fall, and Permian marine sediments were restricted largely to continental margins and a few incomplete suture zones. This sealevel drop presumably was caused by expansion of the Gondwana ice cap, which had been undergoing pulsating advances and retreats throughout the Carboniferous and Early Permian.

Areas around the glaciated region of Pangea (Gondwana) developed an intricate network of rift basins. The exact age relationships among the beginning of rifting, maximum sealevel retreat, maximum extent of glaciers, and closest fusion of Pangea are not known, although all of these events clearly fit within the Permian. Regardless of exact timing, the rift basins preserve most of the terrestrial sediments deposited in the region during the Permian and, locally, into the Triassic. They commonly are referred to as 'Gondwana' or 'Karoo' sequences in all continents, with the latter name derived from the Karoo basin of South Africa. Karoo sedimentation ended as Pangea began to rift and interior areas were submerged under the sealevel advance brought about by glacial retreat. Many of the Karoo basins are now preserved on land, but a few became the locus of oceanic opening and are preserved only as fragments on the rifted margins of continents (Section 8.1).

Karoo sedimentary sequences are subaerial, including fluvial, lacustrine, and aeolian environments. Typical rock types are: arkosic and other red sandstones; green to gray marls and siltstones; micaceous shales; and coals and carbonaceous shaly sediments. Coals are abundant as a result of the lake and fluvial swamps formed in the rift basins. The characteristic fossil is *Glossopteris* (Section 6.5), which is preserved in coaly material. Transport directions in the Karoo sediments are broadly away from the presumed center of the Gondwana ice sheet.

[**References** – General problems of eustasy are discussed by Pittman (1978) and Leggett *et al.* (1981) and papers in volumes edited by Schlee (1984), Kleinspehn and Paola (1988), and Wilgus *et al.* (1988). Information for specific areas is from: Sloss (1963, 1988), Feldmann (1987), and Bond, Kominz and Grotzinger (1988) for North America; Dennison (1989) for the Appalachian basin; Ricketts (1989) for western Canada; Shakhnovskiy (1988) for the Russian platform; and Veevers (1988) for Australia. Worldwide relationships are in Soares, Landim and Fulfaro (1978) and Ross and Ross (1985). Gondwana (Karoo) sedimentation is discussed by Martin (1981), Veevers and Powell (1987) and Dypvik *et al.* (1990).]

7.2 Assembly of Eurasia

HOW can we describe briefly the geologic history of the seemingly endless forests and grasslands and deserts and anastomosing mountain chains of Asia? The answer is simple – we will generalize.

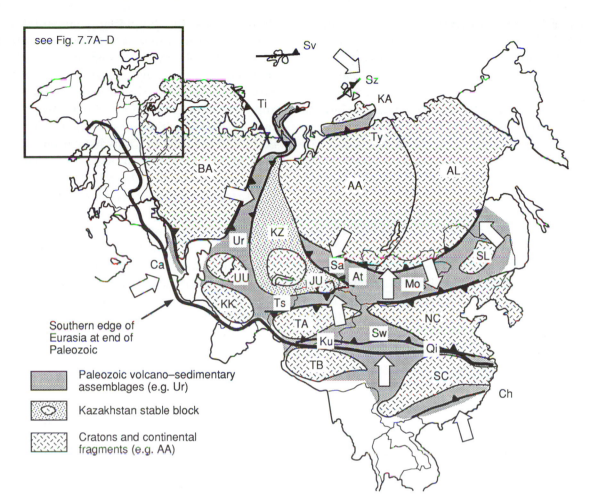

Fig. 7.5. Stable blocks and orogenic belts in Eurasia. This very complicated diagram depicts only major terranes and orogenic zones involved in the assembly of Asia. Most of the accretion occurred during the late Paleozoic, but some continued into the Mesozoic and Cenozoic. The location of many of these individual terranes before accretion is unknown. Thrust vergences are shown by conventional saw-tooth symbols. Inferred direction of movement of the downgoing slab in subduction zones is shown by the large white arrows. The stable blocks are: AA, Anabar–Angara; AL, Aldan; BA, Baltica, also known as the East European platform; JU, Junggar; KA, Kara; KK, Karakum; KZ, Kazakhstan, shown with a different pattern for reasons explained in the text; NC, North China, containing the Sino-Korean craton; SC, South China, containing the Yangtze craton; SL, Songliao; TA, Tarim; TB, Tibet, containing both the Qiangtang and Lhasa blocks; and UU, Ust–Urt. Orogenic belts include: At, Altai; Ca, Caucasus; Ch, southern China; Ku, Kunlun; Mo, Mongolian, also known as Gobi; Qi, Qinling; Sa, Sayan; Sv, Svalbard; Sw, southwestern China flysch basin; Sz, Severnaya Zemlya; Ti, Timan; Ts, Tienshan; Ty, Taimyr; and Ur, Ural.   □

The Eurasian landmass (Fig. 7.5) was assembled largely during the Paleozoic from a motley assortment of blocks of old and young continental crust, microcontinents, island-arc complexes and fragments of oceanic lithosphere (including ophiolites). The two largest (Precambrian) continental plates are Baltica (also referred to as the East European platform) and Siberia. Siberia may have been two plates, with the part containing the Anabar and Angara shields separated from the part containing the Aldan shield

(Section 5.2), but the join between them is mostly covered by flat-lying Phanerozoic sediments. With one exception (see below), most of the other plates shown in Fig. 7.5 probably also consist largely of Precambrian continental crust.

One plate, Kazakhstan, is depicted in Fig. 7.5 with a different pattern from all others. The central part of the Kazakhstan plate contains deformed lower- to middle-Paleozoic orogenic sedimentary assemblages and magmatic rocks that surround (and cover?) limited outcrops of Precambrian crystalline rocks. The northern part of the plate extends an unknown distance beneath the flat-lying sediments of the West Siberian basin, and the southeastern part of the plate may extend into a poorly exposed zone of collision between Siberia and Tarim. The limited data available indicate that the orogenic sedimentary–magmatic suites are continental-margin accretionary complexes, and vergence patterns suggest that early and middle Paleozoic subduction occurred simultaneously on all four sides of an expanding Kazakhstan plate. Simultaneous subduction around all sides of a continental fragment as small as the original Kazakhstan seems unlikely, and more complete data may indicate alternation of activity on different sides at different times. The final docking of Kazakhstan with Eurasia occurred as Kazakhstan was inserted between Siberia and Baltica in the Late Carboniferous, thus causing the Uralian orogeny.

The terrane boundaries shown in Fig. 7.5 are not as accurately known as the diagram implies (i.e. this is a normal geologic map). Part of the difficulty in outlining terranes is cover by younger sediments. Another uncertainty is the amount of craton covered by rocks thrust out of the intervening orogenic belts. Possibly continental blocks on either side of the volcano–sedimentary belts now join beneath these belts, and the oceanic (including ophiolite) suites of the orogenic belts are completely detached from their source and lie on decollements over the old continental crust.

The source of the various plates in Asia is uncertain. Although many of the microcontinents caught in the North Atlantic orogenic system (Section 7.3) appear to have been derived from belts of Pan-African activity in Africa, the terranes assembled into Asia contain ages as old as early Archean and may have been fragments that were never attached to Gondwana. Paleomagnetic data suggest, but do not require, that most of the Asiatic terranes migrated away from Gondwana.

The orogenic belts shown in Fig. 7.5 commonly contain multiple ophiolite/blueschist zones. Some of these zones are within the volcano–sedimentary belts, and some separate the accretionary wedges from the adjacent stable cratons. This complexity could have resulted from two processes. One is intra-oceanic subduction within the closing oceanic basins; in effect, this process would cause the creation of numerous, small, oceanic plates that moved, expanded, and were destroyed as the bounding continental fragments approached each other. A second reason for the complexity of ophiolite occurrences is post-collisional transport by thrusting or strike-slip faulting. This process most likely would move rootless ophiolite zones up and over continental margins but could also move materials within the ensimatic belts themselves. Several of the belts in China have been shown to contain ophiolite/blueschist suites ranging from late Proterozoic through Paleozoic, indicating the sweeping together of lithosphere formed during multiple periods of oceanic spreading and subduction.

The northern margin of the assembling Eurasia (present orientation) exhibits Paleozoic subduction at two places, the Timan belt of northern Russia (and Norway?) and the Taimyr and Severnaya Zemlya belts (with the intervening Kara continental fragment) of northern Siberia. These preserved orogenic belts may be parts of a continental-margin subduction complex that extends along the entire northern edge of the continent, with most of the belt now submerged offshore.

Blocks accreting onto the southern margin of Eurasia from Gondwana were between two ocean basins. The northern one, between these fragments and Eurasia, was a closing ocean with subduction zones on at least one, and possibly both, margins. The ocean between the northward-moving blocks and the northern margin of Gondwana can be regarded as a Paleozoic Tethys. With the exception of southeastern China and the Caucasus, no evidence exists to show subduction of oceanic lithosphere northward under any of the accreting blocks, and their southern margins were apparently sites of passive-margin sedimentation. The lack of subduction under these blocks and along the northern margin of Gondwana indicates that the Paleozoic Tethyan Ocean was a spreading ocean perhaps similar to the modern Atlantic.

The simple pattern of collision of Baltica with Kazakhstan (and Siberia?) seems to have imparted a structural pattern to the Urals that is less complex than that of other central Asian belts. The Urals contain an assemblage of rocks formed throughout most of the Paleozoic along both active and passive continental margins and in intra-oceanic areas (island arcs, ophiolites, etc.). These rocks were swept together and finally compressed during the Carboniferous. Synorogenic magmatism (calcalkaline batholiths) occurs almost exclusively along the eastern side of the belt, consistent with subduction of Baltica and its attached oceanic lithosphere beneath the continental blocks to the east. Thrust complexes without associated magmatism moved up and over the Baltica (East European) platform, and the magmatically active zone to the east contains east-vergent thrusts onto Kazakhstan (see Section 4.1 for further discussion of varieties of orogenic belts).

The largest Paleozoic orogenic belt in Asia extends from the inferred eastern tip of Kazakhstan to near the eastern edge of the continent (Fig. 7.5). The major part of it is between the North China and Siberian (Anabar–Angara plus Aldan) plates, where it is referred to variously as the Mongolian or Gobi orogenic belt. Farther west, the belt is correlatable with the Tienshan, Sayan, and Altai mountains. The belt is virtually a melange of microcontinental and island-arc fragments and contains numerous ophiolite/ blueschist suites. Calcalkaline intrusive suites occur both north and south of the orogenic belt. Magmatism probably occurred throughout much of the early and middle Paleozoic on the south and extended into the late Paleozoic on the north. Post-orogenic granites are abundant throughout the entire deformed region. Closure of the Siberian and North China blocks apparently was rotational, starting in the west in the Permian and extending eastward into the middle Mesozoic. The thrust orientations shown on Fig. 7.5 indicate that volcano–sedimentary suites were thrust out of the belt onto both marginal cratons.

Belts of questionable deformation age south of the Mongolian belts include the Qinling (between North and South China), the Kunlun (between Tarim and Tibet), and a broad basin of deformed flysch northwest of the South China block. The Qinling belt contains ophiolites of Silurian age and Paleozoic sediments caught in a collisional belt of Late Paleozoic to Mesozoic age. Most of the deformation in the Kunlun belt is Mesozoic and Cenozoic, possibly obscuring Paleozoic activity. The flysch basin of western China was affected by continued north-directed subduction of oceanic lithosphere from the Paleozoic into the Mesozoic.

The line shown on Fig. 7.5 as the southern margin of Paleozoic Eurasia is highly questionable. If major blocks such as South China, Tibet and possibly even North China did not collide with Eurasia until the Mesozoic, then the southern edge would have been much farther north. All of the orogenic belts south of

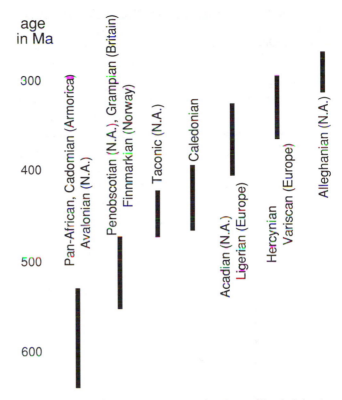

Fig. 7.6. Ages of orogenic events in the circum-North Atlantic orogenic system. □

the Mongolian–Tienshan trend have undergone post-Paleozoic deformation, but whether this activity was an intra-cratonic readjustment or a response to the new docking of South China and Tibet is unclear.

[**References** – Information for Russia, Siberia, and adjoining areas is from Nalivkin (1960, 1973) and Khain (1985). Information for China is from Zhang, Liou and Coleman (1984) and Coleman (1989). General reconstructions are presented by Zonenshayn, Kuz'Min and Natalov (1987) and, for the Carboniferous, by Rowley *et al.* (1985). Revyakin (1987) discusses the Urals.]

7.3 North Atlantic orogenic system

THE DENSE FORESTS and fertile fields of eastern North America and northern Europe hide a geologic pot-pourri of island arcs, microcontinents, continental-margin wedges, oceanic crust and successor basins mashed together by the collision of Baltica, North America, and the African part of Gondwana. This

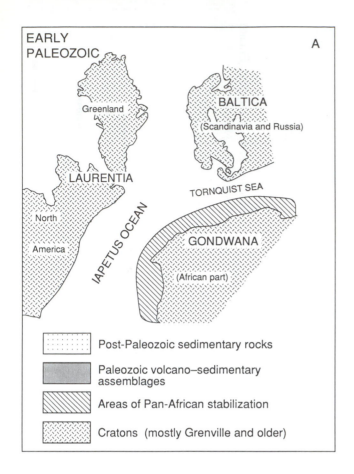

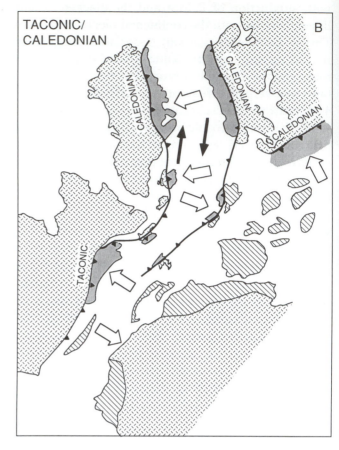

process occupied approximately 300 million years, virtually the entire Paleozoic. It started with rifting along eastern North America, either from Africa, South America, or some continental fragment not yet recognized (see Section 7.0). It continued with collisions of exotic terranes, at least partly from Gondwana, against North America and the southern edge of Baltica and culminated in the middle to late Paleozoic with amalgamation of all surrounding continents and ultimate incorporation into Pangea. This 'North Atlantic orogenic system' underwent almost continual deformation at some place throughout most of the Paleozoic, with a somewhat arbitrary subdivision of events shown in Fig. 7.6. In this section we describe its general evolution and provide somewhat more specific information about the Norwegian Caledonides.

The major events in the development of the North Atlantic orogenic system are summarized in Fig. 7.7A–D. In these diagrams, the widths of the ocean basins between blocks are shown only diagrammatically (largely because they are not known very well).

Laurentia, Baltica, and Gondwana began the Paleozoic separated (Fig. 7.7A) and approached and receded from each other at different times during the era. They also moved laterally past each other, and Fig. 7.7B and C show at least one reversal of major shear directions along Iapetus during the Paleozoic (shear between Baltica and Gondwana is not well constrained). Fig. 7.7D shows the continents in their inferred positions at the end of the Paleozoic, with the 'ocean' between them being the area where present continental shelves fit together; in reference to Mesozoic plate tectonics, Fig. 7.7D is the 'pre-drift configuration' (Chapter 8).

The microcontinental terranes that accreted against North America (Laurentia) and Baltica commonly show the imprint of a Pan-African event (Section 5.2), although the dating is not well established for many of them. (Some of these terranes may be older, which might indicate derivation from some margin other than North Africa; e.g. South America – see Section 7.0.) Much of the African margin appears to have been affected by the Pan-African orogeny, although the

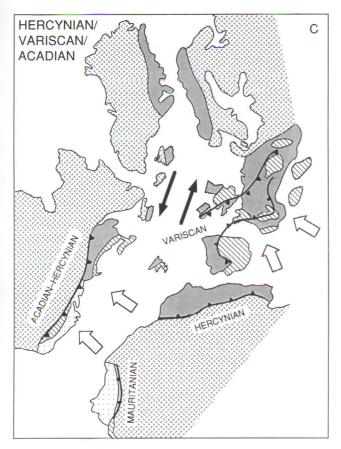

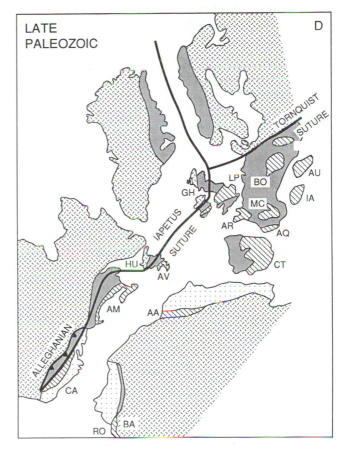

Fig. 7.7 (A–D).   Sketches of events in the circum-North Atlantic orogenic system. This set of four diagrams illustrates the gradual closure of the Iapetus and Tornquist Oceans during the Paleozoic, culminating near the end of the era. This closure fused the original North America, Baltica, and Gondwana and involved the docking of numerous terranes with Europe and the eastern margin of North America. These various accretions led to numerous orogenic pulses at different times and places (see Fig. 7.6). The major terranes also moved laterally past each other in different directions at different times.

   Widths of ocean basins, sizes of blocks, and shapes of oceans and land masses are diagrammatic. Vergences of major thrust belts are shown by saw-tooth patterns. Inferred directions of movement of the downgoing slabs in subduction zones are shown by white arrows. Principal senses of lateral shear are shown by paired black arrows.

   The late-Paleozoic diagram (D) shows closure following the end of Hercynian deformation and development of the Iapetus and Tornquist 'sutures'. In this diagram, allochthonous Pan-African blocks and Pan-African areas east of the Iapetus suture are: AA, Anti-Atlas; AM, Avalon–Meguma; AQ, Aquitaine; AR, Armorica; AU, Austro-Alpine; AV, Avalon; BA, Bassaride; BO, Bohemia; CA, Carolina; CT, Cantabria; IA, Intra-Alpine; LP, London platform; MC, Massif Central; RO, Rokelide. Detached Proterozoic blocks to the west of the Iapetus suture are: GH, Grampian–Highlands; and HU, Humber.   □

shape of the deformed belt is not well known. Apparently either an isolated Africa or the African part of Gondwana shed marginal fragments that drifted across the Iapetus or Tornquist Oceans and were sutured against North America and Baltica. Some geologists use the term 'Rheic Ocean' for the body of water between Africa and its retreating microcontinents.

   The sizes and shapes of Pan-African terranes now in North America or Europe are highly inferential, partly because of poor exposure, and are only diagrammatically represented in Fig. 7.7. For example, Pan-African terranes are: 1) inferred as the base of the London platform, now covered by great thicknesses of Paleozoic and younger rocks; and 2) overprinted by Paleozoic orogeny in the exposed crystalline rocks of the Bohemia massif. Problems of this type prevent an accurate assessment of the amount of European crust that consists of allochthonous terranes and the amount formed by autochthonous accumulation of volcano–sedimentary assemblages and intrusive magmas.

The oldest extensively exposed belt in the North Atlantic orogenic system is commonly referred to as the Caledonian in Europe and the Taconic in North America. Caledonian deformation occurred from the Middle Ordovician through the Silurian, but the Taconic may have ended slightly earlier (Figs 7.6 and 7.7B). A major unconformity occurs between the Silurian and Devonian (about 400 Ma) in most of Europe, and because of later tectonism and development of successor basins, much of the Caledonian (pre-Devonian) history of Europe is obscure. Caledonian-age subduction under the southwestern margin of Baltica apparently caused closure of the Tornquist Sea and formation of the 'Tornquist suture'.

The Caledonides of Norway (see description below) are a complex stack of nappes and thrust sheets that moved eastward onto the Baltic shield as the Iapetus lithosphere and, ultimately, the western part of the Baltic shield descended under eastern Greenland. Subduction under eastern Greenland caused westward-vergent thrusts and abundant Caledonian-age plutonic rocks. The Caledonian orogeny brought Greenland and Baltica together and terminated subduction and compression between them. With the exception of later right-lateral movement, this suturing completed the assembly of North America and Baltica into a single continent (Laurussia).

The Taconic area of North America was affected by a variety of crustal movements (Fig. 7.7B). The dominant direction of subduction was toward the northwest, but some of the Avalonian-age terranes approached North America because of subduction of Iapetus lithosphere under the advancing terranes rather than, or in addition to, subduction under North America. Docking (suturing) of many of these terranes with North America, however, probably did not occur until after the Caledonian. The different directions of subduction shown in Fig. 7.7B imply: 1) subduction in different directions at different times, with the map showing events that occurred over a time range of at least 50 million years; 2) symmetrical subduction on both sides of Iapetus at the same time; and/or 3) the effect of transform faults (Section 4.2) that separated regions of different subduction directions.

Right-lateral movement along the entire Iapetus region can be inferred for at least late Caledonian time. This movement is consistent with motion along the Great Glen fault of Scotland and faults in the eastern part of the allochthonous terranes of southeastern North America. Paleomagnetic data, however, require left-lateral movement along the Iapetus region during the Hercynian orogeny. This movement assembled the major land masses and smaller fragments into substantially the positions that they occupied at the end of the Paleozoic.

The Hercynian orogeny affected the entire North Atlantic orogenic system south of the Greenland/Baltica area (Fig. 7.7C). Subduction was almost exclusively down toward the north. In Europe, where the orogeny is commonly referred to as 'Variscan', the deformation affected pre- and post-Devonian volcano-sedimentary complexes, Pan-African allochthonous fragments, sediments of successor basins on both pre- and post-Devonian terranes, and coeval foredeep and molasse deposits. The line of northward-vergent thrusting shown in Fig. 7.7C represents the northernmost limit of known Variscan deformation; the eastern part of the deformed zone, however, is covered by post-Paleozoic sediments. Complex southward-vergent thrusting appears as closely spaced faults in continental-margin or island-arc accretionary wedges.

In Africa, Hercynian and older Paleozoic deformation occurred along the northern margin of the continent and in the Mauritanides to the west (Fig. 7.7C). The Hercynian in the north is heavily overprinted by Alpine deformation in the Atlas Mountains and, consequently, is little understood. The Mauritanides developed by the scraping of continental-margin and platform sediments eastward onto the West African craton, and they contain very little coeval magmatic rock.

The Acadian–Hercynian orogeny in eastern North America occurred as Africa and microplates in Iapetus collided with the continental margin. The orogeny was characterized by widespread magmatism (with uncertain subduction directions), docking of Pan-African fragments and younger island-arc complexes, and metamorphism up to high grades. Although the orogenic front is shown with west vergence, simple west-moving thrust stacks apparently did not accumulate on the North American continental platform at this time.

Hercynian suturing essentially completed the destruction of the Iapetus (and Rheic) Ocean.

Compression terminated throughout the entire North Atlantic orogenic system except for westward thrusting of sediments onto the North American craton during Alleghanian deformation (Fig. 7.7D; Section 7.4). The basal detachment of the deformed rocks extends far to the east of the exposed Alleghanian belt, and much of the eastern seaboard of North America has been moved westward along this zone. The Alleghanian event was not accompanied by magmatism, and it appears to represent a final compressive adjustment, without subduction, between already-joined Africa and North America.

Norwegian Caledonides

The eastern front of the Caledonian Mountain chain abruptly terminates the flat landscape of the Baltic shield. On the surface, this front is the border between the thick glacial drift that covers the shield and the intricately deformed nappe structures of the mountains. Deep Quaternary glaciation has left excellent exposures in the Caledonides, and the area has been intensely studied. This section contains a brief description of the major features of the various thrust/nappe complexes (e.g. Fig. 7.8) that compose the Caledonides and the tectonic inferences that can be drawn from them. A generalized cross-section is given in Fig. 7.9.

The Baltic shield became a stabilized continental crust during a series of episodes culminating in the middle Proterozoic (Section 5.1). The western part of the shield, underlying the Caledonian front, is covered by flat-lying to slightly deformed, upper Proterozoic to Cambrian, platform sediments. In some areas, Caledonian nappe complexes lie on thrust-faulted blocks of shield and overlying sediments. The shield, presumably unmoved, is referred to as the 'autochthon', and the slightly displaced shield is referred to as the 'parautochthon'. Erosional windows of exposed shield occur at various places through the irregularly shaped allochthons. Exposures of shield-type rocks west of the main Caledonian Mountains (in the 'Western Gneiss terrane') are regarded by some geologists as autochthonous extensions of the Baltic shield and by others as sutured fragments of a North American/Greenland plate. Cambrian and younger Paleozoic sediments are shown in Fig. 7.9 thickening toward the west, where the Baltic shield terminated against the northern Iapetus Ocean.

Numerous classifications have been proposed for the various thrust 'allochthons' in the Caledonian belt, and the one given in Fig. 7.9 is a compromise. As in other mountain belts (Section 5.1), many thrusts carried overlying thrust sheets 'piggy-back', implying at least local younging of the thrust sheets toward the east and downward in the assemblage. All of the Caledonide allochthons are multiple units containing

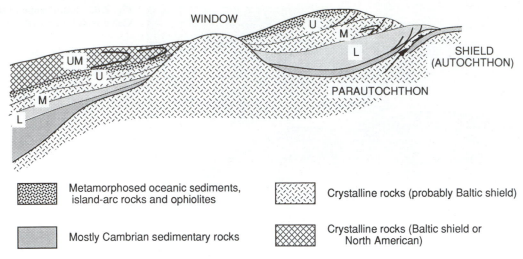

Fig. 7.9. Diagrammatic cross-section of the Norwegian Caledonides. Nappes have ridden upward from the west over the Baltic shield (autochthon), with minor movement of some of the shield (parautochthon). The basement underlying the nappe sequence is exposed in places through windows created by erosion after the thrusting. The location of these windows is partly the result of irregularities in the lower surface of the nappe as it formed and partly the result of minor vertical movement in the crust after thrusting. L, M, U and UM refer to Lower, Middle, Upper and Uppermost Allochthons, respectively.

The Lower Allochthon consists primarily of upper Proterozoic to Lower Ordovician sediments, with an increase in the abundance of intermingled crystalline rocks toward the west. These rocks represent sedimentary suites accumulated on the shield prior to the onset of Caledonian thrusting. The Middle Allochthon consists mostly of crystalline rocks (gneisses and other components of the shield) with intermixed suites of unfossiliferous upper Proterozoic to Cambrian sandstones. The Upper Allochthon contains a lower suite of dominantly crystalline (shield) rocks and an overlying 'oceanic' unit consisting of ophiolites, deep-water and other types of flysch sediments, and ocean-floor and island-arc volcanic rocks; metamorphism is highly variable, with all grades from greenschist to granulite and eclogite. The Uppermost Allochthon is characterized by nappes with crystalline cores and a variety of admixed metasediments at high metamorphic grades, mostly amphibolite facies. ☐

numerous individual thrust sheets. The thrusts are imbricated, intensely mylonitized, and rise eastward through the section, particularly near the fronts of the allochthons. Recumbent folds verge eastward, and many are cut by thrusts. True nappes, representing extensive overfolding and reversal of metamorphic isograds over large areas, are most important in the Uppermost Allochthon.

Details of the various allochthons are given in the caption for Fig. 7.9. As a very great generalization, we can discern a tendency of higher allochthons to sample rocks originally formed farther west in the orogenic belt. The Lower and Middle Allochthons primarily contain rocks of the Baltic shield and overlying platformal or miogeoclinal sediments. The Upper Allochthon contains some 'shield' rocks and also 'oceanic' rocks presumably formed on lithosphere of the Iapetus Ocean. The Uppermost Allochthon (not to be confused with the Upper Allochthon) may contain fragments of the continental crust of Greenland, west of the Iapetus Ocean.

[**References** – The general evolution of the North Atlantic orogenic system is discussed in the volumes edited by Harris and Fettes (1988), Sougy and Rodgers (1988) and Dallmeyer (1989). Information on the evolution of Europe is in Ziegler (1982, 1984, 1988) and Matte and Zwart (1989). The evolution of the British Isles is discussed by Anderton *et al.* (1979) and McKerrow and Soper (1989). Information on the Variscan orogeny is from volumes edited by Hutton and Sanderson (1984) and Matte (1990). The Armorican massif is discussed by Noblet and Lefort (1990). The Appalachians are discussed by papers in the volume edited by Hatcher, Thomas and Viele (1989). The discussion of the Caledonides is from the volume edited by Gee and Sturt (1985) and from D. Roberts and Gee (1985) and Hossack and Cooper (1986).]

7.4 Circum-North American orogens

THE FIGURATIVE DIN of crashing plates that characterized the eastern margin of North America throughout virtually the entire Paleozoic seems to have been more muted on the other three sides of the continent. Continental collisions did not occur on the southern margin of North America until the Late

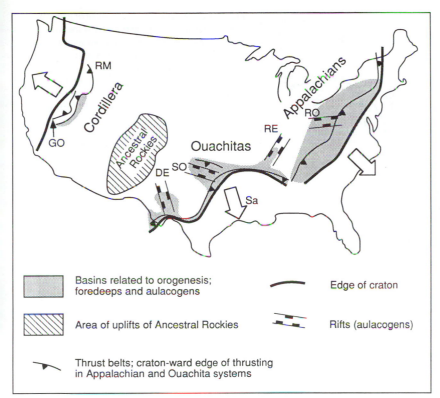

Fig. 7.10. Middle- and late-Paleozoic orogenic belts of the southern and central United States. The major belts are the Appalachians, Ouachitas and Cordillera. The Appalachian and Ouachita orogens created large foredeep areas on the adjacent craton. Intense Mesozoic and Cenozoic deformation in the Cordillera has made its Paleozoic history more difficult to reconstruct than for the other two belts, which did not undergo post-Paleozoic deformation. The Ancestral Rockies are now exposed as a set of basement uplifts and may have been formed in response to long-distance stresses generated by the terrane whose collision caused the Ouachita orogeny. Inferred directions of movement of the downgoing slabs in subduction zones are shown by white arrows. All major Paleozoic subduction was directed away from the North American craton; thus, collisional orogeny was created by subduction under the approaching terranes. Designated features include: DE, Delaware rift; GO, Golconda allochthon; RE, Reelfoot rift; RM, Roberts Mountain allochthon; RO, Rome trough; and SO, southern Oklahoma (Wichita) rift. Sa is the Sabine uplift.   □

Carboniferous (Pennsylvanian). On the northern margin, where the only evidence for collision is a microcontinental fragment, compression did not occur until the Late Silurian or Early Devonian. Apparently continental collision did not occur at all on the western margin of the continent, where middle- and late-Paleozoic orogenic activity was caused by docking of an island-arc complex(es). In this section, we discuss the evolution of the southern (Ouachita), western (Cordilleran), and northern (Innuitian) sides of North America and make some brief comments on the Appalachian foredeep basin (Figs 7.10 and 7.11). In addition to these regions, Paleozoic orogenic activity in western Canada is indicated by fragmentary records of magmatism and metamorphism extensively overprinted by post-Paleozoic orogeny.

Although different in detail, the orogenic suites discussed here have several common characteristics. All of them formed by deformation of thick sequences of sedimentary rocks formed along continental margins. Depositional patterns show a transition from thin sequences of cratonic platforms (Section 7.1) to somewhat thicker miogeoclinal suites and, farther seaward, to very thick sequences probably formed on oceanic lithosphere. All of the belts show intense lateral compression, but no significant evidence exists for subduction of any type of lithosphere under any of the Paleozoic margins of North America except in the North Atlantic orogenic system (Section 7.3). Thus, orogeny along the other margins was caused by subduction of North America away from the continent (i.e. under approaching continents or arcs).

The continental margin of Paleozoic North America is difficult to determine but can be inferred from several lines of investigation: 1) along the eastern and southern margins of the continent, the craton edge must be placed seaward of autochthonous Grenville-age terranes, either exposed or encountered during drilling (Section 5.2) – the most seaward parts of the craton in the west tend to be older than Grenville; 2) the craton edge must be placed seaward of platformal/miogeoclinal sedimentary rocks, including shallow-water carbonates, shales and quartzitic sands; 3) thrust faults that bring basinal/eugeoclinal sedimentary and volcanic rocks up and over the cratonic platforms can be traced seismically backward toward the overridden edge; 4) the seaward edge of the craton is also shown by exploration seismic studies that show deformed wedges of apparent oceanic lithosphere between the craton and

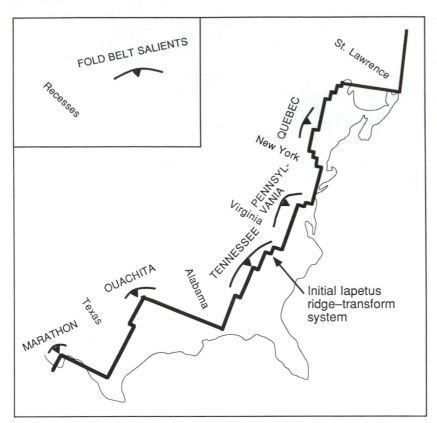

Fig. 7.11. Diagram of promontories and reentrants in the eastern and southern continental margins of North America caused by irregular opening of Iapetus and a southern ocean. The Iapetus ridge–transform system created major compression where the spreading center was 'indented' into the craton. Conversely, recesses (reentrants) formed where transform movement left part of the craton without significant compression. (Modified from Thomas, 1977, 1991.)   ☐

collided blocks; and 5) in the western US, both Rb–Sr and Sm–Nd isotopic data (Section 2.3) show a distinct line that divides intrusive rocks with radiogenic isotopic sources on the east from similar magmatic rocks with non-radiogenic sources on the west – this line presumably is the Paleozoic boundary between continental lithosphere and oceanic lithosphere.

Because of the lack of subduction underneath North America, subduction-related (calcalkaline) igneous rocks occur in the orogenic belts only where they have been thrust onto the continental margin from offshore arc complexes. For example, oceanic and/or back-arc tholeiitic basalts are present in thrust slices throughout the Cordilleran belt, and the subsurface Sabine uplift (Fig. 7.10) contains upper Paleozoic rhyolites and tuffs that are postulated to represent magmatism above a slab descending toward the south during the Ouachita orogeny.

Rifts (aulacogens) extend into the North American craton at three places along the Ouachita belt, one in the Appalachians, and one in the Innuitian belt (Figs. 7.10; also see Fig. 7.2A). The Ouachita and Appalachian aulacogens presumably are related to rifting that formed the original Iapetus Ocean and its southern continuation around the Ouachita trend; as

we discussed in Section 7.0, exactly what rifted away is not clear. The rifting occurred either in the late Proterozoic or earliest Paleozoic, causing all troughs to be filled by Early Cambrian or older sediments. The Reelfoot and Rome rifts may join to form virtually a separate continental plate southeast of the main North American craton. Paleozoic rifts do not extend into the craton of the western United States, although miogeoclinal sequences have been attributed to subsidence of a continental margin perhaps formed by rifting from Antarctica (aulacogen formation along this margin occurred during mid-Proterozoic rifting, Section 5.3).

Ouachita orogenic belt and Ancestral Rockies

The irregular shape of the Ouachita belt, and its continuation into the Appalachians, probably developed as a result of rift–transform interaction during early Paleozoic opening of the marginal ocean basins (Fig. 7.11). The receding continent (Africa?; South America?), therefore, also contained a highly irregular margin. The pattern shown in Fig. 7.11 resulted in two salients, one at the join between the Appalachians and Ouachitas in Alabama and one in Texas between the

Ouachita and Marathon regions. The approximately right-angle junction between the Appalachians and the Ouachitas is covered by Mesozoic and Cenozoic sediments of the Gulf coastal plain and is difficult to study.

Once the continental margin was established, sedimentation in the Ouachita region developed two quite different sequences of rocks on the cratonic platform and in the offshore basins. Platformal sediments (foreland facies) consist of three major suites: 1) basal, transgressive, mostly Cambrian arkoses and quartzitic sands, commonly containing glauconite; 2) Lower Cambrian to Middle Ordovician carbonate rocks; and 3) Upper Ordovician to Upper Devonian carbonate rocks, shales and quartzitic sands derived from cratonic sources. Basinal sediments throughout the early and middle Paleozoic consist of gray to black shales and silty shales, siliceous sandstones, detrital and cherty limestones, and cherts. The basinal sediments contain abundant evidence of slumping, turbidity currents, and other slope-related features.

Sedimentation along the margin changed with the onset of deformation in the Mississippian. Basinal areas accumulated great thicknesses of flysch (mostly gravelly to muddy turbidites) that prograded from the emerging orogenic belt into foreland basins on the craton as the deformation front caused downbuckling of the foreland. Because of the irregular shape of the cratonic margin, collision with a southern plate (Gondwana) was accompanied by lateral movements as jagged pieces tried to fit together. Deformation and sedimentation continued synchronously from the Mississippian through much of the Permian, converting the flysch sediments into a suite of pervasively sheared slaty rocks that imbricated northward onto the foreland in a series of closely spaced thrusts.

The Ancestral Rockies are a mountainous region inferred to have existed almost solely because of the large quantity of clastic sediment that they provided to the adjacent late Paleozoic platform. The ranges are presumed to have been horst-and-graben type block uplifts formed largely in Pennsylvanian time. Their origin is not clear but must be related to the synchronous events occurring in the Ouachitas and along the western margin of the continent.

Appalachian foreland basin

The miogeoclinal basin of the Appalachians is one of the largest foreland basins in the world. Early Paleozoic sediments formed along a typical continental passive margin, but later the repeated orogenic activity to the east (Section 7.3) provided an eastern source for clastic material as the load of the orogenic highlands caused downbuckling of the basin. Sedimentation, however, was sufficiently rapid that water depths rarely exceeded 100 m to 200 m even in deeper parts of the basin, and most of the sediments are shallow-water, deltaic or subaerial.

Sedimentation in the Appalachian basin extended throughout the Paleozoic. Generally, Cambrian sands and gravels are overlain by Upper Cambrian through Ordovician carbonate rocks. Silurian and younger rocks are a more mixed sequence of sands, shales, and carbonates. The two major pre-Alleghanian orogenic pulses (Taconic and Acadian; Section 7.3) provided an eastern highland province that caused progradation of major deltas and general shallowing of water. Reentrants in the initial craton margin are now the sites of thicker sequences of sediments and greater encroachment of Alleghanian thrust faults onto the craton.

The shallow-water Appalachian sedimentary sequence has greater mechanical coherence than the Ouachita deep-water flysch. This coherence caused the development of larger, more laterally extensive thrust slices in the Appalachians than in the Ouachitas. A major decollement zone, apparently of Alleghanian age, extends downward to the east from the Appalachian deformed basin and has been seismically traced eastward under much of the present seaboard of the southeastern US. This thrust may reach depths of 20 km and represent movement of several hundred kilometers, thus causing most crystalline rocks in the southeastern US to be allochthonous.

Cordillera

Along most of the western margins of both North and South America, much of the record of Paleozoic deposition, magmatism, and tectonism has been so overprinted by Mesozoic and Cenozoic activity that it is undecipherable. In the western United States, however, the Phanerozoic orogenic belt is much wider than elsewhere, partly because of extension in the Basin and Range province (Section 4.2), and younger events appear to have left a Paleozoic record that can be ascertained (Fig. 7.10).

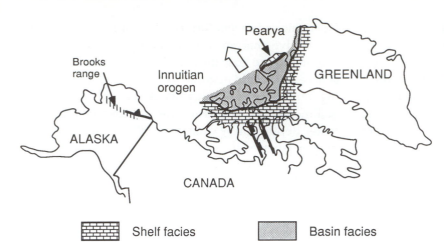

Fig. 7.12. Geology of Innuitian orogenic belt and possible correlative deformational system in Brooks range. The Innuitian volcano-sedimentary suite thickens into a basin north of the craton margin. Orogeny was caused by collision with the Pearya terrane, of which only a small fragment remains after rift-generated opening of the Arctic Ocean.   □

Shelf facies   Basin facies

The western edge of the North American craton in the Paleozoic may have been formed by rifting (see Section 7.0), but the only evidence for such an origin is the rapid initiation of miogeoclinal sedimentation at some time in the late Proterozoic. At approximately the beginning of the Cambrian, deposition of mostly carbonate rocks replaced deposition of Proterozoic quartz sandstones in the miogeocline east of the craton edge. This carbonate deposition continued throughout much of the Paleozoic, with pulses of craton-derived quartz sands and silty shales. Total sedimentary thicknesses of late Proterozoic through Paleozoic rocks range from 1 to 2 km at the transition between the miogeocline and the stable craton (central Utah and eastern Idaho) to nearly 10 km at the craton edge. The miogeocline–craton transition zone is an abrupt hingeline (Wasatch line) that localized some of the Mesozoic/Cenozoic deformation (Section 9.4)

Basinal rocks were deposited west of the thickest parts of the miogeoclinal sequence and consist of shales, cherts, quartzites (derived from a western source), minor limestones and tholeiitic basalts. These suites, plus calcalkaline island-arc assemblages, were thrust eastward over the miogeoclinal sequences in two orogenic pulses, presumably caused by approach and docking of one or more large western island-arc complexes. The older event (Antler orogeny) occurred in the Late Devonian to Early Mississippian, with basinal suites carried eastward on the basal Roberts Mountain allochthon. The highlands created along the craton edge became a western source for clastic debris in the miogeocline during the later part of the Paleozoic. The younger event (Sonoma orogeny)

occurred in the Late Pennsylvanian to Middle Permian, forming the Golconda allochthon and effectively closing the basin between the craton and the island arc by subduction of the basin under the approaching arc.

Innuitian orogen and Brooks range

The major preservation of Paleozoic rocks along the northern margin of North America is in the Innuitian orogen, primarily exposed on islands north of Canada (Fig. 7.12). Rift-related basaltic rocks indicate that downwarp and sedimentation in this area probably started in the late Proterozoic, perhaps at ~750 Ma. Beginning in the latest Proterozoic to earliest Cambrian, two suites of sedimentary rocks were formed in the Franklin shelf and basin that formed over the rifted terrane. South, toward the craton, typical platformal/miogeoclinal carbonates, shales, and quartz sandstones accumulated with thicknesses up to 9 km. North, toward and off of the shelf edge, flysch sedimentation and basaltic volcanism formed a basinal facies of uncertain total thickness. A rift trough (aulacogen) extending southward into the craton was also a site of early to middle Paleozoic sedimentation.

Accumulation of sedimentary rocks continued until the Late Devonian, at which time the entire suite was deformed in a series of orogenic pulses that extended from the Devonian through the Pennsylvanian. The major deformation was in the Late Devonian/Early Mississippian Ellesmere orogeny. Deformation was accompanied by docking of the Pearya terrane, with a crystalline basement of middle Proterozoic or older

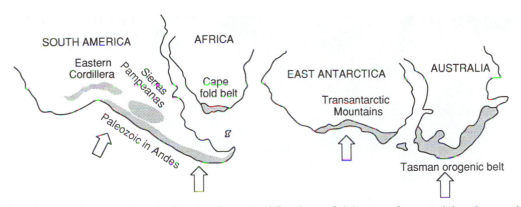

Fig. 7.13. Paleozoic orogenic zones along the 'southern' margin of Gondwana. Subduction of oceanic lithosphere under the margin is shown by white arrows.   ☐

rocks. Because of northward-directed subduction, the Ellesmerian orogeny was not accompanied by magmatism in the Innuitian belt. Beginning in the Pennsylvanian, and extending through the Mesozoic, the deformed rocks of the Ellesmerian orogenic belt were the site of a successor basin (Sverdrup basin), which accumulated several kilometers of undeformed carbonates, evaporites, shales, and minor basalts.

The Brooks range of Alaska (Fig. 7.12) may be correlative with the Innuitian orogen, but younger events have obscured the Paleozoic history. If the Brooks range and the Innuitian belt are correlatable, then a continental-margin orogenic belt may have existed along the entire northern edge of North America in the middle Paleozoic. This belt also may be continuous with orogenic belts along the northern margin of Eurasia (Section 7.1)

[**References** – General information on the United States is provided by the volume edited by Bally and Palmer (1989) and on Canada by the volume edited by Price and Douglas (1972). The volume edited by Hatcher *et al.* (1989) provides information on the Appalachian-Ouachita system, and the Ouachitas are discussed by Lillie (1985) and Handschy, Keller and Smith (1987). Thomas (1977, 1985, 1991) discusses the rifting of the Appalachian-Ouachita margin and the development of the orogenic belts. Western Canada is discussed by Monger, Price and Tempelman-Kluit (1982) and Okulitch (1983). Information on the western United States is in Ernst (1988a, b).]

7.5 Outer margin of Gondwana

SOME PART of the southern margin of Gondwana, as it existed at the end of the Paleozoic, was affected by subduction of surrounding oceanic lithosphere throughout the Paleozoic (Fig. 7.13). A continuous orogenic belt as shown in Fig. 7.13 has been referred to as the 'Samfrau geosyncline', but we must recognize the possibility that parts of the belt may have originated on separate blocks before suturing. Our discussion of the Gondwana orogenic belts proceeds from 'west' to 'east' (in present orientation).

The construction of the Andes along the Precambrian cratonic margin of South America in the Mesozoic and Cenozoic has caused extreme overprinting of Paleozoic activity in the area (Section 9.6). Although not everywhere discernible, Precambrian crust appears to extend westward to, or very near, the present western coast of the continent. The principal coherent areas of pre-Mesozoic outcrop are in the Eastern Cordillera of Bolivia and Peru and the Sierras Pampeanas of Argentina (Fig. 7.13).

Rocks of the Eastern Cordillera are fine-grained, flysch-type, sediments possibly deposited in a foredeep caused by construction of a western orogenic belt. The depositional basin may have developed mostly in mid-Paleozoic time between the Brazilian shield and Precambrian massifs farther west in the present Andes. Rocks of the basin were deformed in an early 'Hercynian' time, ~350 to ~330 Ma, and the belt has been penetratively sheared by eastward-vergent thrusting during the Andean orogeny.

The principal record of Paleozoic activity in the Sierras Pampeanas is shown by the development of a series of late-Proterozoic to the Early-Carboniferous granites intrusive into a metasedimentary/metavol-

Fig. 7.14. Granite of the Sierras Pampeanas of Argentina, showing large 'xenolith' of pre-existing crust.   □

canic terrane of probably late-Precambrian age, possibly with younger components (Fig. 7.14). The granites probably were caused by subduction under the South American margin, although the area is sufficiently far inland from the present margin that they may represent closure of a back-arc zone. Subduction along this margin was associated with outward building of the coastline, which had reached nearly its present position by the start of the Carboniferous either as a result of creation of new crust and/or by docking of one or more continental fragments.

East of the Paleozoic outcrops of southern South America (present orientation) are lower- to middle-Paleozoic rocks in the Cape fold belt of South Africa, which may be correlatable with limited exposures in the Falkland Islands. Sediments in this belt accumulated in shallow-water to terrestrial environments, with the clastic debris supplied from a northern source (the southern African shield). Deformation in the belt is regarded as Triassic because the Paleozoic sediments are deformed with the marginal facies in the adjacent Karoo basin (Section 8.1).

The Transantarctic Mountains form the border between East and West Antarctica (Fig. 7.13; Section 3.0; Section 5.2). The mountains appear partly as scattered exposures through ice fields, and this lack of exposure and the logistical problems of working in the area permit only limited conclusions to be drawn about their origin. Upper Proterozoic through Silurian continental-margin sedimentary and volcanic rocks were deformed in the Late Silurian or Early Devonian and overlain by undeformed Devonian and younger sediments. Thus, subduction apparently occurred along the entire margin during the early to middle Paleozoic but not later. Although West Antarctica is mostly a Mesozoic/Cenozoic orogenic belt (Section 9.6), scattered isotopic evidence indicates subduction-zone magmatism in the Antarctic Peninsula perhaps as old as the late Proterozoic.

Eastern Australia

The most widespread Paleozoic orogenic belts of Gondwana are in eastern Australia (Fig. 7.15). Subduction of oceanic lithosphere occurred under some part of the eastern margin of Australia (present orientation) through much of the Paleozoic, forming the 'Tasman orogenic belt' that is customarily divided into the Lachlan, New England, and Yarrol belts. Some geologists group the New England and Yarrol belts as parts of the New England orogen. The diagonal line on Fig. 7.15 divides the Lachlan belt, with its dominantly early- and middle-Paleozoic activity, from the later Paleozoic belts farther east.

The Lachlan fold belt is dominated by Ordovician (to Silurian) flysch, derived from the west, and

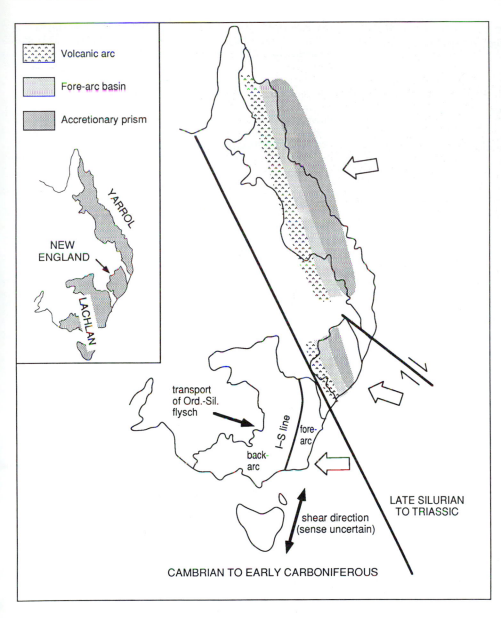

Fig. 7.15. Diagram of orogenic activity along the eastern margin of Australia in the Paleozoic. The diagonal line separates the dominantly early-Paleozoic Lachlan belt from the dominantly late Paleozoic New England and Yarrol belts. White arrows show the direction of descent of the downgoing slabs in the subduction zones.

The Lachlan orogen contains fore-arc and back-arc suites generated by subduction of oceanic lithosphere beneath the Australian continent. This transition is related to the generation of two suites of granites: an 'I-type' (igneous) suite formed seaward of the I-S line by partial melting of material that had not passed through a sedimentary cycle; and an 'S-type' suite that formed landward of the I-S line and incorporated significant amounts of old sediment in the partial melts. The transport direction of sediment during the Ordovician and Silurian is off of the craton into the Lachlan belt.

The Yarrol and New England belts can be divided into three facies: arc, fore-arc basin and accretionary prism. The two belts are offset by a major transform fault.   □

younger granites. Approximately 20% of the outcrop area is granite, with the proportion increasing toward the east. The I-S line is the eastern edge of outcrop of 'S-type' granites, which have compositional properties that indicate their derivation from metasedimentary protoliths. East of that line, and in lesser amounts west of it, are 'I-type' granites, apparently derived from protoliths that have never undergone sedimentary compositional recycling. The I-S line may represent the eastern margin of continental crust in the early Paleozoic.

The basement of the Lachlan belt is commonly regarded as an accumulation of separate blocks, but the time of suturing is controversial. Blocks may have been swept together at various times during the Late-Silurian to Early-Carboniferous orogeny that terminated activity in the belt. Conversely, the terranes may have sutured together near the end of the Proterozoic, forming a platform (back-arc basin) under which continental crust became progressively thinner toward the east. True subduction-zone magmatism, forming calcalkaline granite suites, occurred only in the east, indicating an absence of plate closure within the belt during the development of the granite suites (Late Silurian and Devonian). Many of the granites are post-oro-genic, as shown by their diapiric intrusion of the deformed sediments.

The New England and Yarrol fold belts (Fig. 7.15) began to develop in the middle Paleozoic. The present apparent offset between the New England and Yarrol orogens may signify that they formed in separate belts or, conversely, that they formed as one belt that was later faulted apart. A combined New England and Yarrol orogenic system can be modeled as, from west to east, an eastward-vergent arc, a fore-arc basin, an accretionary wedge, and a trench; these features are consistent with subduction of oceanic lithosphere under the Australian continental margin. Compressive deformation continued into the Triassic and developed westward-vergent thrusts over large areas.

[**References** – General references for this section include the volumes edited by Howell (1985) and Leitch and Scheibner (1987). Information for Australia is from the book by Veevers (1984), papers on the Lachlan fold belt by Fergusson, Gray and Cas (1986) and Chappell, White and Hine (1988), and papers on the New England orogen by Murray *et al.* (1987) and J. Roberts and Engel (1987). The Andes are discussed by James (1971), Dalmayrac *et al.* (1980), and Ramos *et al.* (1986). Truswell (1977) discusses South Africa. Information for the Antarctic is in the volumes edited by Oliver, James and Jago (1983) and Thomson, Crame and Thomson (1991) and the paper by Flottmann and Kleinschmidt (1991).]

References

Anderton, R., Bridges, P. H., Leeder, M. R. & Sellwood, B. W. (1979). *A Dynamic Stratigraphy of the British Isles*. London: George Allen and Unwin, 301 pp.

Bally, A. W. & Palmer, A. R., eds. (1989). *The Geology of North America – an Overview: vol A*. Boulder, Colorado: Geological Society of America, 619 pp.

Bond, G. C., Kominz, M. A. & Grotzinger, J. P. (1988). Cambro-Ordovician eustasy: evidence from geophysical modelling of subsidence in Cordilleran and Appalachian passive margins. In *New Perspectives in Basin Analysis*, ed. K. L. Kleinspehn & C. Paola, pp. 128–60. New York: Springer Verlag.

Chappell, B. W., White, A. J. R. & Hine, R. (1988). Granite provinces and basement terranes in the Lachlan fold belt, southeastern Australia. *Australian Journal of Earth Sciences*, **35**, 505–21.

Coleman, R. G. (1989). Continental growth of northwest China. *Tectonics*, **8**, 621–35.

Dallmeyer, R. D., ed. (1989). *Terranes in the Circum-Atlantic Paleozoic orogens*. Geological Society of America Special Paper 230, 277 pp.

Dalmayrac, B., Laubacher, G., Marocco, R., Martinez, C. & Tomasi, P. (1980). La chaine hercynienne d'amerique du sud – structure et evolution d'un orogene intracratonique. *Geologische Rundschau*, **69**, 1–21.

Dalziel, I. W. D. (1991). Pacific margins of Laurentia and East Antarctica-Australia as a conjugate rift pair: evidence and implications for an Eocambrian supercontinent. *Geology*, **19**, 598–601.

Dalziel, I. W. D. (1992). Antarctica; a tale of two supercontinents? *Annual Reviews of Earth and Planetary Sciences*, **20**, 501–26.

Dennison, J. M. (1989). *Paleozoic Sea-Level Changes in the Appalachian Basin*. 28th International Geological Congress Field Trip Guidebook T354, 56 pp.

Dypvik, H., Nesteby, H., Ruden, F., Aagaard, P., Johansson, T., Msindai, J. & Massay, C. (1990). Upper Paleozoic and Mesozoic sedimentation in the Rukwa–Tukuyu region, Tanzania. *Journal of African Earth Sciences*, **11**, 437–56.

Ernst, W. G. (1988a). Metamorphic terranes, isotopic provinces, and implications for crustal growth of the western United States. *Journal of Geophysical Research*, **93**, 7634–42.

Ernst, W. G., ed. (1988b). *Metamorphism and Crustal Evolution of the Western United States, Rubey Volume VII*. Englewood Cliffs, New Jersey: Prentice Hall, 1153 pp.

Feldmann, R. M., ed. (1987). Paleoceanography of Paleozoic midcontinental seaways. Special section of *Paleoceanography*, **2**, 119–248.

Fergusson, C. L., Gray, D. R. & Cas, R. A. F. (1986). Overthrust terranes in the Lachlan ford belt, southeastern Australia. *Geology*, **14**, 519–22.

Flottmann, T. & Kleinschmidt, G. (1991). Opposite thrust systems in northern Victoria Land, Antarctica: imprints of Gondwana's Paleozoic accretion. *Geology*, **19**, 45–7.

Gee, D. G. & Sturt, B. A., eds. (1985). *The Caledonide Orogen – Scandinavia and Related Areas*. Chichester, U.K.: John Wiley and Sons, 1266 pp.

Handschy, J. W., Keller, G. R. & Smith, K.J. (1987). The Ouachita system in northern Mexico. *Tectonics*, **6**, 325–30.

Harris, A. L. & Fettes, D.J., eds. (1988). *The Caledonian–Appalachian Orogen*. Geological Society of London Special Publication 38, 643 pp.

Hatcher, R. D., Jr., Thomas, W. A. & Viele, G. W., eds. (1989). *The Appalachian–Ouachita Orogen in the United States, The Geology of North America vol. F-2*. Boulder, Colorado: Geological Society of America, 767 pp.

Hoffman, P. F. (1991). Did the breakout of Laurentia turn Gondwanaland inside-out? *Science*, **252**, 1409–12.

Hossack, J. R. & Cooper, M. A. (1986). Collision tectonics in the Scandinavian Caledonides. In *Collision Tectonics*, ed. M. P. Coward & A. C. Ries, pp. 287–304. Geological Society of London Special Publication 19.

Howell, D. G., ed. (1985). *Tectonostratigraphic Terranes of the Circum-Pacific Region. Earth Science Series Number 1*. Houston, Texas: Circum-Pacific Council for Energy and Mineral Resources, 581 pp.

Hutton, D. H. W. & Sanderson, D. J., eds. (1984). *Variscan Tectonics of the North Atlantic Region*. Geological Society of London Special Publication 14, 270 pp.

James, D. E. (1971). Plate tectonic model for the evolution of the central Andes. *Geological Society of America Bulletin*, **82**, 3325–46.

Khain, V. E. (1985). *Geology of the USSR; First Part – Old Cratons and Paleozoic Fold Belts – Beitrage zur Regionalen Geologie der Erde*. Berlin: Gebruder Borntraeger, 272 pp.

Kleinspehn, K. L. & Paola, C., eds. (1988). *New Perspectives in Basin Analysis*. New York: Springer Verlag, 453 pp.

Leggett, J. K., McKerrow, W. S., Cocks, L. R. M. & Rickards, R. B. (1981). Periodicity in the early Paleozoic marine realm. *Geological Society of London Journal*, **138**, 167–76.

Leitch, E. C. & Scheibner, E. (1987). *Terrane accretion and Orogenic Belts*: American Geophysical Union Geodynamics Series, 19, 343 pp.

Lillie, R. J. (1985). Tectonically buried continent/ocean boundary, Ouachita Mountains, Arkansas. *Geology*, **13**, 18–21.

Martin, H. (1981). The late Palaeozoic Gondwana glaciation. *Geologische Rundschau*, **70**, 480–96.

Matte, Ph., ed. (1990). Terranes in the Variscan belt of Europe and circum-Atlantic Paleozoic orogens. Special Issue of *Tectonophysics*, **177**, no. 1/3, 323 pp.

Matte, Ph. & Zwart, H. J., eds. (1989). Paleozoic plate tectonics with emphasis on the European Caledonian and Variscan belts. Special Issue of *Tectonophysics*, **169**, no. 4, 131 pp.

McKerrow, W. S. & Soper, N. J. (1989). The Iapetus suture in the British Isles. *Geological Magazine*, **126**, 1–8.

Monger, J. W. H., Price, R. A. & Tempelman-Kluit, D. J. (1982). Tectonic accretion and origin of the two major metamorphic and plutonic welts in the Canadian Cordillera. *Geology*, **10**, 70–5.

Moores, E. M. (1991). Southwest U.S.–East Antarctica (SWEAT) connection: a hypothesis. *Geology*, **19**, 425–8.

Murray, C. G., Fergusson, C. L., Flood, P. G., Whitaker, W. G. & Korsch, R. J. (1987). Plate tectonic model for the Carboniferous evolution of the New England fold belt. *Australian Journal of Earth Sciences*, **34**, 213–36.

Nalivkin, D. V. (translated by S.I. Tomkeieff) (1960). *The Geology of the U.S.S.R. – A Short Outline: International Series of Monographs on Earth Sciences*. New York: Pergamon Press, 170 pp. plus geologic map of U.S.S.R.

Nalivkin, D. V. (translated by N. Rast) (1973). *Geology of the USSR*. Edingurgh: Oliver & Boyd, 855 pp.

Noblet, Ch. & Lefort, J. P. (1990). Sedimentological evidence for a limited separation between Armorica and Gondwana during the Early Ordovician. *Geology*, **18**, 303–5.

Okulitch, A. V. (1983). Paleozoic plutonism in southeastern British Columbia. *Canadian Journal of Earth Sciences*, **22**, 1409–24.

Oliver, R. L., James, P. R. & Jago, J. B., eds. (1983). *Antarctic Earth Science*. Cambridge: Cambridge University Press, 697 pp.

Pittman, W. C. III (1978). Relationship between eustacy and stratigraphic sequences of passive margins. *Geological Society of America Bulletin*, **89**, 1389–403.

Price, R. A. & Douglas, R. J. W., eds. (1972). *Variations in Tectonic Styles in Canada*. Geological Association of Canada Special Paper 11, 688 pp.

Ramos, V. A., Jordan, T. E., Allmendinger, R. W., Mpodozis, C., Kay, S. M., Cortes, J. M. & Palma, M. (1986). Paleozoic terranes of the central Argentine-Chilean Andes. *Tectonics*, **6**, 855–80.

Revyakin, P. S. (1987). The earth's crust in the eugeosynclinal zones of the Urals. *International Geology Review*, **29**, 1021–34.

Ricketts, B. D., (1989). *The Western Canada Sedimentary Basin – A Case History*. Canadian Society of Petroleum Geologists Special Publication 30.

Roberts, D. & Gee, D. G. (1985). An introduction to the structure of the Scandinavian Caledonides, In *The Caledonide orogen –Scandinavia and Related Areas*, ed. D. G. Gee & B. A. Sturt. Chichester: John Wiley and Sons, 1266 pp.

Roberts, J. & Engel, B. A. (1987). Depositional and tectonic history of the southern New England orogen. *Australian Journal of Earth Sciences*, **34**, 1–20.

Ross, C. A. & Ross, J. R. P. (1985). Late Paleozoic depositional sequences are synchronous and worldwide. *Geology*, **13**, 194–7.

Rowley, D. B., Raymond, R., Totman-Parrish, J., Lottes, A. L., Scotese, C. R. & Ziegler, A. M. (1985). Carboniferous paleogeographic, phytogeographic, and paleoclimatic reconstructions. *International Journal of Coal Geology*, **5**, 7–42.

Schlee, J. S., ed. (1984). *Interregional Unconformities and Hydrocarbon Accumulation*. American Association of Petroleum Geologists Memoir 36, 184 pp.

Shakhnovskiy, I. M. (1988). Aulacogens of the East European craton: their structure and oil and gas potential. *International Geology Review*, **30**, 1313–23.

Sloss, L. L. (1963). Sequences in the cratonic interior or North America. *Geological Society of America Bulletin*, **74**, 93–114.

Sloss, L. L., ed. (1988). *Sedimentary Cover – North American Craton: U.S.: The Geology of North America, vol. D-2*. Geological Society of America, , 506 pp.

Soares, P. C., Landim, P. M. B. & Fulfaro, V. J. (1978). Tectonic cycles and sedimentary sequences in the Brazilian intracratonic basins. *Geological Society of America Bulletin*, **89**, 181–91.

Sougy, J. & Rodgers, J., eds. (1988). The West African connection: evolution of the central Atlantic Ocean and its continental margins. Special issue of *Journal of African Earth Sciences*, **7**, 205 pp.

Thomas, W. A. (1977). Evolution of Appalachian–Ouachita salients and recesses from reentrants and promontories in the continental margin. *American Journal of Science*, **277**, 1233–78.

Thomas, W. A. (1985). The Appalachian–Ouachita connection: Paleozoic orogenic belt at the southern margin of North America. *Annual Reviews of Earth Sciences*, **13**, 175–99.

Thomas, W. A. (1991). The Appalachian–Ouachita rifted margin of southeastern North America. *Geological Society of America Bulletin*, **103**, 415–31.

Thomson, M. R. A., Crame, J. A. & Thomson, J. W. (1991). *Geological Evolution of Antarctica*. Cambridge: Cambridge University Press, 722 pp.

Truswell, J. F. (1977). *The Geological Evolution of South Africa*. Cape Town, South Africa: Purnell, 218 pp.

Unrug, R. (1993). The supercontinent cycle and Gondwanaland assembly: component cratons and suturing events timing. *Journal of Geodynamics*, **16**, 215–40.

Veevers, J. J., ed. (1984). *Phanerozoic Earth History of Australia*. Oxford: Oxford University Press, 418 pp.

Veevers, J. J. (1988). Gondwana facies started when Gondwanaland merged into Pangea. *Geology*, **16**, 732-734.

Veevers, J. J. & Powell, C. McA. (1987). Late Paleozoic glacial episodes in Gondwanaland reflected in transgressive-regressive sequences in Euramerica. *Geological Society of America Bulletin*, **98**, 475–87.

Wilgus, C. K., Hastings, B. S., Kendall, C. G. St.-C., Posamentier, H. W., Ross, C. A. & Van Wagoner, J. C., eds. (1988). *Sea-Level Changes: An Integrated Approach*. Society of Economic Paleontologists and Mineralogists Special Publication 42, 407 pp.

Zhang, M., Liou, J. G. & Coleman, R. G. (1984). An outline of the plate tectonics of China. *Geological Society of America Bulletin*, **95**, 295–312.

Ziegler, P. A. (1982). *Geological Atlas of Western and Central Europe*. Amsterdam: Shell Internationale Petroleum Maatschappij (distributed by Elsevier Scientific Publishing Co.), 130 pp.+ map enclosures.

Ziegler, P. A. (1984). Caledonian and Hercynian crustal consolidation of western and central Europe – a working hypothesis. *Geologie en Mijnbouw*, **63**, 93–108.

Ziegler, P. A. (1988). *Evolution of the Arctic–North Atlantic and the Western Tethys*. American Association of Petroleum Geologists Memoir 43, 198 pp.

Zonenshayn, L. P., Kuz'Min, M. I. & Natalov, L. M. (1987). Phanerozoic palinspastic reconstructions for the USSR. *Geotectonics*, **21**, 487–502.

Cretaceous/Tertiary boundary clay at Gubbio, Italy (courtesy of K. Stewart).

8 THE MESOZOIC AND CENOZOIC – PART I. OCEANS, ATMOSPHERE, CLIMATES AND LIFE

8.0 Introduction

MORE than four billion years after the earth formed, it reached a condition that would appear almost 'normal' to us. By the start of the Mesozoic, the air would have been almost certainly breathable by humans, the oceans had compositions similar to those of the present, atmospheric temperatures varied within the range on the present earth, and the earth's rotation had slowed to a rate that made the day nearly the length of the present one. To be sure, the biotic realm contained some organisms quite different from those with which we are familiar, but the differences were largely of degree rather than kind. The modern world contains no dinosaurs, but their close reptilian relatives are with us. The only major groups of organisms currently on the earth that were not present near the start of the Mesozoic are calcareous plankton (foraminifera and coccolithophores), diatoms, mammals, and angiosperms.

This similarity of the Mesozoic and Cenozoic world to the modern earth provides us with an opportunity to study the interaction between geologic process and result with far greater accuracy than is possible for older ages. This advantage is enhanced by the fact that many Mesozoic and Cenozoic events are the last ones to affect an area – the uppermost and least-deformed sediments; the volcanic rocks not metamorphosed in one or more orogenies; the fossils not replaced by other minerals; the ocean crust not destroyed by subduction. For these reasons, the amount of information available about the Mesozoic and Cenozoic is greater than for any other period of time, and the ratio of (information) to (time interval covered) is enormous.

The Mesozoic and Cenozoic are clearly separate eras, with profoundly different fauna and flora. Furthermore, they are separated by the well-known Cretaceous/Tertiary (K/T) boundary, the result of a catastrophic event proposed (with vigorous dissent) to have been caused by an impact by a meteorite with a diameter of 10 to 20 km. From a faunal and floral viewpoint, the Mesozoic and Cenozoic should be discussed in wholly separate chapters. We do not make this separation here for the very simple reason that the K/T catastrophe seems to have had no effect on processes in the solid earth. All orogenic activity continued without interruption across the K/T boundary. Seafloor spreading patterns were not affected by it. Long-term climatic trends were disturbed briefly at the K/T boundary but returned to normal within a short time (perhaps thousands or only hundreds of years) afterward. In short, the K/T extinction event permits separation of the earth's biota into 'before' and 'after' groups but otherwise had no effect on the earth. For this reason, both eras are considered together in Chapters 8 and 9.

The tectonic history of the earth is summarized in Chapter 9, and in the present chapter we concentrate on the complex history of events that affected the earth's surface outside of compressional zones. One of the most important was the creation of the modern ocean basins. Beginning at the start of the Mesozoic, the earth probably contained one continent and one ocean. The Pacific is the (puny?) remnant of that ocean (Panthalassa). The past 200 million years is dominated by the fragmentation of the continent (Pangea) and the evolution of ocean basins between the separating fragments. This process has generated the Atlantic, Indian, Arctic and Antarctic (Southern) Oceans by patterns of opening discussed in Section 8.1.

Although most of the world's basalt is formed at oceanic spreading centers, the Mesozoic and Cenozoic record provides numerous examples of basaltic or other anorogenic magmatism away from spreading centers. This magmatic activity is commonly regarded as localized over the heads of plumes (Section 4.4), which may be the cause of large plateau basalt eruptions both in continents and oceans. We discuss the general problem of plateaus and elevations on the seafloor in Section 8.2.

As the oceans spread, sealevel and continental elevations varied to produce complex patterns of sedimentation. Much of the volume of Mesozoic and Cenozoic sedimentary rocks occurs on the passive margins of continents, which began subsiding as soon as initial rifting ended. In addition, the combination of climatic variations that produced glacio-eustatic effects and the changes in volumes of spreading ridges caused repeated transgression and regression onto continental interiors. Whether continental glaciers occurred during the Mesozoic is uncertain, but glacioeustatic effects clearly were important during the Cenozoic, when seafloor spreading had formed oceanic circulation patterns that permitted development of continental ice sheets. We discuss these various topics in Sections 8.3 and 8.4.

Although the dominant aspect of biotic evolution during the Mesozoic and Cenozoic was the extinction event that defines the Cretaceous/Tertiary (K/T) boundary, several other episodes of biotic change can be recognized. In the oceans, evolution appears to

have been closely related to, although not totally controlled by, periods of oceanic anoxia, which represent times when oceanic aeration did not occur and organic-rich muds were precipitated on the ocean floor. These episodes and other events in the history of changes in oceanic circulation patterns are discussed in Sections 8.4 and 8.5.

Most major groups of organisms underwent gradual or episodic evolutionary change throughout the Mesozoic and Cenozoic. The ecology of oceans was permanently changed during the middle of the Mesozoic with the development of calcareous plankton and also the silica-secreting diatoms. We have virtually no knowledge of the types of organisms that occupied these niches before their evolution; possibly varieties similar to non-calcareous plankton were simply more abundant. On land, the change from the early Mesozoic to the present was from reptiles moving through a dominantly gymnosperm flora to mammals moving through a primarily angiosperm flora. We discuss the biotic aspects of the Mesozoic and Cenozoic in Sections 8.6 and 8.7.

8.1 History of major oceans

IF there had been any cartographers living in the Late Permian, they would have had little difficulty in designating the world's oceans – in fact, two names would have sufficed. The major ocean was Panthalassa, which accounted for approximately two thirds of the earth's surface. The other ocean was Tethys, which appeared as an eastward-widening wedge of oceanic lithosphere between the Gondwana and Eurasia parts of Pangea. Because of its age, none of the Panthalassan or Tethyan crust remains in modern ocean basins, and the conversion from a Permian oceanic configuration to the present one was accompanied by complete generation of new oceanic lithosphere.

The conversion from a Permian configuration to the present was accomplished largely by rifting of Pangea. The initial rifting occurred during the Permian and Triassic. Some of the fault troughs opened into oceans, but many are preserved on land, where they are filled by sediments and volcanic rocks of the 'Gondwana' or 'Karoo' facies (Section 7.1).

The simplest oceanic opening was the separation of

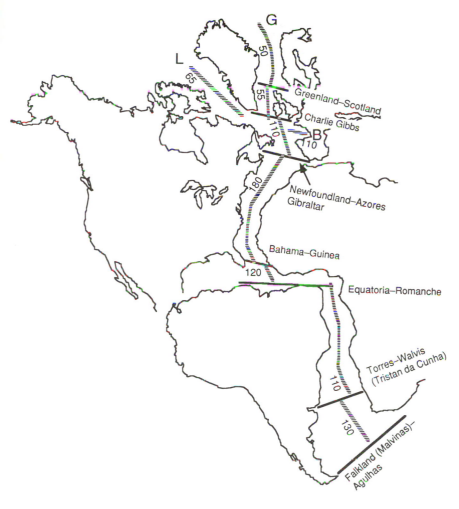

Fig. 8.1. Opening of the Atlantic Ocean. Spreading ridges are depicted as closely spaced horizontal, broken, lines. Various segments of the present ocean began to open at different times. These segments (basins) are separated by fracture zones whose names are shown on the diagram. Ages of initial opening are shown along ridge segments. L, Labrador Sea; G, Greenland Sea; B, Bay of Biscay.   ☐

North and South America from Europe and Africa to create the Atlantic Ocean. Somewhat more complicated was the development of the Indian Ocean, which was caused by fragmentation of the Gondwana part of Pangea by continental stretching and oceanic rifting in various stages and orientations. Formation of the northern Indian Ocean caused the destruction of Tethys by moving of continental fragments from Gondwana onto the Eurasian margin (Sections 9.1 and 9.2). Even more enigmatic was the development of the Arctic Ocean, in which most of the oceanic crust does not show magnetic lineations that could assist in deciphering its origin.

The opening of the Atlantic, Indian, and Arctic Oceans required the closing of the Pacific. The present Pacific Ocean is approximately one half of the area of the Panthalassa that preceded it. This shrinkage was necessarily accompanied by the destruction of oceanic lithosphere around the margins of the Pacific, and nearly 90% of the earth's current subduction occurs along those margins.

An additional aspect of the development of Mesozoic/Cenozoic oceans was the isolation of the Antarctic continent. This separation was not only a tectonic event but also resulted in major modification of ocean currents and the earth's climate. We discuss the tectonics briefly in this section and the climates in Section 8.4.

Atlantic Ocean

The modern Atlantic Ocean is approximately bisected by the mid-Atlantic ridge and its continuation to the north between Greenland and Scandinavia (Fig. 8.1). At present, half spreading rates along the ridge are in the range of 1 to 1.5 cm/yr, and comparison of these

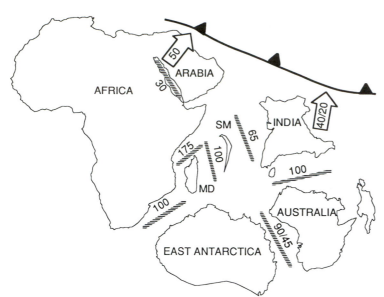

Fig. 8.2. Opening of Indian Ocean. Numbers along the ridge segments (closely spaced horizontal, broken, lines) are ages of initial separation between various fragments of Gondwana that opened to form the Indian Ocean. The separation between Australia and Antarctica probably began at ~90 Ma, with development of the modern spreading ridge at ~45 Ma. Numbers in white arrows are ages of initiation of subduction under the southern margin of Eurasia: 50 Ma for the Arabian peninsula; 40 Ma for initial compression between India and Asia, and 20 Ma for subduction of India beneath Asia.   □

rates with present widths of different sectors of the ocean indicates a comparatively constant spreading since initial opening. Estimated ages of initiation of spreading are shown in Fig. 8.1 for various segments of the ocean separated by transform faults. All spreading ridges are currently active except for those in the Labrador Sea and the Bay of Biscay.

The history of opening of the Atlantic Ocean is dominated by the effects of large transforms, which offset the ridge at intervals of ~100 km. Many of these transforms show small displacement, but some have offsets of several hundred kilometers and have apparently been in existence from the early stages of opening at that location. These large transforms separate regions of the Atlantic Ocean that opened in segments at different times, beginning at ~180 Ma in the main part of the North Atlantic but not starting until the Cenozoic in the northernmost segments (Fig. 8.1). The segmented opening of oceans is discussed in Section 4.3, and Fig. 4.21 is based on the development of the Romanche (Equatoria) fracture zone.

Continental shelves, presumably underlain by thin continental crust, are wider around North America than on its conjugate African margin. This width in North America accommodates a series of rift basins subparallel to the margin, exposed partly on land and partly on the submerged shelf. An age of ~180 Ma for basalts in some of these basins is the principal indication of the time at which the North Atlantic began

opening. The asymmetry of shelf widths on the North American and African margins apparently indicates that a wide zone of stretch developed between the two continents in the latest Paleozoic/earliest Mesozoic, and that final separation and generation of oceanic lithosphere occurred toward the eastern edge (present orientation) of that stretched zone.

Indian Ocean

The Indian Ocean is the result of the progressive fragmentation of the Gondwana part of Pangea (Section 6.7). This opening was necessarily accompanied by the closure of Tethys, which occurred in two stages (Sections 9.1 and 9.2). The earliest blocks (Cimmerian?) accreted onto Eurasia generally during the early and middle Mesozoic, closing a 'Paleotethys' ('Tethys I') and opening 'Neotethys' ('Tethys II') north of India and Arabia, which were part of Gondwana at that time. Later, during the Cenozoic, India and Arabia moved northward, closing Neotethys and leaving the Indian Ocean in the figurative wake of India and the Red Sea behind the Arabian plate.

Comparison of Figs 8.1 and 8.2 show that the pattern of opening of the Indian Ocean is considerably more complicated than that of the Atlantic. Whereas the Atlantic opened along an almost-exclusively north–south spreading ridge, the Indian Ocean resulted from a variety of rifts and consequent spread-

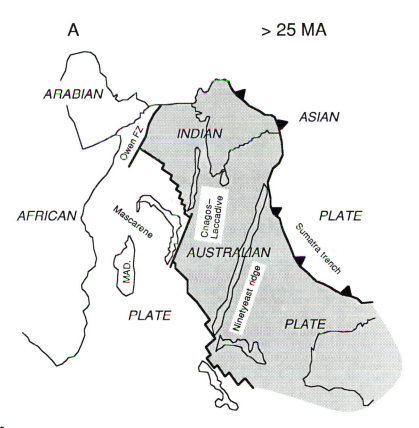

A   > 25 MA

For legend see next page.

ing centers with different orientations and opening ages. A further difference between the two oceans is the long period of continental stretching (perhaps 150 m.y.) that occurred in Gondwana before actual separation occurred, in contrast to a much shorter time (perhaps a few tens of millions of years) in continental areas that separated to form the Atlantic. The long period of stretching formed the Permian-to-Jurassic Karoo troughs and a late-Mesozoic/early-Cenozoic, mostly submerged, stretch zone between Australia and Antarctica (Fig. 8.2). In consequence of these complexities of opening, the Indian Ocean contains numerous plateaus, most of undetermined origin, whereas the Atlantic is dominated by a simple spreading ridge and only a few topographic highs (mostly plume tracks).

Two possibilities can be proposed to explain the contrast between the opening of the Atlantic and Indian Oceans. One is a change in the rotational velocity of Pangea as it moved around the earth. During the early and middle Mesozoic, when the Atlantic

began to open, the earth's continents were moving rapidly with respect to a (fixed?) hotspot reference frame, and the high angular momentum permitted passive pulling apart of some of Pangea but prevented buildup of active embryonic ridge-spreading systems. The part of Pangea that separated to form the Indian Ocean, however, began to split in the middle to late Mesozoic, at a time of low angular rotation of the remaining Pangea and during the long magnetic quiet period in the age range of 125 Ma to 85 Ma (Section 1.2). These two different modes of rifting may have generated very different patterns of separation. A second possibility for the difference between the Atlantic and Indian Oceans is a thermal anomaly under Gondwana. A mantle that was hotter than normal – or perhaps colder? – might cause an extended period of rifting of an overlying crust until separation of continental blocks could begin.

An important plate reorganization occurred in the Indian Ocean region at about 20 to 25 Ma (Fig. 8.3A and B). Prior to that time, India and Australia were

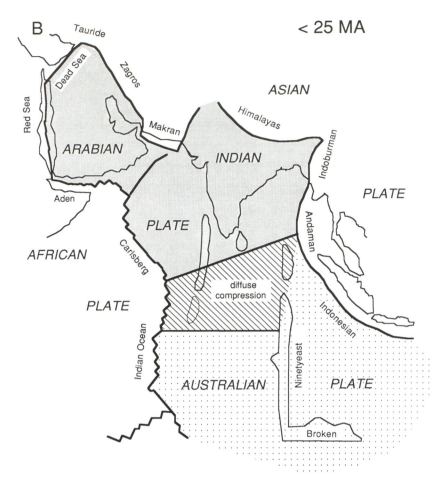

Fig. 8.3. Arrangement of plates in Indian Ocean region. (A) Prior to ~25 Ma, showing an Indian–Australian plate moving coherently and subducting under the southern margin of eastern Asia. The Owen fracture zone was active, permitting movement of the Indian–Australian plate as a unit. (B) Younger than ~25 Ma, showing fusion of Arabian and Indian plates and separation of Australian plate by a zone of diffuse compression in the center of the Indian Ocean. The Owen fracture zone is now locked between the former Arabian and Indian plates. The zone of compression presumably results from the resistance of the continental plate of India to subduction beneath Asia.   □

apparently joined on the same plate, which was bordered on the north and east by subduction zones, on the south by spreading ridges, and on the west by the transform fault of the Owen fracture zone. At 20 to 25 Ma, a broad compressional zone developed south of India, possibly as a result of the resistance to further northward movement of India when the continent finally collided with the southern margin of Asia (Section 9.1). This zone of compression, and transform movement along the Ninetyeast ridge, isolated the Indian plate from the Australian plate. At the same time, and for possibly unrelated reasons, the Owen fracture zone became 'locked' against any more than insignificant further movement. This locking fused the

Indian and Arabian plates, which have been moving northward as a single block since ~20 Ma. The present western boundary of the Indian–Arabian plate is a transform zone along the Dead Sea and Gulf of Aqaba, which has enhanced the opening of the Red Sea by rotation of the Arabian peninsula toward the northeast (Section 4.2).

Arctic Ocean

The Arctic Ocean is a small area of oceanic lithosphere surrounded by very broad continental shelves (Fig. 8.4). Its origin presents several problems. One is the configuration of continental blocks prior to the open-

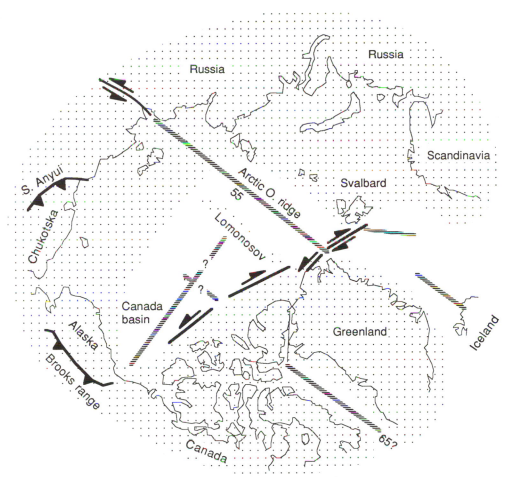

Fig. 8.4. Opening of the Arctic Ocean. Continental crust is shown in the dot pattern and includes the Lomonosov ridge. At some time in the Mesozoic, this continental crust fit together without intervening oceanic lithosphere. Spreading ridges are shown as closely spaced horizontal, broken, lines. Where known, initial ages of opening are shown by numbers along the spreading ridges. The two intersecting ridge segments (with ?) between the Canada basin and Lomonosov ridge diagrammatically indicate controversy over the mechanism, including direction of spreading, and age of opening of this part of the Arctic Ocean. Subduction zones and strike-slip faults are shown with conventional patterns.   □

ing of the Arctic Ocean; how did they fit before oceanic lithosphere was developed? A second problem is the almost total absence of magnetic anomalies in the area underlain by oceanic crust. Their absence, except in the rifts connected with the North Atlantic (Fig. 8.1), leaves the origin of the Arctic ocean virtually unconstrained. A third problem is the number and nature of separately moving plates in the Arctic area during the Mesozoic/Cenozoic and their interactions with each other.

Although normal world maps (Mercator or similar projections) do not show the fit, a Permian Pangea (Fig. 7.1A) is not only closed 'east–west' but also with respect to North America and Asia. This configuration of Pangea places the northern margin of North America, with its broad offshore shelf, against the wide shelf of Russia/Siberia, which extends out to Svalbard (Fig. 8.4). Extension and downwarp, without seafloor spreading, began to develop roughly correla-

tive sedimentary basins around the entire North American–Siberian area in the late Paleozoic, perhaps as early as Mississippian. This extension continued for approximately 150 m.y. until the first separation began in the Early Cretaceous.

Because of the absence of magnetic anomalies in most of the Arctic, Fig. 8.4 shows two questionable spreading centers and no age designations in the major area of oceanic lithosphere. These two orientations correspond to the two major hypotheses for the development of the Arctic. One is spreading along a ridge that is subparallel to the Canadian and Arctic margins, probably with a rotational pole close to its southern end. This orientation created oceanic crust along the rifted margins of North America and Asia and also caused rotation of Alaska away from Canada. A second possible direction of spreading is shown approximately perpendicular to the Canadian shelf and is intended merely as a diagrammatic representation of

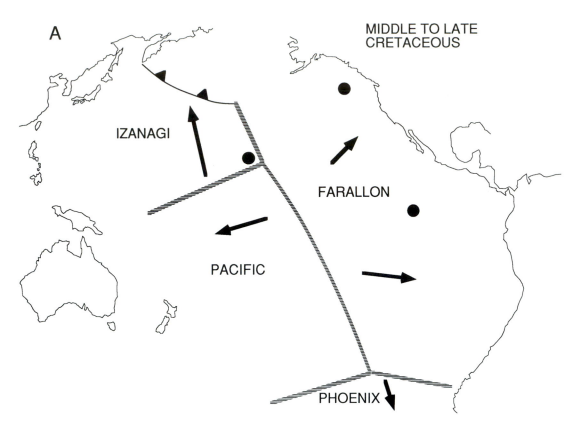

A

MIDDLE TO LATE
CRETACEOUS

IZANAGI

FARALLON

PACIFIC

PHOENIX

some unknown suite of small spreading centers of various orientations. Regardless of the orientation of spreading ridges, patterns of plate movements require most of the Arctic lithosphere to have been formed during the Cretaceous.

The one part of the Arctic Ocean that contains decipherable magnetic stripes is the linear basin along the Arctic Ocean ridge (Fig. 8.4). Opening along this spreading center apparently began in the early Cenozoic and may have been responsible for separation of the continental crust of the Lomonosov ridge from the broad continental shelf extending to the Russian and Scandinavian coastlines.

Exact movement of plates and microplates in the Arctic is virtually impossible to reconstruct with available evidence. At least two small continental fragments appear to have been separately mobile: Alaska north of the Brooks range and the Chukotska block, which constitutes northeastern Siberia and is bordered on the west by the South Anyui fold belt (Fig. 8.4). The South Anyui fold belt is regarded by many geologists as the western margin of the North American plate, which thus extends from Alaska across the Bering Straits into Siberia and southward into northern Japan

(Section 9.3). The validity of this designation may be doubted, but the absence of collision in the Bering Sea area prohibits drawing a plate boundary there.

Pacific Ocean

The Pacific Ocean is the shrunken remnant of Panthalassa. The oldest lithosphere in the Pacific basin is Jurassic, (exactly what part of the Jurassic is disputed), and the absence of older crust prevents us from drawing the configuration of spreading ridges for a Panthalassa before the rifting of Pangea began. Not only has pre-Jurassic crust disappeared down appropriate subduction zones, but a considerable area of younger crust has also been destroyed. Thus, the positions of some ridges that no longer exist must be inferred from relic magnetic stripes that do not correspond to present spreading centers. Fig. 8.5A–C represents an effort to depict the evolution of the Pacific Ocean from the latter part of the Cretaceous to the present relative to fixed positions of three hotspots (Hawaii, Yellowstone and Galapagos); (there is no evidence for a pre-Cenozoic, perhaps pre-Neogene, existence of either Yellowstone or Galapagos; Section 9.4).

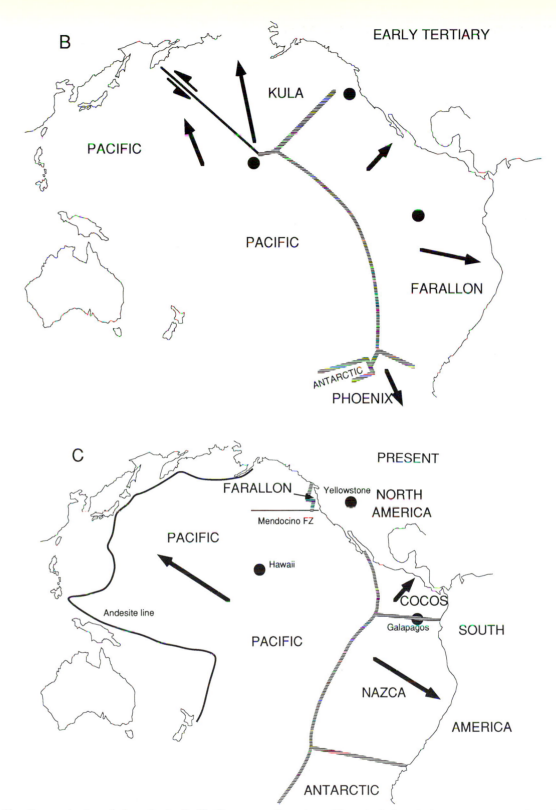

Fig. 8.5(A–C). Reorganization of plates in the Pacific Ocean at various times. The principal evolutionary pattern consists of expansion of the Pacific plate (west of the East Pacific Rise) and destruction of plates around the margin of the ocean by subduction. Arrows show dominant movement of each plate. Spreading centers are shown as closely spaced horizontal, broken, lines. Subduction zones and transforms are indicated by conventional symbols. Large dots (named on the map for the present configuration) show positions of the Hawaiian, Galapagos and Yellowstone hotspots, presumably stationary during the time period shown. The Mendocino fracture zone and Andesite line are also shown on the diagram for the present configuration. ☐

Fig. 8.5A shows a Middle- to Late-Cretaceous arrangement of plates. The Phoenix plate is now represented by lineaments formed north of the Phoenix–Pacific ridge, and the spreading center has disappeared against the northern margin of Antarctica. The Farallon plate is shown east of the precursor of the modern East Pacific rise, with the Izanagi and Pacific plates to the west. North America is shown to the east of the Yellowstone hotspot. The Izanagi plate was being subducted on its northern margin in the Late Cretaceous, and both the Izanagi and Pacific plates were subducted under the orogenic belts and back-arc seas along the western margin of the Pacific (Section 9.3).

By the Early Tertiary (Fig. 8.5B), the Izanagi and Pacific plates had become fused, but a new spreading center had subdivided the old Farallon plate into a new Farallon plate and a Kula plate. The Kula plate moved almost directly northward until the spreading ridges on its southern margin were subducted against the Aleutians; now only relic magnetic stripes formed south of the ridge survive to show its existence. North and South America are shown farther west of their positions in the Cretaceous, with continued active subduction along both margins (Sections 9.4 and 9.6). A complex development of spreading centers in the south began the formation of an Antarctic oceanic lithosphere.

Fig. 8.5C shows the present configuration of plates in the Pacific. The old Farallon plate has been divided into three remnants: a very small plate north of the extremely long Mendocino fracture zone (Section 9.4); a Cocos plate between the Galapagos and East Pacific spreading centers, which combine to push the plate northeastward under Central America; and the Nazca plate, which is being thrust under South America and has a complex ridge–transform border against the Antarctic plate. The main Pacific plate changed from a northward direction to a northeastward one at ~40 Ma and now occupies most of the oceanic basin.

The 'Andesite line' shown on Fig. 8.5C has long been recognized as one of the earth's most important geologic boundaries. In early studies in the Pacific, the line was interpreted as a separation of the basalt/andesite/dacite/rhyolite ('calcalkaline') suite erupted landward of the line from primarily alkaline basaltic and related rocks within the Pacific Ocean basin. This distinction was necessarily modified with

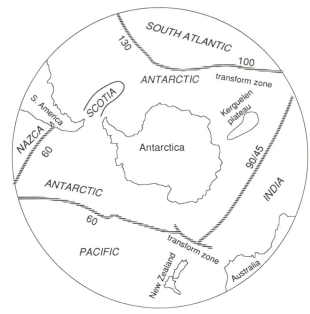

Fig. 8.6. Formation of the Southern Ocean surrounding the Antarctic during the Late Cretaceous and Early Tertiary. Spreading ridges are shown as closely spaced horizontal, broken, lines, with ages of initial opening indicated. Plates are shown in italics.   □

the recognition that both MORB and most of the volume of oceanic islands are types of tholeiites rather than alkaline rocks. A further improvement was the recognition that the Andesite line coincides with Pacific-margin subduction zones. Now, the Andesite line is simply one type of evidence indicating that calcalkaline magmatic suites are related to subduction, whereas magmatism within the Pacific is caused by other varieties of thermal anomaly (Section 4.1).

Antarctic Ocean

The Antarctic (or Southern) Ocean is really the southern extent of the Atlantic, Pacific and Indian Oceans (Fig. 8.6). In the Atlantic, the southern mid-Atlantic ridge terminates eastward in a long zone of ridge–transform segments that extend into the Southwest Indian Ocean ridge and westward in a poorly defined transform to the Scotia arc. Australia and Antarctica are separated by a ridge that began to create oceanic lithosphere at some disputed time in the range of 100 to 50 Ma following a long period of continental stretching. East of Australia, the Antarctic plate is bordered on the north by another long zone of ridge–transform segments that extends to the East Pacific rise. The Nazca–Antarctic boundary is also a ridge–transform margin. At the southern tip of South America, the opening of the Scotia Sea by ridge spreading in the early Tertiary completed the isolation of the Antarctic from all other continental areas. In Section 8.4, we discuss the fact that the opening of the Drake passage between South America and the Antarctic peninsula created sudden onset of glaciation in the Antarctic and contributed to global cooling.

[**References** – The discussion of the *Atlantic Ocean*: is based on the books by Emery and Uchupi (1984) and Tankard and Balkwill (1989), the volume edited by Scotese and Sager (1988), and papers by Rowley and Lottes (1988) and Blundell *et al.* (1991). The development of the *Indian Ocean*: is discussed in an original comprehensive paper by Norton and Sclater (1979) and subsequent publications by Weins *et al.* (1985), Besse and Courtillot (1988), Coffin and Rabinowitz (1988), Powell, Roots and Veevers (1988), Bohannon (1989), Veevers (1989), and Kent (1991); Hynes (1990) discusses the angular momentum of continents in the Mesozoic. Information on the *Arctic Ocean*: is provided by the volume edited by Scotese and Sager (1988) and the papers of Rowley and Lottes (1988) and Jackson and Gunnarsson (1990). An original paper on the opening of the *Pacific Ocean*: is by Larson and Chase (1972); more-recent publications include Barker (1982), Woods and Davies (1982), Whitman, Harrison and Brass (1983), Engebretson, Cox and Gordon (1985), Lonsdale (1988), Mammerickx and Sharman (1988), and Nakanishi, Tamaki and Kobayashi (1989). The *Antarctic (Southern) Ocean*: is discussed in the book by Kennett (1982), the volume edited by Scotese and Sager (1988), and papers by Kennett (1977), Barker (1982), and Barker and Kennett (1988).]

8.2 Continental flood basalts, plume tracks and elevations on the ocean floor

WHAT do plume tracks, other (commonly enigmatic) elevations in the ocean basins, and continental basalt plateaus have in common? The answer is not clear, and that is why these possibly diverse topics are grouped in this chapter. We will investigate oceanic regions first and then synthesize this information in a discussion of continental plateaus.

The oceans

The ideal ocean is one in which depths are controlled solely by the age of oceanic lithosphere formed at a spreading ridge (Section 4.2). In such an ocean, depth steadily increases away from the ridge, forming abyssal plains that are terminated landward by continental rises on passive margins or fore-arc trenches on active margins. Actual oceans, however, are not so simple, and several processes bring ocean floor to higher elevations than would be attained by thermal contraction of ridge-generated lithosphere. One of these processes is plume magmatism, which is reviewed in Section 4.4.

Fig. 8.7A displays the locations of plumes that are currently active or have recently been active. The figure shows somewhat more plumes than a 'conservative' approach would indicate because several 'plumes' are located as the tails of hotspot tracks (Section 4.4) that some investigators would regard as submerged continental plateaus or other elevations of non-hotspot origin (see below). With the exception of Yellowstone and the plume at the southern end of the Red Sea (Afar), all current plumes are in oceanic areas, and all are associated with geoidal highs (Section 4.4). The relative positions of the plumes shown in Fig. 8.7A are proposed to have been stationary with respect to each other throughout their existence, although some absolute polar wander may have occurred (Section 4.4).

Fig. 8.7B rather confidently shows oceanic plateaus and other high elevations classified into three categories (locations of a few plumes are duplicated from Fig. 8.7A). The actual origin of many of the features shown in Fig. 8.7B is extremely controversial; nevertheless, the three categories are as follows.

Continental fragments. These plateaus represent thin continental crust, mostly fragments of rifted margins. The difficulty in designating plateaus as 'continental' is the standard problem of identical densities, and hence other geophysical properties, of lower continental crust and layer 3 of oceanic crust (Section 4.2). The only definitive evidence for a continental origin of a submerged plateau is recovery of continental sialic rocks (gneiss, granite, etc.) by drilling or dredging.

Plume tracks. In oceanic areas, plume tracks are represented by chains of islands and submerged seamounts (Section 4.4). Proof that a linear chain of islands was formed by plate motion over a plume rather than by some other process requires that ages of volcanism decrease steadily toward the active plume, which has been found for some tracks (e.g.

A   HOTSPOTS

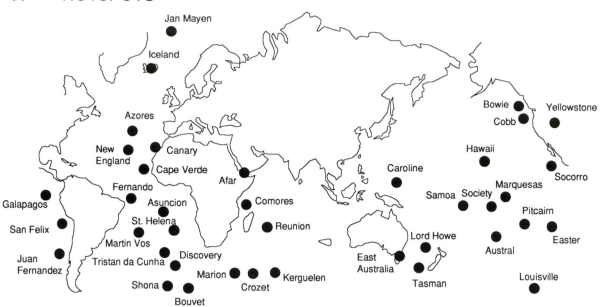

B   PLATEAUS

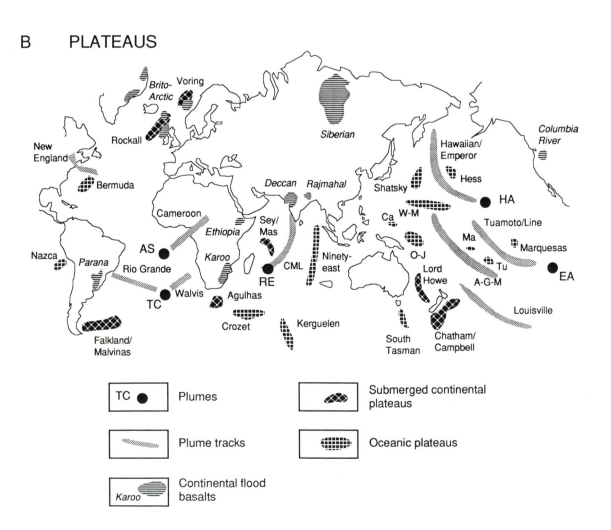

Hawaii–Emperor) but not for others that have been proposed. Some plume tracks can be traced across continental margins (Cameroon line, probably the New England track). Recognition of plume tracks no longer headed by an active hotspot is very difficult, but several (e.g. Louisville, Austral–Gilbert–Marshall) are shown in Fig. 8.7B. In addition to plumes at the end of linear island chains, some 'plumes' are recognized by long-continued volcanism at plate boundaries, where the plume does not appear to move (e.g. Iceland).

Oceanic plateaus. This category includes elevated areas of probably very diverse origins (some may be continental fragments – see above). At least three origins have been proposed. One is formation of a plateau in areas of higher-than-normal volcanism (perhaps the heads of plumes) at ridge crests and movement of symmetric fragments of the plateau away from the ridges. A second origin is off-ridge volcanism, perhaps caused by shallow convection cells; for example, linear ridges that do not show age trends may represent volcanism above fracture zones (possibly the Ninetyeast ridge, although it has been proposed to be a plume track from the Rajmahal basalts). A third possible origin is synchronous formation of different basalt plateaus above a large area of diffuse plume activity; this proposal might particularly apply to regions containing numerous islands and seamounts, such as the west-central Pacific.

Fig. 8.7B shows an Atlantic Ocean relatively free of enigmatic plateaus. Several linear plateaus are present east of the mid-Atlantic ridge, one (Iceland) on the ridge, and a former track (New England) west of the

ridge. The Cameroon line is clearly a plume track, but the Walvis and Rio Grande rises are controversial. The Rockall plateau is a continental fragment formed by rifting from northern Europe, but the origin of the Voring plateau is less certain. The most controversial Atlantic high area is the Bermuda rise, which carries the island of Bermuda. Apparently it represents some type of off-ridge volcanism.

As we discussed in Section 8.1, the Indian Ocean contains a large number of strange plateaus, most of which do not have a readily specifiable origin (Fig. 8.7B). The Chagos–Maldive–Laccadive ridge may (or may not, depending on interpretation) head into the current Reunion Island hotspot; some geologists have regarded the Deccan basalts of India, which formed at the Cretaceous/Tertiary boundary, as the initial location of the plume (Section 8.7). The Crozet and Kerguelen plateaus have been interpreted as tracks from the Crozet and Kerguelen 'plumes,' respectively (Fig. 8.7A), but no definitive proof is available. The Seychelles Islands contain continental granites, but the origin of the remainder of the Mascarene plateau is uncertain. This diversity of uncertain plateaus in the Indian Ocean may be related to the long period of rifting that the eastern part of Gondwana underwent before it finally separated into oceanic basins (Section 8.1), with the implication that more of the plateaus are floored by thin continental crust than is shown in Fig. 8.7B.

Except for shallow areas around New Zealand (Section 9.3), the plateaus and islands of the Pacific cannot be attributed to continental fragmentation (Fig. 8.7B). Many areas can be attributed to plumes or to some other type of intraplate volcanism that was not active for a sufficient period of time to leave a track. The prime example of a plume is the Hawaii–Emperor seamount chain, with the oldest part (80 Ma) now disappearing down a subduction zone at the northern margin of the Pacific, a knickpoint at about 40 Ma (Midway Island) when the Pacific plate changed spreading direction (Section 8.1), and a modern plume at the Hawaiian Islands; tracks emanating from Easter Island and the Austral chain are also reasonably well confirmed by age progressions of volcanism. Conversely, both the Shatsky rise and the Ontong–Java plateaus cover areas of several hundreds of thousands of square kilometers but do not have associated tracks (the collision of the Ontong–Java plateau with the

Fig. 8.7. (A) Proposed location of hotspots (plumes) that are active or have recently been active (adapted from Duncan and Richards, 1991).

(B) Oceanic and continental plateaus and hotspot tracks in present configuration. Plumes are shown in Fig. 8.7A, and only a few are shown here: AS, Asuncion; EA, Easter; HA, Hawaii; RE, Reunion; and TC, Tristan da Cunha. Plume tracks are labeled but use two abbreviations: CML, Chagos–Maldive–Laccadive; and A-G-M, Austral–Gilbert–Marshall. Some areas simply designated as oceanic plateaus have been proposed to be plume tracks (see Fig. 8.7A for possible sources). Abbreviations used for oceanic plateaus are: Ca, Caroline; Ma, Manihiki; O-J, Ontong–Java; Tu, Tubuai; and W-M, Wake–Marshall. Presumed submerged continental plateaus are all labeled without abbreviation except Sey/Mas, which indicates the combined Seychelles and Mascarene plateaus. All continental flood-basalt areas are labeled (in italics) without abbreviation.   □

Fig. 8.8. Horizontal basalt flows forming 'steps' along the Columbia River, Oregon and Washington.   □

island arcs of the southwestern Pacific has been responsible for much of the orogeny in the region – Section 9.3).

The most controversial of the elevated regions of the Pacific is the broad region of seamounts, islands and swells in the central-western part of the ocean (Fig. 8.7B). Both the Tuamoto–Line and Austral–Gilbert–Marshall plume tracks pass through the area, but unless they are very wide, they cannot account for all of the excess volcanism in the area. At one time, the area was regarded as the result of a temporary thermal anomaly that created both uplift and broad-scale volcanism. This 'Darwin rise' was thought to have formed in the Cretaceous and to have subsided in the Cenozoic, drowning most of the islands and creating seamounts capped with Cretaceous reef limestones. Because thermal subsidence alone generally is not sufficiently rapid to drown active limestone growth, the formation of reef-capped seamounts requires some marine environmental change in addition to subsidence, but this environmental effect has not been identified.

A more-recently proposed origin of the mid-Pacific mountains and surrounding areas is based on the observation that a broad region of elevated oceanic crust now occurs west of the East Pacific rise in the general area of the Marquesas and Tuamoto islands. This region has been referred to as a 'superswell' and is regarded as the present position of a long-lived mantle uplift resulting from excess heat. Pacific crust originating at the ridge and moving westward would pass

through this region and retain both the excess volcanic products and the high temperature that would maintain a relatively high elevation, and then subside to form seamounts as it moved westward .

A third hypothesized origin of the western-central Pacific seamount/island area is sudden appearance of an enormous mantle plume ('superplume'?) at some time during the period of 140 Ma to 100 Ma. This plume might have created nearly synchronous island formation over the entire area, which has not yet been verified by dating.

Continental basalt plateaus (flood basalts)

While all of this activity was occurring in the oceans, a number of thick sequences of basalt flows were constructed on land. Their broad, flat surfaces have given them the name 'plateaus', although some are not particularly elevated; the Columbia River basalt, for example, is actually at a lower elevation than most of its surroundings. Edges of the plateaus, or steep river gorges through them, commonly display a 'stairstep' topography of benches and cliffs controlled by the flat-lying flows (Fig. 8.8), and the German *treppen* (steps) has been converted into the word 'traps' to refer to some suites (e.g. the Deccan Traps; Fig. 8.7B).

The plateaus all have several features in common. One is that some type of basalt constitutes almost all of the rock suite, although thin layers of siliceous pyroclastics are present locally; these pyroclastic layers distinguish the continental plateaus from possible correlatives on the ocean floor. A second characteristic of the continental plateaus is that the eruptions were mostly subaerial (e.g. showing columnar joints), with only local subaqueous conditions. A third characteristic is that each plateau was apparently formed within a very short period of time, possibly less than one million years. This rapid eruption has been recognized only recently, mostly by $^{40}Ar/^{39}Ar$ dating.

The apparently simple characteristics of the plateau basalts belie the complexity of the questions that they pose. For example, what is their source? Clearly the basalts originated in the mantle, but can we be more specific about the properties of the source volume? Plateau basalts range from tholeiites to alkali–olivine basalts, but they are all richer in K_2O than comparable oceanic basalts, and the continental tholeiites contain more SiO_2 than oceanic counterparts. Does this compositional difference signify a difference in the composition of the upper mantle in subcontinental and suboceanic areas? Does the composition of the basalts represent partial melting of a plume consisting of peridotite from the deep mantle with a composition different from that of 'normal' upper mantle? The issue has been nearly drowned under a blizzard of isotopic and other geochemical data without resolution.

Regardless of their source, can we attach a tectonic significance to the basalt plateaus? One possibility is that they represent the sudden initiation of plume activity, with nearly instantaneous release of large volumes of partial melt. If the plateaus are the older ends of plumes, then continental movement should provide tracks leading to recognizable modern hotspots. This demonstration may be possible for some plateaus, with the Deccan basalts possibly originating at the Reunion hotspot and the Columbia River basalts and basalts of the Snake River plain forming a plume track originating at the Yellowstone hotspot (Fig. 8.7A and B; Section 9.4). Most of the other plateau basalts, however, are not easily assigned to definable plume tracks, raising the question of whether or not they represent plumes at all.

The ages of plateau basalt eruption have been correlated with the ages of two other processes. One is continental rifting, with the basalts representing eruption through continental lithosphere just as it becomes weak enough to rupture and start drifting. Many of the correlations seem quite good, as shown in the following short table.

	Eruption age	Drifting age of adjoining ocean
Ethiopian	25 Ma	25–30 Ma (Red Sea)
Brito-Arctic	55–60 Ma	55 Ma (North Atlantic)
Deccan	65 Ma	65 Ma (NW Indian Ocean)
Parana and Namibia	130?	110–130 Ma (South Atlantic)
Karoo	175–195 Ma	175 Ma (early Indian Ocean)

The only basalt plateaus that do not appear to be related to continental splitting are the Columbia River basalts (16 Ma) and the Siberian basalts (250 Ma).

Another correlation that has been attempted is between the ages of plateau eruptions and paleontologic extinction events (or other transitions). A relationship between massive volcanism, with its attendant environmental effects, and worldwide extinction is certainly plausible. The correlation between Deccan eruption and extinction at the Cretaceous/Tertiary boundary has been well established, although cause-and-effect relationships are unclear (Section 8.7). Similarly, the Siberian basalts, with an age of ~250 Ma, are approximately synchronous with the Permo-Triassic extinction event. Data for other basalts and extinctions, however, are difficult to interpret, partly because of uncertainties in the exact ages of paleontologic events and the continued debate over their abruptness.

A final problem to be considered is the relationship between plateau eruptions and meteorite impacts. The evidence for meteorite impact at the Cretaceous/ Tertiary boundary and the occurrence of the Deccan basalts at the same time has led some investigators to propose that the basalts were initiated by the shock of impact. Conversely, impact structures of the same or different age have been located in several places where no basalts occur. Furthermore, there is little reason to believe that impact can develop a long-lived plume, which the Deccan may represent.

[**References** – General information for this section is in volumes edited by Thompson and Morgan (1984) and Keating *et al.* (1987) plus the review by Duncan and Richards (1991). The Pacific Ocean is discussed by Schlanger, Jenkyns and Premoli-Silva (1981), Abers, Parsons and Weissel (1988), and Hekinian *et al.* (1991). Larson (1991) proposed a

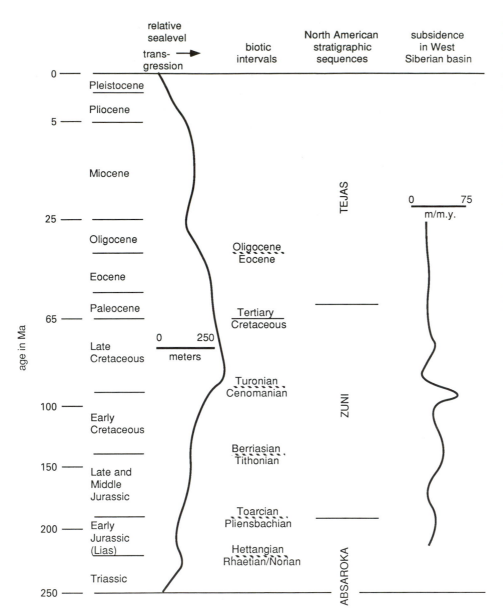

Fig. 8.9. Stratigraphic relationships in continental sediments. The major Cretaceous/Tertiary boundary is shown as a solid line, and all other extinction events as dashed lines. Relative sealevel is generalized from Haq et al. (1987), with a scale of ~250 m shown in order to provide limited quantification for otherwise qualitative data. Stratigraphic intervals are from Sloss (1988), and the major Zuni/Tejas boundary does not correlate with the Cretaceous/Tertiary extinction event. Subsidence in the West Siberian basin is shown in terms of meters per million years of accumulated sediment and indicates rapid tectonic downwarp; information from Rudkevich and Maksimov (1988).   ☐

broad 'superplume' in the Pacific. A general discussion of plateaus is in Schubert and Sandwell (1989). The volume edited by Johnson (1989) provides information on the Australia–New Zealand region. Mid-Eocene events are discussed by Schwan (1985). Periodicities of basalt eruptions are considered by Rampino and Stothers (1988) and Baksi (1990). The data on ages of continental plateaus are from Baksi (1990) and Renne and Basu (1991). Vogt (1991) discusses Bermuda.]

8.3 Platform and continental-margin cover

THE RISING sealevels of the Mesozoic crept over subsiding continental margins and into the interiors of most of the continents that were dispersing from the rapidly disintegrating Pangea. Punctuated by numerous episodes of minor regression and transgression, mean sealevel rose until the Late Cretaceous and then began a decline to the lowstands of the Pleistocene. At the height of the Cretaceous marine transgression, global sealevels were some 250 m to 300 m above their present position, raising the proportion of the earth's surface under water to about 80% (Figs 8.9 and 8.10). This sealevel may have been the highest in earth history (although possibly exceeded in the Ordovician). In this section we discuss the processes that controlled the complex interaction of sealevel movement and subsidence during the Mesozoic and Cenozoic and then concentrate on the Cretaceous event.

Relationship between sealevel and tectonics

The changes in sealevel shown on the left side of Fig. 8.9 are 'first-order' changes that represent the major trends in eustatic sealevel over the approximately 250-m.y. period of the Mesozoic and Cenozoic. The most likely, although perhaps not total, cause of such long-term variations is change in volume of oceanic spreading ridges. We can calculate the volume as follows.

Assume that no ice was held above sealevel on continents during the Cretaceous and that present icecaps in the Antarctic and Greenland hold approximately 100 m of water. Thus, Cretaceous sealevels 250 m above present would require 150 m of sealevel rise (250 − 100) by expansion of oceanic ridges. With an average depth of 3.8 km, a rise of 150 m in sealevel within the present area of the oceans would require about 4% (0.15/3.8) of the present ocean volume to be displaced by newly created crust. Expansion of the oceans over one half of the area of continental crust (about 30% of the earth) would require additional displacement of perhaps 1% to 2% of ocean volume. Thus, without any compensating changes in 'absolute' continental elevation, spreading ridges that filled approximately 5% more of the ocean basin than at present could account for the Cretaceous marine transgression. This simple estimate is affected by several problems that do not lend themselves to accurate calculation, the major one being the failure to consider isostatic adjustment of both continents and oceans to new sealevels and to the readjusted patterns of sediments on the earth's surface.

Variations in ridge volume are presumably dependent on changes in the rate of ridge spreading, which is the same as the rate of creation of new lithosphere. A more rapid spreading rate creates oceans with crust of younger average age, thus 'floating' at a higher level and causing sea water to lap onto land areas. Although periods of continental rifting, such as the early part of the Mesozoic, have traditionally been regarded as the result of rapid spreading, the evidence shown in Fig. 8.9 does not substantiate this conclusion.

The complexity of interactions between tectonics and sealevel is illustrated in Fig. 8.9 with information from North America and from western Siberia. In North America, marine transgressions and regressions in the Mesozoic and Cenozoic exhibit the same pattern of sequence-bounding unconformities in the sedimentary suites of cratonic platforms that we discussed for the Paleozoic in Section 7.1. The oldest boundary in the Mesozoic is the base of the Zuni sequence, which follows the Carboniferous-to-Jurassic Absaroka suite (Fig. 8.9). The pre-Zuni unconformity is controversial, with some workers proposing that marine sediments covered most of the Canadian shield before being stripped during later marine regression and continental uplift, and other workers postulating that Cretaceous seas encroached only to areas now underlain by rocks of that age. Fig. 8.9 shows that sequence-bounding unconformities have the same lack of relationship to geologic age boundaries in the Mesozoic and Cenozoic that they have in the Paleozoic (Section 7.1).

The West Siberian basin provides a sharp contrast to cratonic North America (Fig. 8.9). The West Siberian basin is one of the largest sedimentary basins in the world (nearly as large as the cratonic interior of North America) and contains one of the world's most continuous records of marine sedimentation throughout the Mesozoic and Cenozoic. The basin is underlain by a complex basement of arcs, microcontinents, and oceanic slices apparently swept together when Kazakhstan collided with Baltica near the end of the Paleozoic (Section 7.2). Thus, the West Siberian basin is underlain by a crust considerably younger (less 'mature'?) than that of the North American interior. Presumably this newly accreted crust permitted the rifting near the beginning of the Mesozoic and the extensive downwarp that led to the accumulation of 10 km to 15 km of Mesozoic–Cenozoic sediment. This thickness indicates that sedimentation must have been controlled almost entirely by tectonic downwarp rather than by sealevel variation. Further indication of this tectonic control is provided by the observation that the periods of maximum sediment accumulation (and presumably maximum subsidence) are Middle Jurassic, Early Cretaceous and Oligocene (Fig. 8.9), which do not coincide with times of maximum eustatic sealevel rise.

The Cretaceous marine highstand

The extreme height of Cretaceous sealevel (at ~90 Ma) probably was the result of a variety of processes. Some

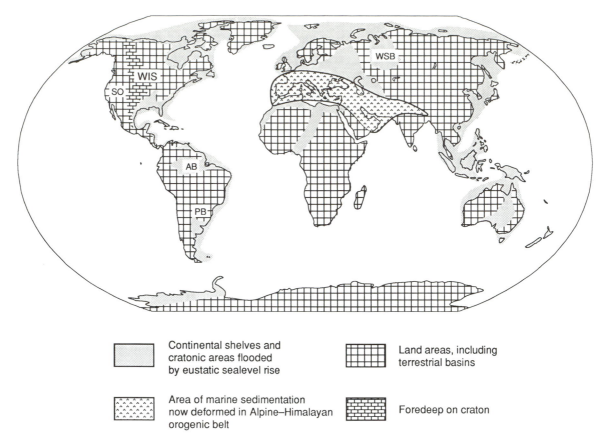

| | Continental shelves and cratonic areas flooded by eustatic sealevel rise | | Land areas, including terrestrial basins |
| | Area of marine sedimentation now deformed in Alpine–Himalayan orogenic belt | | Foredeep on craton |

Fig. 8.10.   Cretaceous oceans and platform cover. Relationships are particularly generalized in the area of the Alpine–Himalayan orogeny. The Western Interior Seaway (WIS) of North America is shown as a foredeep caused by loading by the Sevier orogen (SO) to the west. Other areas of marine incursion represent combinations of eustatic sealevel rise (to ~250 m above the present) and tectonic downwarp. They include: AB, Amazon basin; PB, Parana basin; WSB, West Siberian basin.

of the sealevel rise undoubtedly was associated with activity along several spreading centers in the (formerly much larger) Pacific basin that have now been destroyed (Section 8.1). Another possibility is the accelerated formation of oceanic plateaus throughout the Pacific early in the time interval of high sealevel (Section 8.2).

The interaction between the Cretaceous sealevel highstand and tectonic processes produced a variety of depositional environments, summarized in Fig. 8.10. On continents, sedimentation occurred in flooded platforms and foredeeps of orogenic belts. Along rifted continental margins, sediments accumulated from the time of rifting to the present. Accumulation rates on these margins diminished toward the present as thermal contraction rates became slower. Thick sequences of sediment also accumulated along the margins and

interior of Tethys as it closed to develop the Alpine/Himalayan mountain chain (Sections 9.1 and 9.2). We discuss the various sedimentary environments briefly below.

The rifting of Pangea formed stretched continental margins with ages ranging from ~180 Ma (e.g. eastern North America) to nearly the present (northern Red Sea). In some places, such as the North Sea, networks of rift valleys created similar broad areas of thin crust that now form shallow seas extending into the continents. Sediments on the continental margins are dominantly clastic, but in some areas the early deposits are shallow-water carbonates and evaporites.

A second environment shown on Fig. 8.10 is areas of actual (or in some places potential) encroachment of marine water onto unthinned continental crust. The limit of this transgression is approximately the 250- to

300-m elevation contour, which would have been reached by the equivalent sealevel rise. The simplest sedimentary pattern is shoreline and deltaic complexes overlapped transgressively by offshore sediments, including carbonate rocks. Onlap near the 250-m contour is found in continental areas that have not undergone orogenic activity during or since the Cretaceous, such as narrow marginal zones of Africa. Onlap to similar elevations occurred across much of eastern Australia, but the major transgression was in the Middle Cretaceous, and the seas had withdrawn almost completely from the continent at the time of the Late-Cretaceous sealevel maximum. Coincidence of transgressive shoreline facies with present contours indicates that a continental area has not undergone tilting since the Cretaceous.

Broad areas of Cretaceous marine sediments occur in some continents at elevations considerably above the level that could have been reached by simple transgression. The largest area of both uplift and downwarp is the collision zone along the Alpine/Himalayan orogen (Fig. 8.10). A simpler example is the rocks of the Western Interior Seaway of North America (Fig. 8.10), where some of the present elevation is the result of upward arching of the entire western part of North America during the Cenozoic.

The Western Interior Seaway formed over a period of several tens of millions of years and records at least five major transgressive–regressive pulses. The creation of the marine basin in the seaway occurred partly because the general elevation of western North America was closer to sealevel than it is now and partly because of foreland downwarp caused by loading of the Sevier orogenic belt to the west (Section 9.4). Repeated orogenic pulses were apparently responsible for the transgressive–regressive cycles rather than some complex pattern of eustatic sealevel variations. The influx of fresh water from the Sevier highlands was so large that parts of the seaway remained brackish for long periods of time.

A potential foreland basin east of the Andes did not develop in the same fashion as the Western Interior Seaway in North America. The reason is probably the infilling of the area by the enormous volume of clastic debris washed eastward from the developing Andes, thus preventing marine encroachment. The supply of sediment was so large that terrestrial deposition (mostly fluvial) occurred along nearly the entire extent of the Amazon basin to the eastern edge of the continent, despite the fact that the basin is now below the 250-m contour. The absence of marine transgression in the area may also indicate some eastward tilting of South America during the Cenozoic, yielding lower elevations in the eastern interior now than were present in the Cretaceous.

One of the largest areas of marine Cretaceous sedimentation was the Middle East and southwestern Asia (Fig. 8.10). During the Cretaceous, the Arabian peninsula was separated by Tethys from Asia, and similar carbonate–clastic sequences developed on the continental platforms of both sides. Closure of the ocean basin by mostly Cenozoic collision has apparently raised the area well above its former elevation near sealevel.

[**References** – General information on the Mesozoic is provided by Moullade and Nairn (1978a, 1978b) and on the positions and shapes of plates by Barron *et al.* (1981). Worldwide sealevel variations are discussed by Haq, Hardenbol and Vail (1987) and the concept of a superplume by Larson (1991). Information on sequences in North America is from volumes edited by Sloss (1988) and Bally and Palmer (1989); on the West Siberian basin from Nalivkin (1960, 1973), Rudkevich and Maksimov (1988), and Aplonov (1989); on Africa from Petters (1981) and the volume edited by Kogbe and Lang (1990); and on Australia from Veevers (1984).]

8.4 Climates and atmosphere

THE WATER of the Atlantic Ocean near the equator is heated by the sun and begins to evaporate as it moves slowly westward under the influence of the earth's rotation. Passing through the Caribbean into the Gulf of Mexico, it is rerouted around Florida and passes northward along the western margin of the Atlantic, where it is known as the Gulf Stream. In the North Atlantic, this warm, salty water cools, mixes with Arctic meltwater, sinks to mid-ocean depths, and is referred to as 'North Atlantic deep water' (NADW). At that point, it begins its subsurface journey southward, collecting saline effluent from the Mediterranean, and ultimately works around the southern tip of Africa, through the Indian Ocean, and into the North Pacific. Here, at the northernmost possible limit of movement, the last remnants of NADW rise to the surface and begin their return trek on the surface, ultimately reaching the equatorial Atlantic and completing the cycle.

A   CRETACEOUS AND EARLY TERTIARY

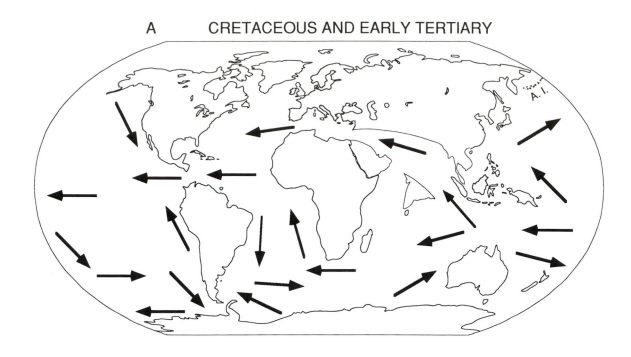

B   LATE TERTIARY

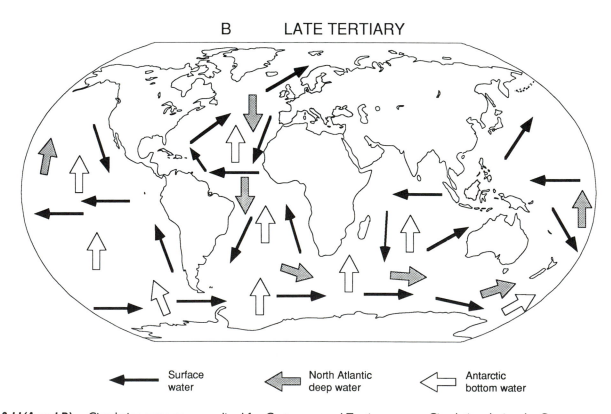

Fig. 8.11(A and B).   Circulation patterns generalized for Cretaceous and Tertiary oceans. Circulation during the Cretaceous and Early Tertiary shows a dominantly equatorial pattern without circum-Antarctic flow. Circulation for most of the Cenozoic shows interruption of equatorial circulation and establishment of a circum-Antarctic current.   □

The present journey of NADW has not always been possible, and its history is closely related to changes in the earth's climate through time. These differences may result from variations in the output of energy from the sun, from variations in the earth's orbit and tilt, and from movements of continental masses that force redirection of oceanic currents. We discuss the broad-scale evolution of Mesozoic/Cenozoic climates first, followed by more specific consideration of Pleistocene climatic fluctuations.

Icehouse to greenhouse to icehouse

In the spectrum of Phanerozoic climates, the two end-member conditions can be identified as the 'icehouse' and the 'greenhouse'. Periods of major accumulation of ice on continents, such as the Gondwana glaciation, are icehouses. The Quaternary is also an icehouse, and we do not know whether the earth has passed out of that condition or is merely in another 'interglacial' stage. The type greenhouse is represented by the Cretaceous, when the earth was either ice-free or, at least, had very little land surface covered by ice.

The concept of 'average' worldwide climates is difficult. In particular, glacial climates have little effect on equatorial areas, and equatorial climates may remain the same through both glacial and interglacial stages. Another problem is the distinction between worldwide climate changes and plate movements. For example, the early and middle parts of the Mesozoic not only witnessed the disintegration of Gondwana and its ice sheets but also dispersal of continents from generally equatorial positions to more northerly ones (Section 8.1). North America moved northward at the same time as climatic amelioration was occurring, and the relative importance of these two factors is not clear. Our best interpretation of climatic fluctuations is as follows.

The breakup of Gondwana and apparent movement of Antarctica somewhat off of the South Pole melted the widespread Permian glaciers, caused sealevel rise, and presumably brought the earth to a more equable climate at the start of the Mesozoic. The climatic history of the Triassic and Jurassic is not particularly reliable, but the Cretaceous clearly was a time when fauna and flora were similar over much of the earth, latitudinal temperature gradients were low, and sealevel may have been the highest in the Phanerozoic (Section 8.3).

By the Cretaceous, the Antarctic apparently had reoccupied a position over the South Pole, but probably did not contain any permanent ice cover, and North America and Eurasia had not reached latitudes far enough north to develop icecaps. This lack of ice cover means that small-scale fluctuations in Cretaceous sealevels must have had some origin other than glacioeustacy. Sealevel began to fall before the Cretaceous/Tertiary boundary, but the rate of fall is more consistent with decrease in volume of spreading ridges than with more rapid development of continental ice cover.

Major change in oceanic circulation and climate began in the Paleogene, with the pace particularly rapid in the Oligocene (Fig. 8.11A and B). The principal changes were caused by northward movement of the Gondwana continents, which terminated circum-equatorial currents by closing the Caribbean (Section 9.5), Tethys (Sections 9.1 and 9.2), and the ocean between Australia and Indonesia (Section 9.3). These equatorial currents had established various other small current cells within the Atlantic, Pacific and Indian Oceans and kept the Antarctic continent relatively warm. The small size of these cells also permitted repeated periods of near anoxia to develop in the deep oceans (Section 8.5). The early-Tertiary change in oceanic circulation patterns may coincide with, and be responsible for, the biotic change at the Eocene/Oligocene boundary, although exact timing is not adequate for certainty.

The closing of equatorial seaways necessarily coincided with the separation of Antarctica from other continents, and by the start of the Miocene the Antarctic was surrounded by an 'Antarctic (Southern) Ocean' with a circumpolar current (Fig. 8.11B). The isolation of the Antarctic is shown more comprehensibly in Fig. 8.6, but the ages of initiation of spreading on the ridges are considerably older than the ages at which the various passages opened to deep-ocean currents. Neither the Scotia Sea (Drake passage) nor the Australia–Antarctic seaway were free of obstructing ridges until the Oligocene. As soon as isolation was obtained, warm water was diverted from the Antarctic and continental glaciers began to form; they expanded episodically into the Pleistocene.

The circumpolar flow of the Southern Ocean had a profound effect on the rest of the world's oceans. The reason is shown diagrammatically in Fig. 8.12. Ice

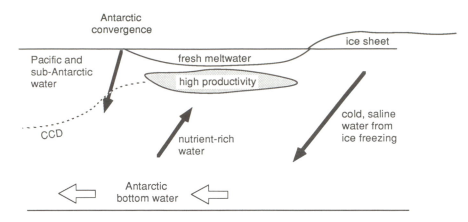

Fig. 8.12. Diagram showing development of Antarctic bottom water by density-driven sinking of cold water and movement northward toward the world's major oceans. The zone of high productivity surrounding the Antarctic continent results from rise of nutrient-rich water. The calcite compensation depth (CCD) is shallow near the Antarctic and deepens toward the north. □

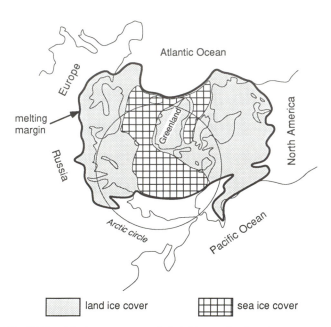

Fig. 8.13. Maximum extent of glacial ice on land and as ocean cover during the Pleistocene. □

shelves around the Antarctic continent yield fresh meltwater during Spring and Summer, and this low-density water moves northward on the ocean surface from the continent. Conversely, the freezing of ice shelves over water leaves highly saline, cold water that sinks because of its high density. The meltwater moves northward until it meets more normal sub-Antarctic and Pacific water at the 'Antarctic convergence', where the two water masses form a downgoing cell. South of the convergence, nutrient-rich water from intermedi-

ate depths is displaced upward, causing the extraordinary biologic productivity of circum-Antarctic seas.

The cold water that sinks around the Antarctic forms an 'Antarctic bottom water' (AABW), with slightly lower salinity than the overlying oceans, that flows northward throughout virtually the entire Pacific and Atlantic Oceans. The movement of AABW contributes to the continued ventilation of the present world's oceans, ensuring that oxygen is available throughout the entire water column down to the ocean floors. Partly because of low productivity of calcareous organisms in Antarctic waters, the calcite compensation depth (CCD) increases from relatively shallow around the Antarctic to normal depths in the rest of the world's oceans (Section 8.5).

After the formation of circum-Antarctic flow, another feature of modern ocean currents was established by the separation of the Atlantic/Caribbean and the Pacific Oceans by construction of the Isthmus of Panama, which probably occurred in the Pliocene (about 3 to 4 Ma). This isolation permitted the formation of a nearly worldwide current of North Atlantic deep water (NADW) (Fig. 8.11B), whose movement was described at the start of this section. The flow of the NADW is largely responsible for the ventilation of the world's oceans, and the continual circulation across latitudes reduces temperature gradients. Apparently, this circulation pattern occurs only during non-glacial periods.

The Neogene was a time of increasing ice cover on the Antarctic and probably continued reduction of worldwide temperatures. As many as 60 transgressive–regressive marine cycles have been proposed for the Cenozoic, and many of those in the Neogene may

Fig. 8.14. Toe of the Rhone Glacier, Switzerland, which has retreated up its valley following the last major glaciation.   □

glacial
toe

have been caused by fluctuation in the amount of glacial ice on the Antarctic. The major development in the late Neogene, however, was glaciation in the northern hemisphere (see below).

Pleistocene glaciation

The toe of a glacier is a messy place. The ice front is trenched by melting, and relic debris of boulders, sand and rock powder coat the exposed surface. Away from the glacier, braided streams distribute this load as quickly as possible through piles of till and outwash. On multiple occasions during the Pleistocene (roughly the past one million years), this view could be seen along ice fronts up to several tens of meters high that snaked for some 10 000 km across North America and Eurasia (Fig. 8.13). What caused them, and what effect did they have on the earth? We explore these issues in this section.

The progressive isolation of ocean basins and formation of deep, saline currents during the Neogene was associated with increasing latitudinal climatic gradients. The gradual change, however, was not sufficient to cause Arctic glaciation without an additional triggering event. That event apparently occurred in the Pliocene (about 3 to 4 Ma), shortly before the first indications of permanent glaciers are found in North America and in deep-sea sediments (about 2 to 3 Ma).

One possible trigger was the closing of the Isthmus of Panama by continued construction of the volcanic arc between the Nazca and Caribbean plates (Section 9.5). This closure isolated the Pacific Ocean from the Atlantic (and Caribbean) Ocean, thus causing increased provinciality and possibly providing a source for the NADW as it starts on its circuitous modern route (see above). The resultant modification of ocean currents should have had a major effect on the earth's climate that may have led to glaciation.

Regardless of the cause, northern-latitude ice sheets have advanced and retreated for approximately the past 2 to 3 million years (Fig. 8.14). Their extent in the most recent major glacial advance (Wisconsin/ Weichsel) is shown in Fig. 8.13, which is close to the maximum range of ice cover throughout the Pleistocene.

One of the most puzzling aspects of Pleistocene glaciation is the periodicity of glacial advance and retreat, which seems to have affected both the Arctic and Antarctic at nearly the same times. Exactly how many glacial and interglacial stages occurred in the Pleistocene is uncertain, but clearly six to seven glacial advances and interglacial retreats have occurred in the past 700 000 to 800 000 years. Fig. 8.15 shows oxygen-isotope variations with age in deep-sea sediments and also provides a rough indication of time-dependent variation in other parameters that measure climate and glacial development. The $\delta^{18}O$ variation results from

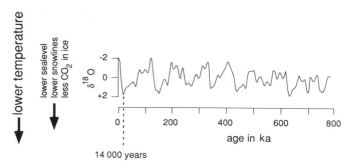

Fig. 8.15. $\delta^{18}O$ variations during the past 800 000 years, showing approximate 100 000-year periodicity. $\delta^{18}O$ is $[(^{18}O/^{16}O)_{sample} - (^{18}O/^{16}O)_{standard}]/[(^{18}O/^{16}O)_{standard}] \times 1000$.
Increase in ^{18}O content of sea water indicates increase in volume of ice sheets, which are impoverished in the heavy ^{18}O isotope. Correlative variations that indicate lower temperatures in polar regions include lower sealevel, lower snowlines and less CO_2 preserved in gas trapped in ice.

the fact that ice on continents is frozen from fresh rain water that has low $\delta^{18}O$ values because of preferential evaporation of water enriched in ^{16}O from the oceans. Thus, increasing cover by land ice results in increasing $\delta^{18}O$ in ocean water, and the record of this variation is preserved by the tests of planktic organisms (particularly calcareous foraminifera) that settle on the seafloor. Therefore, a plot of $\delta^{18}O$ in calcareous microfossils in deep-sea sediments vs. age measured by some combination of radiometric and paleontologic techniques provides a measure of variation in extent of ice cover on the continents.

Two other indicators of glaciation are also shown in Fig. 8.15. One is sealevel (see below), which can be estimated both from terraces above present sealevel, when the world contained even less land ice than it does today, and by the extent of subaerial exposure on continental shelves during glacial advance. Another indicator is the altitude of the 'snowline', which is the line separating net accumulation of snow at higher elevations from net loss (ablation) at lower elevations. The snowline in a mountainous area is commonly estimated as the average elevation of valley glaciers, which can be determined by mapping of glacial debris and erosional patterns.

How do we interpret the periodicity. An old, qualitative, and formerly popular, explanation for glacial/interglacial alternation was based solely on the effect of CO_2 in the atmosphere. The cyclicity, according to this theory, was as follows: high atmospheric CO_2 concentration causes high temperature ('greenhouse effect'); the high temperature causes glacial

melting; the increased ocean volume absorbs CO_2, thus lowering atmospheric CO_2; the resultant cooling causes ice caps to form, thus reducing ocean volume; to complete the cycle, the loss of ocean water places more CO_2 into the atmosphere. Some support for this concept is shown by ice cores in Greenland and the Antarctic, which record episodic changes in atmospheric CO_2 content, with higher abundances during interglacial stages (Fig. 8.15). Fluctuations in concentrations of other 'greenhouse gases', such as methane, that absorb radiation reflected from the earth probably are also important in controlling climatic variations.

The qualitative explanation offered above does not explain the precise cyclicity of the glacial/interglacial indicators. One line of investigation that has proved very useful is spectral analysis of the data. Merely by inspection, the curve in Fig. 8.15 shows six to eight periods of glacial advance. This interpretation can be greatly improved by a Fourier analysis, and when we do one, we find at least four important periodic components. The largest component has a period of ~100 000 years. Smaller components have periods of ~41 000, ~23 000, and ~19 000 years. These periodicities correspond approximately to Croll–Milankovitch periodicities (Section 2.1) – 100 000 years for orbital eccentricity, 41 000 years for obliquity variations, and 23 000 and 19 000 years for precession of the equinoxes. Presumably, the variations in solar energy reaching the earth as a result of Croll–Milankovitch cycling are major controlling factors in the alternation of glacial and interglacial stages.

Despite   the   correspondences   outlined   above,

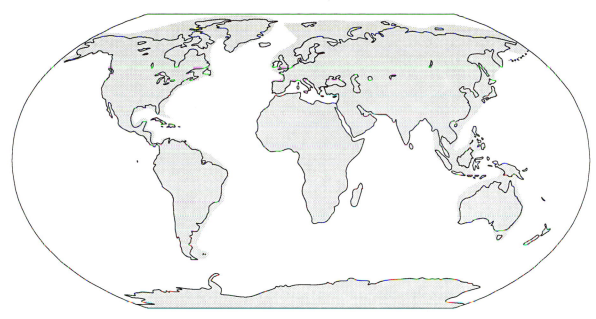

Fig. 8.16.   Extent of continental emergence during maximum Pleistocene glaciation, with shaded area showing land.

Croll–Milankovitch cycles do not appear to be solely responsible for glacial cycles. One of the largest problems is that the apparent 100 000-year periodicity of major glacial advances and retreats is only approximate and, thus, corresponds only approximately to the 100 000-year Croll–Milankovitch cycle. Furthermore, the 100 000-year orbital eccentricity cycle is the weakest of the Croll–Milankovitch variations, causing only ~1% variation in the amount of solar radiation on the earth, which seems far too small to cause (force) glacial alternations. Therefore, Croll–Milankovitch orbital variations presumably are important only in their ability to act as small triggers for ocean/atmosphere processes that can cause major climatic variations.

The periodicities exemplified in Fig. 8.15 reveal another aspect of the glacial cycle – the asymmetry of temperature variations. The onset of glaciation is commonly a slow process, with gradual increase in continental ice over some 75 000 years, whereas deglaciation is almost instantaneous, with time spans of only a few thousand years. For example, the peak of Wisconsin glaciation occurred about 20 000 to 15 000 years ago, but melting was well started by 14 000 years, and sealevel had reached almost its present position by 9000 to 8000 years ago.

The surface of the glacial earth was a far different place than the one with which we are familiar. Climatic zones were generally compressed toward the equator, with forests and grasslands occupying areas that have become deserts in this interglacial stage. Loess plains were major periglacial features in some areas. Sealevel was in the range of 100 to 150 m lower than the present (Fig. 8.16), nearly exposing the continental shelves. This lowering caused entrenchment of river valleys and exposure of former marine sediments to subaerial erosion and oxidation. Holocene sediments now lie on this oxidized surface.

Lowered sealevels had profound effects on Pleistocene biotic trends. One of the major effects was the exposure of the Bering Sea as a land bridge between Asia and North America. Although at high latitude, this area was mostly beyond glacial cover, allowing animals to move freely between continents. One result was the human occupation of North America. Another area of increased migration was Indonesia and Southeast Asia, where the Sunda shelf (Section 9.3) became a land bridge.

The effects of Pleistocene glaciation are still very evident today. Large parts of North America and Eurasia are covered by subglacial and glacier-front tills and periglacial outwash and loess, a fact lamented by many 'bedrock' geologists who find their outcrops

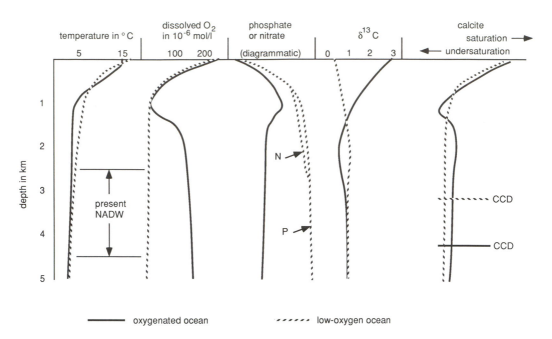

Fig. 8.17. Typical variations in compositional parameters in idealized, oxygenated, ocean (solid lines) and anoxic, or low-oxygen, ocean (dashed lines). In oxygenated oceans, the thermocline, showing rapid decrease in temperature downward, is approximately the location of the minimum concentration of oxygen and maximum of phosphate (P) and nitrate (N). $\delta^{13}C$ is

$$[({}^{13}C/{}^{12}C)_{sample} - ({}^{13}C/{}^{12}C)_{standard}]/[({}^{13}C/{}^{12}C)_{standard}] \times 1000.$$

The $\delta^{13}C$ is high in surface waters of oxygenated oceans because of preferential incorporation of ^{12}C in organisms. Sea water is saturated in calcite at the surface but becomes undersaturated at greater depths; CCD is the calcite compensation depth, below which calcite cannot be preserved.   □

buried. In northern areas of continents, river and lake patterns are largely inherited from the glaciers and processes during the first stages of deglaciation. Exposure of the continental shelves during glaciation permitted vast amounts of sediment that normally would have remained on the shelves to be remobilized by turbidity currents down to abyssal depths, where they contributed to the formation of present continental rises (Section 4.2). These currents are still active, partly controlled by submarine valleys cut during the interglacial stages. Many of the world's shorelines show either glacial fjords or, outside of the glaciated region, estuaries that attest to the inability of fluvial deposition to keep up with the flooding of valleys incised during times of lower sealevel.

[**References** – General aspects of paleoclimatology are discussed in the book by Crowley and North (1991). Information on long-range Mesozoic/Cenozoic climatic variations is from the book by Kennett (1982), a volume edited by Kennett (1985), and a paper by Holland, Lazar and McCaffrey (1986). Berner (1991) discusses the effects of CO_2 variations during the Phanerozoic. The climatic effects of Croll–Milankovitch cycles are discussed by Hays, Imbrie and

Shackleton (1976) and papers in the volume edited by Berger *et al.* (1984). A review paper by Broecker and Denton (1991) discusses the causes of Pleistocene climatic fluctuations. Characteristics of the earth during the Pleistocene are summarized from the work of CLIMAP Project Members (1976), Imbrie (1985), Denton and Hughes (1986), and Ruddiman *et al.* (1989). The importance of the Antarctic in controlling climatic patterns is discussed in the volume edited by Olausson (1988) and the paper by Kennett (1977). The Vostok ice core is described by Barnola *et al.* (1987) and Genthon *et al.* (1987). Paull, Ussler and Dillon (1991) discuss climatic effects of methane. Porter (1975) discusses the concept of snowlines.]

8.5 Changes in ocean compositions

MODERN OCEANS are an ecologically 'friendly' environment for the organisms that live in them. They provide abundant oxygen and a copious supply of microflora and microfauna as a base of the food chain. Hostility is mostly restricted to the fact that so many animals try to eat each other. Ancient oceans were not always so agreeable, and in this section we explore the changes in oceanic composition through time and the associated (resultant?) changes in the sedimentary and biotic record. We proceed in the following order: 1) a discussion of the compositional parameters that can be used to characterize oceans; 2) an outline of the types of sediment formed in the oceans; and 3) the changes in types of oceans and their sediments through the Mesozoic and Cenozoic.

Types of oceans

The chemical parameters that we use to characterize oceans are intimately related to, and largely controlled by, organic activity. The upper, photic, layers of the ocean extend to depths of not more than ~100 m but are responsible for the photosynthetic activity that provides a food chain for the animal life. Organisms that live in this zone drift downward after death, and the organic carbon consumes oxygen in the middle and lower parts of the ocean. These synthesis and decay reactions, and accompanying chemical changes, interact with oceanic currents to control much of the composition of sea water.

We use five parameters to divide oceans into two categories (Fig. 8.17). The first parameter is the range of temperature. In present oceans, a shallow (~70-m) layer of warm water is separated from the colder deep ocean (below ~1000 m) by a 'thermocline' across which the temperature gradient is steep. The steep gradient is maintained by the combination of warming of the photic zone by sunlight and insertion of cold water from polar regions at depth; this Arctic and Antarctic water constitutes most of the volume of modern ocean basins. At present, the earth's climate is still generally 'glacial', and a steep latitudinal temperature gradient exists in surface water. In non-glacial times, the absence of polar cold water and small latitudinal temperature gradients may yield a more gradual vertical temperature gradient (Fig. 8.17).

A second parameter that characterizes oceans is the vertical gradient in dissolved oxygen (Fig. 8.17). Much of the world's ocean shows an 'oxygen minimum' zone. This minimum is established partly by decay of organic carbon falling from the overlying photic zone and partly by organic consumption of oxygen in the moving belt of North Atlantic deep water (NADW; Section 8.4). A combination of factors makes the oxy-gen-minimum zone particularly evident in the Pacific. Water below the oxygen-depleted zone may contain higher oxygen concentrations because of its origin in oxygenated Antarctic seas (AABW, Section 8.4). In well-ventilated modern oceans, the dissolved oxygen of the deep oceans is not significantly affected by the organic carbon that reaches it. In oceans of restricted circulation, however, oxygen supply at depth may be so small that the oxygen minimum is maintained all the way down to the seafloor. In extreme situations, the entire ocean may be effectively 'dead'.

A third (combined) parameter in Fig. 8.17 is the concentrations of dissolved nitrate and phosphate. These two ions are important because they are the major, required, nutrients for all life. In fact, almost all marine plants and animals have a P:N:C ratio of 1:15:105 (the 'Redfield Ratio'). Most modern sea water also has a N:P ratio of 1:15, but the total carbon is much greater than 105. Thus, nitrate and phosphate are generally the limiting factors in the growth of marine plants and animals. The N:P ratio in organisms now controls the N:P ratio of dissolved solutes, but an interesting (shall we say 'philosophical'?) question is whether the N:P ratio in organisms was established by ocean compositions during the development of primordial life (Section 3.6).

Consumption of oxygen during decay of organic matter necessarily releases N and P to the sea water, and Fig. 8.17 shows the observed inverse relationship between nitrate/phosphate and dissolved oxygen. This relationship probably also is found in oceans of more restricted circulation. In water of very limited (or no) oxygen concentration, the tendency of some organisms to obtain energy by nitrate reduction may cause the N:P ratio to decrease somewhat, particularly in restricted water bodies.

A fourth parameter for characterizing ocean composition is the $^{13}C/^{12}C$ ratio, normally described by the value of $\delta^{13}C$ (Fig. 8.17). Plants (and some bacteria) concentrate ^{12}C during metabolic processes, thus reducing the $\delta^{13}C$ in the organic carbon relative to carbon in the environment. In an ocean in which microorganisms live in the upper layers and drift downward after death, their progressive decay gives bottom water a lower $\delta^{13}C$ than the near-surface water. Thus, when the calcareous tests of benthic organisms (e.g. foraminifera) equilibrate with the water, they have a $\delta^{13}C$ that is two to three parts per thousand lower than the tests of planktic organisms. In a less-productive ocean (warmer, less bottom circulation), this $\delta^{13}C$ difference is not maintained, and the tests of both planktic and benthic forms tend to have similar $\delta^{13}C$ values. This parameter must be used with great care because of the tendency of calcareous material to undergo diagenesis or otherwise reequilibrate with its environment.

The fifth parameter shown in Fig. 8.17 is the degree of saturation in $CaCO_3$. The surface water of almost all oceans is supersaturated in both calcite and aragonite, and any organism capable of synthesizing a calcareous test has an ample supply of raw material. After death, calcareous plankton sink, and the decay of their organic matter places CO_2 into the oceans. This dissolved CO_2 causes the dissolution of both calcite and aragonite, primarily to form HCO_3^-. The solubility of $CaCO_3$ is also enhanced by increasing pressure and decreasing temperature downward in the ocean. In low-O_2 oceans, microorganisms can flourish only at very shallow depths, and their decay causes the calcite to dissolve at shallower levels than in oceans with greater primary productivity.

One measure of the solubility of $CaCO_3$ at depth is the 'calcite compensation depth' (CCD). The CCD is the depth at which dissolution of calcite equals its supply, and below it virtually no calcite is added to seafloor sediment. The factors that control the CCD are not well understood, but they must include the increased solubility of CO_2 as pressure increases with depth. Other measures of carbonate dissolution are the 'aragonite compensation depth', below which aragonite does not survive, and the 'lysocline', a shallower level below which the rate of solution of carbonate in the water column increases sharply. In modern oceans, the CCD varies between depths of about 5000 m (Atlantic) to 3500 m (Pacific), approximately half of the distance between Pacific ridge crests and abyssal depths. Thus, deposition of calcareous sediments on the seafloor is mainly in the higher levels of ridges.

Based on the differences shown in Fig. 8.17, we can classify oceans into two 'endmembers' of biologic and chemical activity. One type is similar to oceans now in existence. It is well ventilated by interchange of water, driven either by temperature or salinity differences, and thus contains oxygen throughout the water column. The currents accompanying interchange cause erosion of the bottom, and the oxidative

sediment–water interface permits bioturbation. A deep CCD causes accumulation of calcareous sediment on much of the elevated parts of spreading ridges or perhaps at greater depths. By contrast, an ocean that is less well ventilated contains lower dissolved oxygen and less organic activity throughout the water column. The seafloor may be a dead place without currents or bottom-dwelling fauna, and the high CCD restricts the precipitation and preservation of calcareous sediments.

What controls the establishment of restricted oceans? One factor is size, with circulation more easily restricted in small basins, such as the Mediterranean or the Black Sea, than in major oceans. Another factor is the establishment of stable vertical density gradients, commonly in one of two ways: 1) excess discharge of low-density fresh water onto the top of the ocean; or 2) discharge of high-density saline water from an evaporative basin into the bottom of an ocean (an 'injection event'). Some investigators have speculated that entire large ocean basins can be made anoxic by a worldwide process (increase in atmospheric CO_2 concentrations?) that causes such extreme organic productivity in surface waters that the downward-moving, decaying, material would convert the ocean into an anoxic, lifeless, water mass.

Oceanic sediments

Oceans with high organic productivity generate two major types of biogenic sediment. One is siliceous, formed mostly by the tests of radiolaria and diatoms, and the other is calcareous, formed mostly by the tests of planktic foraminifera and nannofossils. Productivity is highest in the Antarctic divergence, south of the convergence (Fig. 8.12), and in the equatorial divergence, where trade winds form a zone of upwelling of nutrient-rich deeper water. Siliceous deposits accumulate in both places, but calcareous deposits are concentrated in equatorial regions.

Non-biogenic material in the abyssal areas of the ocean is largely classified as 'red clay'. The clay consists of clay minerals, with iron oxides carried down with the clay (and perhaps oxidized in place). The clay is obviously terrigenous and is distributed in the oceans by some combination of river processes, ocean currents and wind transport. The margins of the abyssal areas (continental rises) have slopes intermedi-

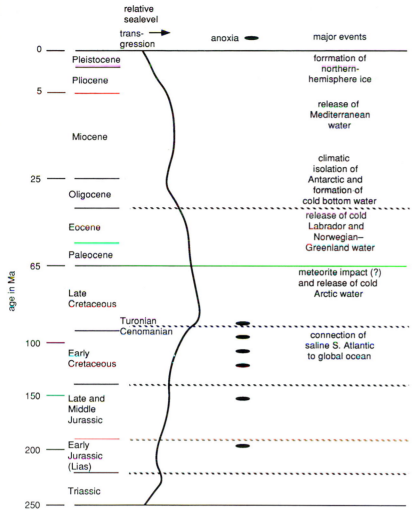

Fig. 8.18. Major events in ocean history. The sealevel curve is from Fig. 8.9. Times of biotic extinction are shown as dashed lines plus the solid line for the Cretaceous/Tertiary event (from Fig. 8.20A). Episodes of deposition of deep-ocean black shales are mostly from Fischer and Arthur (1977), Thierstein and Berger (1978) and Stein *et al.* (1986). (NOTE – the time scale is not linear; see caption for Fig. 8.20.)

ate between those of the seafloor and those of the continental slope and apparently consist largely of turbidity-current debris swept off of the continental shelves.

Two types of sediment represent conditions of 'abnormal' productivity. One is organic-rich (euxinic) muds that become black shales when lithified. Their formation requires deposition of organic carbon in excess of its destruction, either by increased productivity that overwhelms the decay process or by enhancement of the oxygen-minimum zone to prevent decay. Enlargement of the oxygen-minimum zone could be effected in deep oceans by a variety of methods of density stratification (see above), but in shallow (epeiric) seas the only feasible method is formation of a fresh-water cap.

A second, but closely related, 'unusual' sediment is phosphorite, which consists largely of the mineral apatite. Numerous proposals have been made for the mode of formation of phosphorites, and we can only mention a few salient observations. Phosphorites must represent a time and place of high supply of nutrients, and most investigators assume that this supply is provided by upwelling of deep water at topographically favorable areas along continental margins. The phosphate in these waters is then precipitated biogenically, but some of the apatite may have formed by diagenesis. Almost all phosphorites were deposited in shallow water. Some phosphates are closely associated with black shales, perhaps indicating organic-carbon preservation because of high productivity.

History of oceanic changes

Based on the preceding discussions, we are ready to outline the history of compositional changes in sea

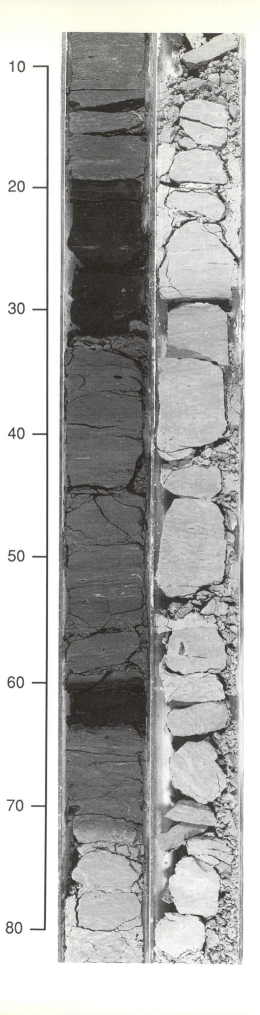

water during the Mesozoic and Cenozoic. The major events are shown in Fig. 8.18, which uses the same chronology that we use in Section 8.6 to show the patterns of organic evolution.

Fig. 8.18 shows four types of information. One is a very generalized curve of variation in sealevel. This curve is from Fig. 8.9 and shows very high levels during the Late-Cretaceous transgression, smaller highstands during the Early Jurassic and Miocene, and regression during times of initiation of continental glaciation (Antarctic in the Oligocene and Arctic in the Pliocene). This generalized curve of sealevel changes does not correspond to times of biotic crises or other oceanic events.

A second type of information shown in Fig. 8.18 is the times of major extinction events on the earth. These events are recorded mostly in the oceans, but several have also been proposed for terrestrial fauna and flora (Section 8.6). The typical pattern of most periods of major biologic change is for organisms to show great diversity for perhaps 5 to 10 million years, following which many of the groups (species, genera, families) become extinct. (We discuss in Section 8.7 the possibility that the massive K/T extinction event was much more abrupt.) The new fauna and flora that follow the extinction event may contain some varieties that 'bloom' very quickly to large numbers of individuals ('opportunists'), and then normal patterns of diversity are reestablished.

The third type of information shown in Fig. 8.18 is the times of development of oceanic black shales, presumably indicative of the times of oceanic anoxia (pre-Late-Jurassic black shales are preserved only in sections structurally emplaced on land). The anoxic oceans represent times of low O_2 content, current-free bottoms, shallow CCD, and accumulation of organic-rich muds (Fig. 8.19).

The fourth set of information shown in Fig. 8.18 is a list of special events in the history of ocean basins.

Many of the oceanic events shown in Fig. 8.18 were probably the consequence of plate reorganizations and their effects on water movements. The extinction/diversification between the Early and Late Cretaceous (Section 8.6) occurred shortly after the

Fig. 8.19. Oceanic black shale at the Cenomanian/Turonian boundary, Indian Ocean. DSDP Site 763. (Courtesy of T. Bralower.) □

centimeters depth

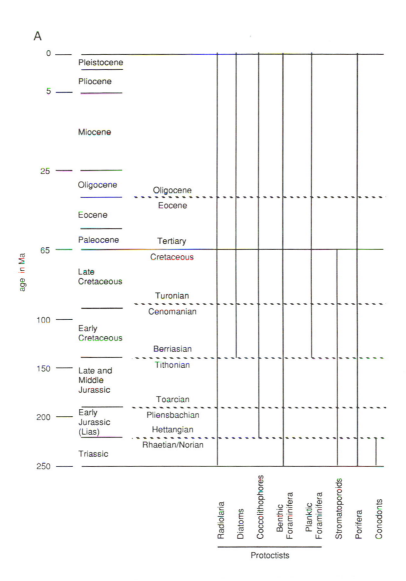

For legend see page 241.

South Atlantic Ocean opened sufficiently that it was no longer a set of isolated, possibly saline, basins. Release of saline water into the global oceans could have caused a biotic crisis as eustatic sealevel was rising to its Late-Cretaceous maximum (see discussion of injection events above). Similar saline injections may have accompanied the repeated development of hypersaline water in the Mediterranean in the Late Miocene (Section 9.2) provided that the sill at the Straits of Gibraltar could have been lowered sufficiently that some of that water was released periodically into the Atlantic. Injections of cold, dense, water may have occurred by: 1) reorganization in the Arctic shortly before the end of the Cretaceous; 2) opening of the Labrador and Norwegian–Greenland Seas shortly before the Eocene–Oligocene transition; 3) isolation of the Antarctic in the Late Eocene or Oligocene; and 4) formation of northern-hemisphere ice sheets after closure of the Central American Isthmus in the Pliocene (Section 8.4).

[**References** – General information for this section is from books by Broecker and Peng (1982) and Kennett (1982), a volume edited by Sundquist and Broecker (1985), and papers by Fischer and Arthur (1977) and McElroy (1983). Oceanic black shales and anoxic events are discussed in the volume edited by Brooks and Fleet (1987) and papers by Stein, Rullkotter and Welte (1986), Bralower (1988), and Pedersen and Calvert (1990). The Black Sea as an anoxic basin is dis-

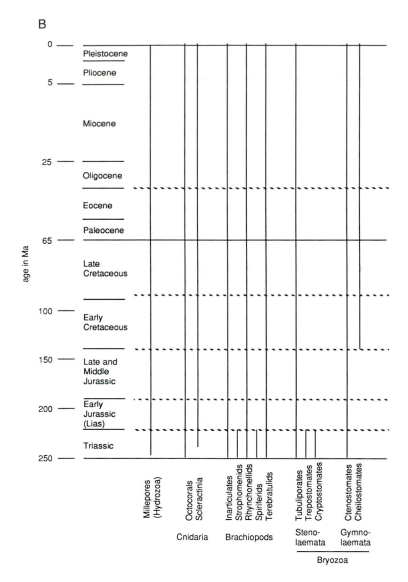

B

cussed by Glenn and Arthur (1985). Information on injection events is from Thierstein and Berger (1978) and Brass, Southam and Peterson (1982). Phosphate deposition is discussed by Piper and Codispoti (1975) and Riggs (1984). Sheridan and Grow (1987) provide an edited volume on the North Atlantic continental margin, and Young and Taylor (1989) provide an edited volume on Phanerozoic ironstones.]

8.6 Evolution

A LARGE NUMBER of animals and plants died over a short time interval – how short is not clear – at the end of the Paleozoic, bringing an end to most of the Paleozoic species and genera. The effect on high taxo-nomic levels, however, was very small. Among invertebrates, the only major extinctions in the phyla to class range (depending on the classification used) are trilobites, some varieties of corals and echinoderms, and fusulinids (Section 6.9). Most of the major invertebrate lineages passed through the crisis and began diversifying rapidly in the early Mesozoic. They established the principal pattern of invertebrate life in the Mesozoic and Cenozoic, with only a few higher taxa (particularly calcareous plankton) evolving during the Mesozoic and none in the Cenozoic (see below). Similarly, most plant higher taxa had been established in the Paleozoic, and the major younger event was the arrival of angiosperms in the Cretaceous (Section 6.5). Much of the history of evolution of vertebrates, how-

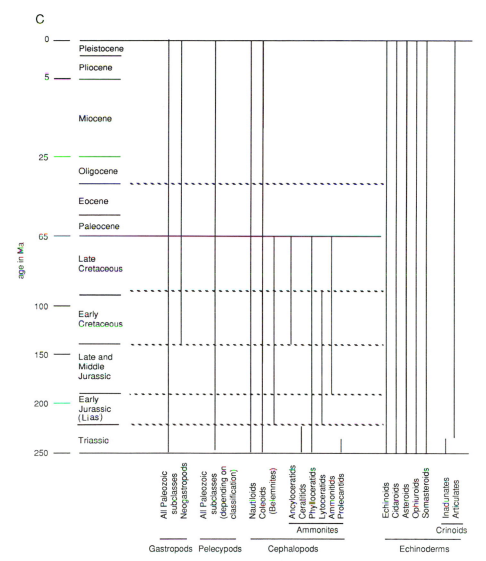

Fig. 8.20(A–C).   Ranges of major invertebrate groups in the Mesozoic and Cenozoic. Principal proposed 'extinction events' are named on Fig. 8.20A and shown on all other parts of Figures 8.20 and 8.23. (NOTE – the time scale is linear in the Mesozoic but expanded in the Cenozoic.) (A) Protoctists, stromatoporoids, porifera and conodonts; (B) hydrozoans, cnidarians, brachiopods and bryozoans; (C) molluscs and echinoderms. □

ever, is during the Mesozoic and Cenozoic, and our discussion is largely concentrated on them.

Despite the absence of new higher taxa, the Mesozoic and Cenozoic represent a time of great diversification of life, with still-uncounted new genera and species. This diversification, however, was punctuated by at least seven episodes of extinction (the last one in the Pleistocene). These episodes are shown in Fig. 8.20A and, without names, in all other diagrams used to discuss Mesozoic/Cenozoic events.

Invertebrates

The typical pattern of invertebrate evolution during the Mesozoic and Cenozoic was diversification along evolutionary lineages into new species and genera until an 'extinction event' decimated the groups and the process started again. The morphological changes permit the establishment of a paleontological time scale, with the organisms most frequently used for chronology including foraminifera, nannofossils,

A

B

Fig. 8.21. (A) Triassic coccolith *Crucirhabdus primulus*; DSDP site 761, Indian Ocean; specimens have average diameters of a few microns (courtesy of T. Bralower). (B) Pliocene globigerinids, with diameters of 0.5 mm to >1 mm; the upper form is *Globigerinoides fistulosis*; the lower photos are an edge view (left) and an umbilical view (right) of *Globorotalia tumida* (courtesy of M. Leckie).   □

ammonites, and pelecypods (Fig. 8.20A–C). The precision in fossil zonation varies with the type of organism and its age, with uncertainties of 1 to 2 m.y. through much of the Cenozoic and less than 5 m.y. in at least the later part of the Mesozoic.

No record of calcareous microplankton exists before the middle Triassic, although the absence of older crust on the modern seafloor and the fragility of the tests (skeletal parts) has led some investigators to suggest that the organisms existed earlier and were not preserved. This occupation of the upper levels of the oceans by generally autotrophic plankton that secreted calcium carbonate had a dramatic effect on oceanic evolution. For the first time in history, a rain of fine-grained calcite and aragonite settled toward the seafloor (Fig. 8.21A and B). The deeper parts of the oceans presumably have always (almost always?) been below the carbonate compensation depth, but after the development of calcareous plankton, vast areas of elevated oceanic regions plus quiet waters of continental shelves accumulated thick sequences of chalks, marls, and related rock types. These organisms clearly helped to control the calcium balance in Mesozoic and younger oceans, and the mechanism for maintaining a Ca-balanced ocean prior to their arrival is not understood.

Plants

The evolution of plants is discussed in Section 6.5, and here we will make only a few comments on the major aspects of Mesozoic/Cenozoic evolution. The history of these eras is essentially one of progressive replacement of less-competitive varieties with more advanced ones. At the start of the Mesozoic, lycopods were still present, pteridosperms were abundant, and gymnosperms were developing. The advanced properties of gymnosperms enabled them to become dominant throughout most of the later Mesozoic. Angiosperms first appeared in the Early Cretaceous, and their ability to nurture seeds within the plant gave them a strong competitive edge. They had become important members of the plant community by the end of the Mesozoic, and the Cretaceous/Tertiary catastrophe cleared the land for them to become dominant in the early part of the Paleogene. The progression from lycopods and pteridosperms in the Paleozoic through gymnosperms in the Mesozoic and angiosperms in the

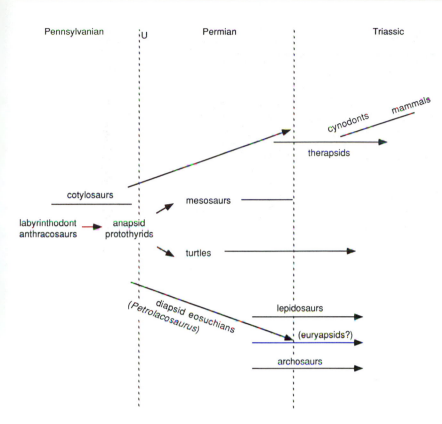

Fig. 8.22. Diversification of vertebrates in the late Paleozoic and early Mesozoic. The terminology is discussed in the text.   □

Cenozoic was partly an evolutionary improvement in reproductive mechanisms. It was also an increase in efficient use of energy, with the more advanced plants secreting stiffening and protective agents that required less energy for synthesis than the primitive plants.

Vertebrates (General evolution)

As with invertebrates and plants, the fundamental characteristics of vertebrate differentiation were established in the Paleozoic (Section 6.4). They include the development of a spinal cord and bony skeletons and the ability to survive on land. Adaptation to land was first made by amphibia, which required water to breed and to keep their skins wet. In the later part of the Carboniferous (Pennsylvanian), true land animals evolved by development of a scaly skin and, perhaps most important, the use of an egg that did not have to be fertilized and hatched in water. This transition apparently occurred within a taxonomic group known as 'cotylosaurs', presumably a misnomer because it refers to both amphibians and reptiles, and led to the development of protothyrid lizards (Fig. 8.22).

In Section 6.4 we discussed the classification of rep-tiles based on the types of temporal openings in the area of their cheekbones. Anapsid reptiles had no temporal openings; euryapsid and synapsid reptiles had one opening, either above or below the cheekbones, respectively; and diapsid reptiles had two openings, both above and below the cheekbones. The earliest lizards (protothyrid) were anapsid, but differentiation into the major synapsid and diapsid vertebrate lineages occurred almost immediately (Fig. 8.22). Anapsid forms continued as two principal lines, both of which returned largely to marine habitats. One was the mesosaurs, shark-like carnivores that have the dubious honor of being the only major vertebrate line that became extinct at the end of the Paleozoic. The other possible anapsid lineage is the turtles.

The earliest significant synapsid reptile was the pelycosaur (sail lizard). The large, spiny, sail on their backs was obviously useful for something (temperature regulation?), for pelycosaurs constitute a high proportion of Permian land animals. Some group of pelycosaurs was apparently ancestral to a group of vertebrates with a more specialized skeleton, known as therapsids (Section 6.4). Therapsids were successful from the later part of the Permian into the Triassic, at

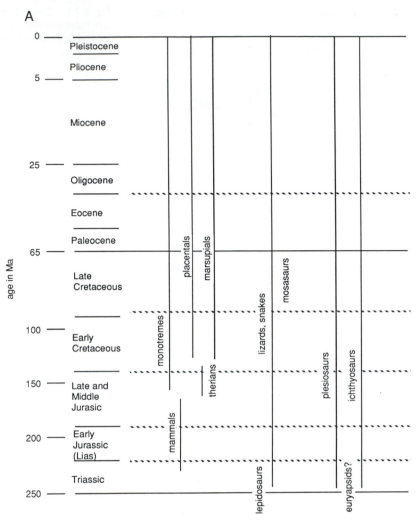

For legend see page 246.

which time the next major change in reproductive processes occurred – the development of mammals.

The obvious difference between mammals and reptiles is that mammals are furry and cuddly, at least when small, and reptiles clearly are not. The more important difference, however, is that mammals nurse their young, whereas reptiles simply lay eggs that hatch the young with nearly the full complement of adult capabilities. Mammals accomplish this nursing in three different ways. Monotremes are oviparous (egg-laying), but the helpless newborn immediately attach themselves to the mother. Marsupials are viviparous (hatching within the body), but the newborn must immediately move to a mother's pouch or other place where they can be nourished and protected during the infant stage. Placentals are viviparous and the most advanced, with newborn being physically unattached but still requiring a long period of nurture in infancy.

In addition to reproductive differences, mammals have an advanced set of physiological characteristics that distinguish them from reptiles (and that are reflected in their fossilized skeletons). In particular, a change of metabolic pathways from reptile to mammal is probably as important as the change in reproductive processes. Mammals regulate their own temperatures, except when young and too small for their metabolic processes to overcome heat loss, and their oxidative metabolism permits them to sustain physical activity for long periods of time, whereas reptiles can move quickly but then must rest for a while. An additional difference is that the brain cases of the earliest mam-

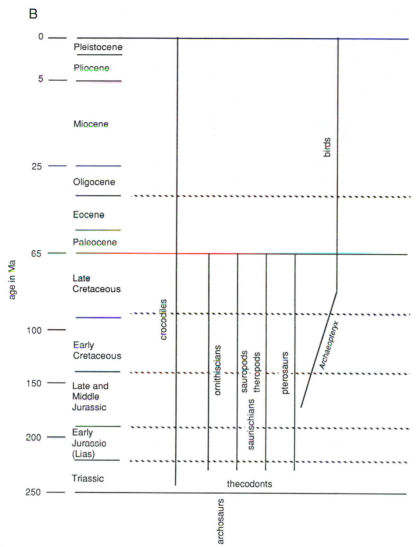

For legend see page 246.

mals were several times larger than the brain cases of their immediate reptilian ancestors, and the presumably advanced mental ability has characterized mammals throughout their history.

The earliest mammals were probably monotremes and evolved via a group of therapsid reptiles known as cynodonts (Fig. 8.22). Differentiation into monotreme and therian (marsupial and placental) lineages occurred early in mammal evolution, certainly by the middle of the Jurassic (Fig. 8.23A). Placentals are now dominant, with marsupials concentrated largely in Australia and South America and monotremes only in Australia.

Vertebrates (Dinosaurs and their relatives)

For some unknown reason, diapsid reptiles never evolved into mammal-like animals. For the diapsids, the days of glory were the Mesozoic, and they are represented in modern biota mostly as crocodiles, snakes, lizards and birds. The earliest true diapsid was probably the Pennsylvanian fossil *Petrolacosaurus*. In the Permian, these primitive forms radiated into two major lineages, lepidosaurs and archosaurs (Fig. 8.23A; see below), and possibly to a related group of marine reptiles. Lepidosaurs are reptiles that never brought their legs under their bodies, either retaining

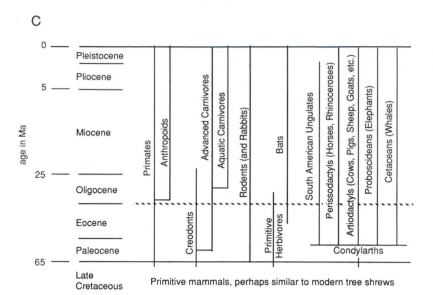

Fig. 8.23. Ranges of major vertebrate groups in the Mesozoic and Cenozoic. This figure continues the evolutionary patterns shown in Fig. 8.22. (A) general groups, (B) archosaurs (dinosaurs), (C) placentals. ☐

the splayed limbs of lizards or abandoning limbs to become modern snakes or snake-like marine reptiles (mosasaurs) in the Cretaceous (Fig. 8.23A).

In contrast to the lepidosaurs, in the Permian the archosaurs began to swing their limbs more directly beneath their bodies. This arrangement enabled them to move faster and, in some groups, to become bipedal. Archosaurs quickly diversified into crocodiles, which have persisted to the present, and a group of primitive 'dinosaurs' classified as thecodonts (named for their characteristic socket teeth; Fig. 8.23B). The thecodonts were important parts of the fauna in the Triassic, but by the end of the Triassic they evolved into the major reptile groups of the Jurassic and Cretaceous, including the dinosaurs (Fig. 8.23B).

The fundamental distinction between dinosaur groups is into ornithischians and saurischians, with the distinction based on the construction of the pelvis. The ornithischians were all herbivores and included both bipedal and quadripedal forms. For protection, some varieties developed a tough armor, as exemplified by the well-known *Stegosaurus* and *Triceratops*. The saurischians include carnivorous bipedal forms (theropods; e.g., *Tyrannosaurus*) and herbivorous bipedal and quadripedal forms (sauropods). Some of the quadripedal saurischians developed into the largest land animals in history, including *Diplodocus* (up to 30 m long) and *Brachiosaurus* (up to 80 000 kg).

In addition to the saurischians and ornithischians, the Triassic thecodonts apparently spawned another

lineage, the pterosaurs (Fig. 8.23B). With the exception of insects, pterosaurs were the first animals to inhabit the air. They did so by essentially stretching membranes out from the body along their front limbs, which gradually evolved into more functional wings. In the Late Jurassic, one branch of pterosaurs (*Archaeopteryx*) developed into true birds by the replacement of reptilian scales with feathers. The feathers gave these early birds, which still retained teeth and other reptilian features, a great advantage in flight, and they were able to survive the Cretaceous/Tertiary extinction event.

The physiology of dinosaurs has attracted both intense investigation and vigorous controversy. Dinosaurs are obviously reptiles, and modern reptiles do not metabolize oxygen rapidly enough to sustain body temperatures or engage in continued muscular activity. In modern reptiles, body temperature is maintained by habits, such as moving between sunlight and shade. Possibly dinosaurs were different. The large size of some of them would have improved their ability to maintain constant body temperature simply because of the difficulty of diffusing heat out of their interiors. Furthermore, the position of limbs underneath the dinosaur body implies rapid and continued motion.

Dinosaurs occupied the land surface between two major Mesozoic extinction events (Fig. 8.23B). At the end-Triassic event (Rhaetian/Norian) the primitive thecodonts were replaced by the more developed dinosaur and pterosaur lineages. Different dinosaur groups began dying out in the Late Cretaceous, the ichthyosaurs possibly as early as the Cenomanian/Turonian extinction event. Although varieties of both major dinosaur groups, plus the pterosaurs and plesiosaurs, were present just before the Cretaceous/Tertiary catastrophe, many families had already become extinct.

Vertebrates (Mammals)

Most of the basic mammalian lineages were in existence before the Cretaceous/Tertiary boundary event (Fig. 8.23C). These Cretaceous forms included primitive varieties of primates, creodonts (which evolved into more advanced carnivores), various unsuccessful groups of primitive herbivores, and condylarths. Recognition of eating habits, and hence classification into major groups, is made primarily by the shape of the teeth. Carnivores require sharp teeth for grasping and cutting their dinners, and herbivores require broad teeth that can be used to grind plants into a digestible form. The condylarths are Late-Cretaceous to Early-Tertiary forms that were ancestral to the major herbivore groups and differentiated at some time in the Paleocene or Eocene into five groups, four of which are now in existence (Fig. 8.23C). These groups include: Cetaceans (whales), which migrated back into the oceans early in the Cenozoic; Proboscideans (elephants); and 3) hoofed animals (ungulates).

The major ungulate groups are classified by their toes. Perissodactyls are 'odd-toed', generally by extreme development of the third toe of the basic mammalian foot into a single hoof. They were dominant forms during the early part of the Tertiary, but many small lineages died out, and the principal living forms are horses and rhinoceroses (rhynoceri?). Artiodactyls are 'even-toed', with the third and fourth toes developing equally and now forming a single 'cloven' hoof in such modern forms as pigs, cows, goats. The artiodactyls are the most numerous of the modern ungulates.

All mammalian evolution has been affected by the isolation of continents from each other during much of the Tertiary. For example, numerous mammalian groups in South America evolved separately from those in the rest of the world until some time in the Pliocene/Pleistocene, when South America was joined to North America by construction of the Central American isthmus (Section 9.5). After connection of the two continents, many of the primitive South American forms were replaced by more-competitive forms from North America.

Summary of biotic episodes

We can divide the Mesozoic and Cenozoic into time intervals separated by periods of apparent rapid transition in fauna and flora. These intervals are shown in Fig. 8.20A and are shown without names in all parts of Figs 8.20 and 8.23. The most important change, at the Cretaceous/Tertiary boundary, is discussed in Section 8.7, and here we summarize the general nature of the other biotic boundaries in the past 250 million years. The problems of recognizing extinction events are discussed in Section 6.6.

Many of the paleontologic transitions appear to occur at times of oceanic anoxia, or at least diminished oxygenation of most of the water column. The causes of this process are discussed in Section 8.5. These anoxic events may cause widespread extinction and reduction in biotic diversity. Following the reestablishment of 'normal' oxygenated conditions, evolutionary diversification occurs rapidly, commonly generating forms that are morphologically simpler than the ones that existed previously. Some of the oceanic anoxic events appear to correlate with biotic transitions on land, but others do not. Not all of the 'extinction events' shown in Figs 8.20 and 8.23 can be related to oceanic anoxia. Other possible causes are injection of cold water into the ocean system, habitat reduction by marine regression, meteorite impact, etc. Regardless of the cause, here are the proposed events and their intervening intervals.

The earliest interval was the Triassic, bounded below by the end-Paleozoic catastrophe and above by the Rhaetian/Norian event very near the end of the Triassic. The Triassic was a period of transition. Plants were gradually changing from forests dominated by pteridosperms and lycopods to ones dominated by gymnosperms, but neither the beginning nor the end of the Triassic seem to have been times of abrupt floral change. Also on land, the vertebrate population was dominated by relatively primitive reptiles (thecodonts) that appear to have died out abruptly at the end of the Triassic. Marine vertebrates were a mixture of reptiles and primitive actinopterygian fish (chondrosts; Section 6.4). Triassic seas gradually developed Mesozoic varieties of invertebrates, such as scleractinian corals, but contained some remnants (in greatly diminished diversity) of Paleozoic forms that disappeared by the end of the Triassic. Comparison of extinction rates among different families of terrestrial and marine organisms indicates that the Rhaetian/Norian 'event' may have been spread out over ten or more million years. By the end of the Triassic, however, nearly one-quarter of all Triassic faunal families on land and in the sea had disappeared. The first calcareous plankton (coccolithophores) appeared near the end of the period,

The Jurassic began with a rapid expansion of new terrestrial and aquatic organisms. The ammonites, which had nearly become extinct at the end of the Triassic, formed an extraordinary abundance of new orders (superfamilies?), and rapid diversification occurred among other molluscs. Fish diversified, and modern teleosts (bony fish) began to appear. On land, forests were dominantly gymnosperms, and the primitive reptilian stock of the Triassic was rapidly superseded by true dinosaurs.

A marine anoxic event may have occurred near the end of the Early Jurassic (Lias), between the Pliensbachian and Toarcian stages (Fig. 8.20A). Planktonic marine organisms, however, did not undergo much change, and although land plants and animals show some diversity decrease at this time, the records are probably not adequate for the recognition of a true 'extinction event'.

During the Late Jurassic (Tithonian and perhaps somewhat earlier), another biotic transition has been correlated with oceanic anoxia. Faunal and floral changes, however, appear to have been mostly at species and generic level, and groups such as brachiopods, foraminifera and coccolithophores show little effect. The end of the Jurassic occurred during a time of rising sealevels and climatic warming that heralded the astonishing diversity of the Cretaceous.

Rapid seafloor spreading and a distribution of continents that created a virtually ice-free world caused the Cretaceous to be characterized by global high sealevels, equable climates, and a diverse and rapidly evolving fauna and flora. Reptiles dominated the land, with some of them evolving huge sizes and bizarre forms. Mammals had differentiated into their three groups (monotreme, marsupial and placental), but they remained a minor part of the fauna. Angiosperms appeared in the Early Cretaceous, and by the middle of the period they had begun to displace gymnosperms and more primitive plants, some of which became extinct.

Cretaceous seas were the home of an abundant population. Ammonites developed sutures of bewildering intricacy. Pelecypods assumed a variety of forms, and some with the approximate shape of corals (rudists) contributed to the formation of large reefs. Coccolithophores were joined by planktic foraminifera (generally referred to as 'globigerinids'; Fig. 8.21B) in the production of copious amounts of fine calcareous sediment. Ichthyosaurs, plesiosaurs, and less advanced reptiles chased teleost fish through broad expanses of epeiric seas, and pterosaurs chased their prey from the sky.

This richness of Cretaceous life was interrupted by at least one, and possibly two, anoxic events recorded both in sections now on land and those in the deep sea. The major one is at the boundary between the Cenomanian and Turonian (Figs 8.19 and 8.20A), and an earlier one is likely at the Aptian/Albian boundary. This event does not represent a qualitative change in the major types of organisms on the earth but simply a quantitative change measured by biotic diversity and extinction and evolution at the species and genera level.

The fine climate and many of the presumably happy organisms of the Cretaceous were terminated abruptly by some event, although major biotic changes had begun several millions of years earlier (see above). This event put an end to the dinosaurs, swimming and flying reptiles, and ammonites. Because of its significance, it is discussed separately in Section 8.7.

Various extinction events have been proposed for the Cenozoic, but except for the Pleistocene, the evidence is not clear. No extinctions or appearances have been documented above the species or genera level, but a major reduction in diversity and rapid evolution at lower levels of taxa are proposed for several times in the Paleogene. Most attention is centered on the transition from Paleocene to Eocene, but several other 'events' can be proposed throughout ten or more million years of the Eocene through Oligocene. A 'Terminal Eocene Event' appears to have affected benthic organisms, both in the deep and shallow seas, but

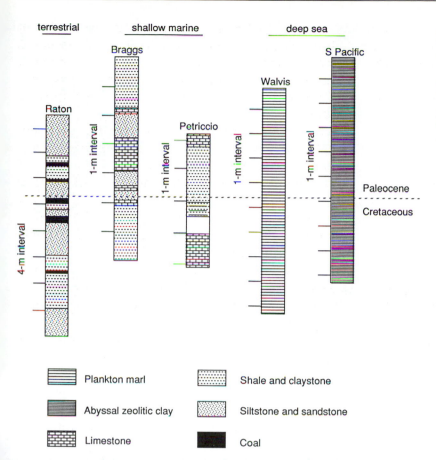

Fig. 8.24. Occurrence of Cretaceous/Tertiary boundary in various stratigraphic sections. Raton, New Mexico from Orth *et al.* (1981); Braggs, Alabama, from Zachos, Arthur and Dean (1989); Petriccio, Italy, from Montanari *et al.* (1983); Walvis Ridge, Atlantic Ocean, from Moore *et al.* (1984), and S. Pacific site from Menard *et al.* (1987). □

not to have had major effect on planktic organisms or terrestrial animals or plants. No anoxic sediments have been found associated with this Eocene/Oligocene transition. Because the Oligocene represents the start of glaciation in the Antarctic, the biotic changes associated with the start of the Oligocene may simply represent increased flux of cold bottom water and general climatic cooling.

The major Cenozoic extinction event may be the Pleistocene, possibly even the present. The extreme development of glacial climates beginning in the Pliocene (Section 8.4) brought general cooling of both the oceans and atmosphere. Extinction affected the mammal population, particularly in the northern hemisphere, where more than half of the mammalian megafauna (>44 kg) died. Plants and oceanic organisms seem not to have suffered extinction but merely to have moved to more equatorial regions. Because no oceanic anoxia developed during the Pleistocene, some investigators have proposed that the effects merely of cold climate and cold ocean water in the

Pleistocene could be used as models for other extinction events.

[**References** – General references include: a book by Brasier (1980) on microfossils; books on invertebrate paleontology by Clarkson (1986) and Boardman, Cheetham and Rowell (1987); a book on vertebrate paleontology by Carroll (1988); and a book on general paleontology by Stearn and Carroll (1989). Extinction events are reviewed in a book by Donovan (1989) and volumes edited by Pomerol and Premoli-Silva (1986), Walliser (1986), Larwood (1988) and Kauffman and Walliser (1990). Papers on extinction events include Hallam (1981, 1986), Beaton (1986), Jarvis *et al.* (1988), Jablonski and Bottjer (1991) and Bice *et al.* (1992). Specific reviews include: the Jurassic, by Hallam (1975); birds, by Martin (1983); molluscs, by Elder (1989); plants, by Robinson (1990); and dinosaurs by Farlow (1989).]

8.7 The Cretaceous/Tertiary boundary

A LAYER OF CLAY, commonly several centimeters thick, separates rocks with undoubted Cretaceous fossils from rocks of undoubted Tertiary fossils at numerous places around the world (Figs 8.24

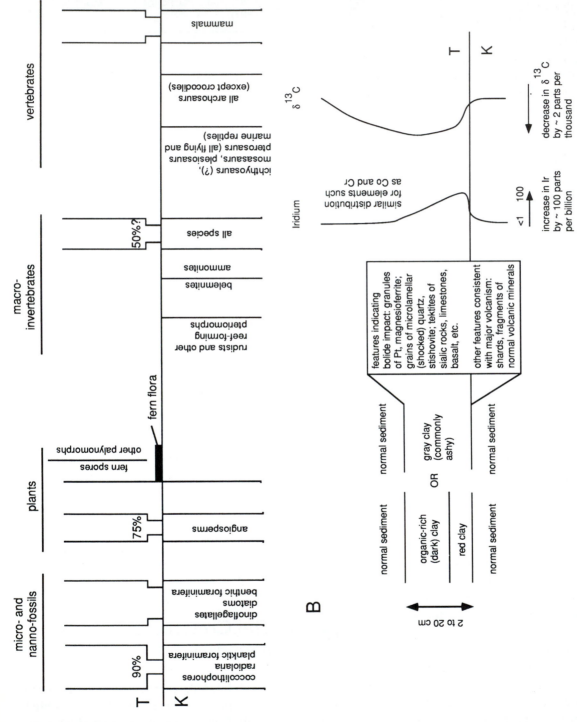

Fig. 8.25. Summary of sedimentary, geochemical and biotic changes across Cretaceous/Tertiary boundary.

(A) Paleontologic information. Five groups of organisms are shown within frames whose widths diagrammatically indicate changes in the degree of biotic diversity across the K/T boundary: coccolithophores, radiolarians and planktic foraminifera show an approximate 90% extinction rate at the boundary; dinoflagellates, diatoms and benthic foraminifera show very little change; angiosperms show an approximate 75% extinction rate; all species of macro-invertebrates show an approximate 50% extinction rate, although the averaging to obtain that figure is very problematic; and mammals show minor extinction. Organisms showing complete extinction before the Tertiary include rudists and related forms, belemnites, ammonites, all flying and marine reptiles and all archosaurs (including dinosaurs) except crocodiles.

(B) Lithologic and compositional information. The boundary clay is commonly from 2 cm to 20 cm thick, with sediment normal for its environment of formation on both sides. Lithologic features have been proposed to support either an impact or a volcanic cause for deposition of the clay. Typical variations in the concentrations of Ir and related elements and in δ^{13}C also show some sudden event at the K/T boundary, and 8.25A and B). At most outcrops it is gray and ashy, but reddish and greenish colors and some dark, organic-rich, horizons are also present. It occurs in subaerial and shallow-water sediments and in some deeper, basinal, accumulations now exposed on land. In cores from the deep oceans, however, the components of the clay appear to be diluted into the normal sedimentary column, making the transition from Mesozoic to Cenozoic less noticeable.

Near the town of Gubbio, in the Italian Appenines, the 'boundary clay' is only a few centimeters thick (see Frontispiece to this chapter), but investigations here have spawned a set of publications that would cover tens of meters (possibly more) of shelf space. The clay at Gubbio, sandwiched between relatively deep-water limestones, was the first to yield high levels of iridium, lending support to the proposal by L. Alvarez *et al.* (1980) that the exceptional paleontologic changes at the Cretaceous/Tertiary (K/T) boundary were caused by impact of a large bolide with the earth (the term 'bolide' refers to a comet or an asteroid that exploded on impact). The general effects of impacts and their frequency are discussed in Section 2.1, and here we will concentrate on the evidence and arguments concerning the rapid transition from Cretaceous to Paleocene.

The first issue is whether the change from Cretaceous to Tertiary fauna and flora was an instantaneous one, requiring a dramatic cause, or a more gradual one that could be explained by conventional processes leading to extinction and evolution. The total effect of the transition is not in dispute; indeed, the exceptional biotic change led early geologists to an easy definition of the Mesozoic and Cenozoic eras. The major aspects of the change are shown in Fig. 8.25A and B. All dinosaurs were extinct before the start of the Tertiary. This extermination ended the diapsid lineage, except for crocodiles (in some classifications) and birds (Section 8.6). All swimming and flying reptiles (plesiosaurs, pterosaurs, etc.) were also extinct before the Tertiary. Among the invertebrates, all ammonites and belemnites disappeared. The flourishing Cretaceous reef communities characterized by rudists and related pteriomorph pelecypods were destroyed. Conservative estimates indicate that more than 50% of all invertebrate species became extinct at or near the end of the Mesozoic. The most extreme carnage was among certain microorganisms, in which coccolithophores, radiolarians, and planktic (calcareous)

foraminifera generally suffered greater than 80% (some more than 90%) extinction at the species level. Among plants, approximately three-quarters of all angiosperm species became extinct. The extinction of angiosperms is responsible for the very high ratio of fern spores to other palynomorphs in earliest Tertiary sediments.

Despite the major casualties at the end of the Cretaceous, extinction was not universal among all types of organisms. Benthic foraminifera were only moderately affected, and the planktic diatoms and dinoflagellates did not undergo the extreme changes shown by other planktic groups. Although the ammonites disappeared, the closely related nautiloids survived as the modern *Nautilus*. Some extinction occurred among primitive mammal groups, but many varieties survived relatively unharmed. Similarly, fish populations underwent only moderate change.

The patterns of evolution and extinction described above do not yield an absolutely defined environmental cause. In general, however, a large reduction in light intensity (perhaps near darkness) seems mandated by the changes in angiosperms and calcareous plankton. Darkness would not have affected diatoms and dinoflagellates because of their ability to form 'resting cysts' that helped them to disappear from surface waters into a vegetative state until environmental conditions permitted regeneration. The extinction of reef communities probably was also related to reduction in light. Similarly, the inability of most reptiles to survive may have been partly caused by their inability to regulate their temperatures at a time when low radiation intensity permitted rapid cooling of the earth's surface.

Major reduction in light intensity at the K/T boundary is supported by carbon-isotope studies. Variation in $\delta^{13}C$ is characteristic of many of the earliest Tertiary sediments immediately overlying the boundary clay (Fig. 8.25B). In marine sediments, the $\delta^{13}C$ value is commonly 2 to 3 parts per thousand lower than in normal parts of the column. This decrease can be attributed to a sharp decrease in primary (photosynthetic) productivity in surface waters, which prevented uptake of the light ^{12}C by organisms and, thus, increased the concentration of ^{12}C throughout the water column and in bottom sediments.

Regardless of the direct cause of the extinctions, one of the principal issues in the investigation of the K/T boundary is the speed with which the biotic changes occurred. Was it instantaneous? That is, one day the dinosaurs ruled the world and a few weeks later they were all dead? Was it a gradual or 'stepwise' extinction spread out over hundreds of thousands or millions of years? Both dinosaur and ammonite diversity have been proposed to have declined greatly within the Late Cretaceous, before its termination. Was a sudden event imposed on an environment that was already undergoing major biotic change? Geologists cannot agree on the answers.

What does the term 'rapid' mean, both with regard to the rate of extermination and the duration of the hostile environment? Almost total darkness lasting for even a few weeks would probably have destroyed much of the calcareous plankton. Thus, extinction may have occurred within a few years, or possibly even less. The time required for the environment and its biota to recover differs with the type of organism. Among microorganisms, Tertiary plankton communities evolving from a few survivors did not become as rich and diverse as the Cretaceous community until the Late Paleocene or possibly Eocene. Conversely, new varieties of angiosperms apparently adapted to the land much more rapidly. The speed of recovery of both terrestrial and marine environments is another line of evidence favoring a sudden cause of the extinctions rather than a gradual environmental change.

If we conclude that the extinction was a rapid process, then we must conclude that it was caused by a sudden event. Since the publication of the data from Gubbio (see above), a leading candidate for that event has been the catastrophic impact of an asteroid or comet. The environmental catastrophe at the end of the Cretaceous would have required collision with a body of at least 20 km diameter, resulting in an impact crater more than 100 km in diameter if it had hit the land surface. A crater of this size could have provided all of the necessary ash for a worldwide layer a few centimeters thick. Impact in the oceans would have formed a smaller crater but should have generated sufficient disturbance of the water mass to cause total reworking of the sediment column by seafloor erosion; evidence of tsunamis has been proposed in some Cretaceous/Tertiary boundary sections. The impact may have occurred in areas that have been subducted during the Cenozoic, in which case no direct evidence of the impact site would remain. Possibly the incoming body broke up and produced multiple synchronous collisions.

An impressive array of properties of the boundary clay have been used to demonstrate that impact was a cause of the extinction event (summarized in Fig. 8.25A and B). The Ir content, which is more than 100 parts per billion in some samples, seems diagnostic of metallic meteorites, and similar 'meteoritic' values are found for related elements such as cobalt. Some boundary clays contain other components of metallic meteorites such as platinum and magnesioferrite. Other materials in the clay include shocked (microlamellar) quartz and the high-pressure silica mineral stishovite, both presumably formed by explosive impact. Very small spherical bodies (tektites) could have formed by quenching of liquids derived from melting of rocks at the impact site.

The evidence cited for meteorite impact has also been used to support the proposal that the K/T event was caused by extraordinary volcanic eruptions instead of impact. The glass shards and fractured minerals in the clays could have been formed by massive volcanic eruptions. Much of the boundary clay at locations worldwide probably consisted largely of volcanic debris before alteration to some suite of clay minerals. Particles regarded as impact-produced tektites by some investigators are regarded as quenched volcanic liquids by others. Detailed stratigraphic studies suggest more than one horizon of Ir-enriched, altered, volcanic material near the K/T boundary, which may be more consistent with a period of major volcanism than with a single impact.

The problem of interpreting components of the K/T boundary layer can be illustrated by studies at Beloc, Haiti. The proposal that this section represents accumulation of debris from an impact near Chixculub, in the Yucatan Peninsula of Mexico, is based partly on the proposal that glassy globules (tektites?) in the boundary layer have compositions of both limestone and sialic (gneissic) rocks. These compositions could have been produced by impact melting of Mesozoic carbonate rocks and their underlying continental basement on the Yucatan Peninsula. Other investigators, however, have proposed that the 'tektites' have typical compositions of volcanic glasses. Distinction between the two proposed origins is, theoretically, simple, but the glasses are altered and difficult to sample without including veinlets and matrix. Thus, the interpretation remains controversial.

The possibility that the K/T event was caused by enormous volcanic eruption is consistent with the observation that the boundary is approximately synchronous with the eruption of the Deccan basalts of India (Section 8.2). In the Deccan plateau, ~2×10^6 km^3 of basalt were erupted within a few million years near the Cretaceous/Tertiary boundary, similar to the apparent synchrony of the Paleozoic/Mesozoic extinction with eruption of the Siberian basalts at ~250 Ma (Sections 6.9 and 8.2). Very detailed dating of the Deccan and K/T events, however, suggests that the major pulse of Deccan eruption was in the range of 67 to 69 Ma, some 2 to 4 m.y. older than the K/T boundary at 65 Ma. Thus, the major eruption would have been completed well before the extinction event and formation of the boundary clay, and the eruption could not have been directly responsible for the extinction.

The environmental picture that emerges from the various studies of the K/T extinction is similar to the feared 'nuclear winter' but without the enhanced radiation. Some combination of impact, explosion, and volcanism placed enormous quantities of ash into the atmosphere. This ash caused reduction in photic intensity and cooling of the earth's surface. Wildfires raged over the entire land surface, destroying the angiosperms and leaving a charred earth that was reseeded by the ferns that are commonly the first replacements of modern burned forests. The soot from the fires and the ash from the eruptions and explosions fell into fresh and salt water, possibly causing heavy-metal poisoning of both animals and marine plankton. Lack of surface photosynthesis created oceans low in dissolved oxygen (but not completely anoxic). Specialized organisms that did not have protection from the cold, the dark, and/or the poisoning died. Organisms that could protect themselves (burrowing, hibernation, etc.) survived, along with chance survivors from groups such as calcareous plankton, and the earth was quickly restocked with a diversity of fauna and flora.

[**References** – The original proposal of meteorite impact at the K/T boundary was made by Alvarez *et al*. (1980). Considerable information is in the volumes edited by Silver and Schultz (1982), Berggren and Van Couvering (1984), Sharpton and Ward (1990) plus reviews by Hallam (1987), Gilmour and Anders (1989) and Smit (1990). Descriptions of on-land and deep-sea sections are from Orth *et al*. (1981), Gilmore *et al*. (1984), Moore *et al*. (1984), Menard *et al*. (1987), Pollastro and Pillmore (1987), Zachos, Arthur and Dean (1989) and Zhou, Kyte and Bohor (1991). Specific discussions include: compositions of spheroids at the K/T boundary, by Montanari *et al*. (1983); paleontology of the boundary, by Keller (1988); and layering in the boundary, by Schmitz (1988). The age of the Deccan basalts is from Vandamme *et al*. (1991) and of the K/T boundary from Izett, Dalrymple and Snee. (1991). The controversy at Beloc, Haiti, is illustrated by the papers of Sigurdsson *et al*. (1991), Jehanno *et al*. (1992) and Lyons and Officer (1992).]

References

Abers, G. A., Parsons, B. & Weissel, J. K. (1988). Seamount abundances and distributions in the southeast Pacific. *Earth and Planetary Science Letters*, **87**, 137–51.

Alvarez, L. W., Alvarez, W., Asaro, F. & Michel, H. V. (1980). Extraterrestrial cause for the Cretaceous–Tertiary extinction. *Science*, **208**, 1095–108.

Aplonov, S. V. (1989). The paleogeodynamics of the West Siberian platform. *International Geology Review*, **31**, 859–67.

Baksi, A. J. (1990). Search for periodicity in global events in the geologic record: quo vadimus? *Geology*, **18**, 983–6.

Bally, A. W. & Palmer, A. R., eds (1989). *The Geology of North America – an Overview: vol A*. Boulder, Colorado: Geological Society of America, 619 pp.

Barker, P. F. (1982). The Cenozoic subduction history of the Pacific margin of the Antarctic Peninsula: ridge crest–trench interactions. *Geological Society of London Journal*, **139**, 787–801.

Barker, P. F. & Kennett, J. P. (1988). Weddell Sea palaeoceanography: preliminary results of ODP Leg 113. *Palaeoceanography, Palaeoclimatology, and Palaeoecology*, **67**, 75–102.

Barnola, J. M., Raynaud, D., Korotkevich, Y. S. & Lorius, C. (1987). Vostok ice core provides 160,000 year record of atmospheric CO_2. *Nature*, **329**, 408–14.

Barron, E. J., Harrison, C. G. A., Sloan, J. L., II, & Hay, W. W. (1981). Paleogeography, 180 million years ago to the present. *Eclogae Geologicae Helvetiae*, **74**, 443–70.

Beaton, M. J. (1986). More than one event in the late Triassic mass extinction. *Nature*, **321**, 857–61.

Berger, A., Imbrie, J., Hays, J., Kukla, G. & Saltzman, B., eds. (1984). *Milankovitch and Climate: Understanding the Response to Astronomical Forcing.*, part 1. Dordrecht: D. Reidel Publ. Co., 510 pp.

Berggren, W. A. & Van Couvering, J. A., eds (1984). *Catastrophes and Earth History*. Princeton, New Jersey: Princeton University Press, 464 pp.

Berner, R. A. (1991). A model for atmospheric CO_2 over Phanerozoic time. *American Journal of Science*, **291**, 339–76.

Besse, J. & Courtillot, V. (1988). Paleogeographic maps of the continents bordering the Indian Ocean since the Early Jurassic. *Journal of Geophysical Research*, **93**, 11 791–808.

Bice, D. M., Newton, C. R., McCauley, S., Reiners, P. W. & McRoberts, C. A. (1992). Shocked quartz at the Triassic-Jurassic boundary in Italy. *Science*, **255**, 443–6.

Blundell, D. J., Hobbs, R. W., Klemperer, S. L., Scott-Robinson, R., Long, R. E., West, T. E. & Duin, E. (1991). Crustal structure of the central and southern North Sea from BIRPS deep seismic reflection profiling. *Geological Society of London Journal*, **148**, 445–57.

Boardman, R. S., Cheetham, A. H. & Rowell, A. J., eds (1987). *Fossil Invertebrates*. Palo Alto, California: Blackwell Scientific Publications, 713 pp.

Bohannon, R. G. (1989). The timing of uplift, volcanism, and rifting peripheral to the Red Sea: a case for passive rifting. *Journal of Geophysical Research*, **94**, 1683–701.

Bralower, T. J. (1988). Calcareous nannofossil biostratigraphy and assemblages of the Cenomanian-Turonian boundary interval: implications for the origin and timing of oceanic anoxia. *Paleoceanography*, **3**, 275–316.

Brasier, M. D. (1980). *Microfossils*. London: George Allen and Unwin, 193 pp.

Brass, G. W., Southam, J. R. & Peterson, W. H. (1982). Warm saline bottom water in the ancient ocean. *Nature*, **296**, 620–3.

Broecker, W. S. & Denton, G. H. (1991). The role of ocean–atmosphere reorganization in glacial cycles. *Geochimica et Cosmochimica Acta*, **53**, 2465–501.

Broecker, W. S. & Peng, T.-H. (1982). *Tracers in the Sea*. Palisades, New York: Lamont-Doherty Geological Observatory, 690 pp.

Brooks, J. & Fleet, A. J., eds. (1987). *Marine Petroleum Source Rocks*. Geological Society of London Special Publication 26, 444 pp.

Carroll, R. L. (1988). *Vertebrate Paleontology and Evolution*. New York: W.H. Freeman and Co., 698 pp.

Clarkson, E. N. K. (1986). *Invertebrate Paleontology and Evolution*, 2nd ed. Boston: Allen and Unwin, 382 pp.

CLIMAP Project Members (1976). The surface of the ice-age earth. *Science*, **191**, 1131–7.

Coffin, M. F. & Rabinowitz, P. D. (1988). *Evolution of the conjugate East African–Madagascan margins and the western Somali basin*. Geological Society of America Special Paper 226, 78 pp.

Crowley, T. J. & North, G. R. (1991). *Paleoclimatology*. Oxford: Oxford University Press, 339 pp.

Denton, G. H. & Hughes, T. J. (1986). Global ice-sheet system interlocked by sea level. *Quaternary Research*, **26**, 3–26.

Donovan, S. K. (1989). *Mass Extinctions*. New York: Columbia University Press, 266 pp.

Duncan, R. A. & Richards, M. A. (1991). Hotspots, mantle plumes, flood basalts, and true polar wander. *Reviews of Geophysics*, **29**, 31–50.

Elder, W. P. (1989). Molluscan extinction patterns across the Cenomanian–Turonian Stage boundary in the western interior of the United States. *Paleobiology*, **15**, 299–320.

Emery, K.O. & Uchupi, E. (1984). *The Geology of the Atlantic Ocean*. New York: Springer Verlag, 1050 pp.

Engebretson, D. C., Cox, A. & Gordon, R. G. (1985). *Relative Motions between Oceanic and Continental Plates in the Pacific Basin*. Geological Society of America Special Paper 206, 59 pp.

Farlow, J. O., ed. (1989). *Paleobiology of the Dinosaurs*. Geological Society of America Special Paper 238, 100 pp.

Fischer, A. G. & Arthur, M. A., (1977). Secular variations in the pelagic realm. In *Deep-Water Carbonate Environments*, ed. H. E. Cook & P. Enos, pp. 19–50. Society of Economic Paleontologists and Mineralogists Special Publication 25.

Genthon, C., Barnola, J. M., Raynaud, D., Lorius, C., Jouzel, J., Barkov, N. I., Korotkevich, Y. S. & Kotlyakov, V. M. (1987). Vostok ice core: climatic response to CO_2 and orbital forcing changes over the last climatic cycle. *Nature*, **329**, 414–18.

Gilmore, J. S., Knight, J. D., Orth, C. J., Pillmore, C. L. & Tschudy, R. H. (1984). Trace element patterns at a non-marine Cretaceous–Tertiary boundary. *Nature*, **307**, 224–8.

Gilmour, I. & Anders, E. (1989). Cretaceous–Tertiary boundary event: Evidence for a short time scale. *Geochimica et Cosmochimica Acta*, **53**, 502–11.

Glenn, C. R. & Arthur, M. A. (1985). Sedimentary and geochemical indicators of productivity and oxygen contents in modern and ancient basins: The Holocene Black Sea as the 'type' anoxic basin. *Chemical Geology*, **48**, 325–54.

Hallam, A. (1975). *Jurassic Environments*. Cambridge: Cambridge University Press, 269 pp.

Hallam, A. (1981). The end-Triassic bivalve extinction event. *Palaeogeography, Palaeoclimatology, Palaeoecology*, **35**, 1–44.

Hallam, A. (1986). The Pliensbachian and Tithonian extinction events. *Nature*, **319**, 765–7.

Hallam, A. (1987). End-Cretaceous mass extinction event: argument for terrestrial causation. *Science*, **238**, 1237–42.

Haq, B. U., Hardenbol, J. & Vail, P. R. (1987). Chronology of fluctuating sea levels since the Triassic. *Science*, **235**, 1156–67.

Hays, J. D., Imbrie, J. & Shackleton, N. J. (1976). Variations in the Earth's orbit: pacemaker of the ice ages. *Science*, **194**, 1121–32.

Hekinian, R., Bideau, D., Stoffers, P., Cheminee, J. L., Muhe, R., Puteanus, D. & Binard, L. (1991). Submarine intraplate volcanism in the South Pacific: geological setting and petrology of the Society and the Austral regions. *Journal of Geophysical Research*, **96**, 2109–38.

Holland, H. D., Lazar, B. & McCaffrey, M. (1986). Evolution of the atmosphere and oceans. *Nature*, **320**, 27–33.

Hynes, A. (1990). Two-stage rifting of Pangea by two different mechanisms. *Geology*, **18**, 323–6.

Imbrie, J. (1985). A theoretical framework for the Pleistocene ice ages. *Geological Society of London Journal*, **142**, 417–32.

Izett, G. A., Dalrymple, G. B. & Snee, L. W. (1991). ^{40}Ar/^{39}Ar age of Cretaceous–Tertiary boundary tektites from Haiti. *Science*, **252**, 1539–42.

Jablonski, D. & Bottjer, D. J. (1991). Environmental patterns in the origins of higher taxa: The post-Paleozoic fossil record. *Science*, **252**, 1831–3.

Jackson, R. H. & Gunnarsson, K. (1990). Reconstructions of the Arctic: Mesozoic to present. *Tectonophysics*, **172**, 303–22.

Jarvis, I., Carson, G. A., Cooper, M. K. E., Hart, M. B., Leary, P. N., Tocher, B. A., Horne, D. & Rosenfeld, A. (1988). Microfossil assemblages and the Cenomanian–Turonian (late Cretaceous) oceanic anoxic event. *Cretaceous Research*, **9**, 3–103.

Jehanno, C., Bocler, D., Froget, L., Lambert, B., Robin, E., Rocchia, R. & Turpin, L. (1992). The Cretaceous–Tertiary boundary at Beloc, Haiti: No evidence for an impact in the Caribbean area. *Earth and Planetary Science Letters*, **109**, 229–41.

Johnson, R. W., ed. (1989). *Intraplate Volcanism in Eastern Australia and New Zealand*. Cambridge: Cambridge University Press, 408 p.

Kauffman, E. G. & Walliser, O. H., eds. (1990). *Extinction Events in Earth History: Lecture Notes in Earth Sciences 30*. Berlin: Springer Verlag, 432 pp.

Keating, B. H., Fryer, P., Batiza, R. & Boehlert, G. W., eds. (1987). *Seamounts, Islands, and Atolls*. American Geophysical Union Geophysical Monograph 43, 405 pp.

Keller, G. (1988). Extinction, survivorship and evolution of planktic foraminifera across the Cretaceous/Tertiary boundary at El Kef, Tunisia. *Marine Micropaleontology*, **13**, 239–63.

Kennett, J. P. (1977). Cenozoic evolution of Antarctic glaciation, the circum-Antarctic Ocean, and their impact on global paleoceanography. *Journal of Geophysical Research*, **82**, 3843–60.

Kennett, J. P. (1982). *Marine Geology*. Englewood Cliffs, New Jersey: Prentice-Hall, 813 pp.

Kennett, J. P., ed. (1985). *The Miocene Ocean: Paleoceanography and Biogeography*. Geological Society of America Memoir 163, 337 pp.

Kent, R. (1991). Lithospheric uplift in eastern Gondwana: evidence for a long-lived mantle plume system. *Geology*, **19**, 19–23.

Kogbe, C.A. & Lang, J., eds. (1990). Major African continental Phanerozoic complexes and dynamics of sedimentation. Special Issue of *Journal of African Earth Sciences*, **10**, 1–408.

Larson, R. L. (1991). Latest pulse of the Earth: evidence for a mid-Cretaceous superplume. *Geology*, **19**, 547–50.

Larson, R. L. & Chase, C. G. (1972). Late Mesozoic evolution of the western Pacific Ocean. *Geological Society of America Bulletin*, **83**, 3627–44.

Larwood, G. P., ed. (1988). *Extinction and Survival in the Fossil Record: Systematics Association Special Volume 34*. Oxford: Oxford University Press, 365 pp.

Lonsdale, P. (1988). Structural pattern of the Galapagos microplate and evolution of the Galapagos triple junctions. *Journal of Geophysical Research*, **93**, 13 551–74.

Lyons, J. B. & Officer, C. B. (1992). Mineralogy and petrology of the Haiti Cretaceous/Tertiary section. *Earth and Planetary Science Letters*, **109**, 205–24.

Mammerickx, J. & Sharman, G. F. (1988). Tectonic evolution of the North Pacific during the Cretaceous quiet period. *Journal of Geophysical Research*, **93**, 3009–24.

Martin, L. (1983). The origin and early radiation of birds, In A. Brush & G. Clark, eds, *Perspectives in Ornithology*. Cambridge University Press, Cambridge, pp. 291–338.

McElroy, M. B. (1983). Marine biological controls on atmospheric CO_2 and climate. *Nature*, **302**, 328–9.

Menard, H. W. & 20 others, eds. (1987). *Leg XCI*. Initial Reports of the Deep Sea Drilling Project, 91, 494 pp.

Montanari, A., Hay, R. L., Alvarez, W., Asaro, F., Michel, H. V. & Alvarez, L. W. (1983). Spheroids at the Cretaceous–Tertiary boundary are altered droplets of basaltic composition. *Geology*, **11**, 668–71.

Moore, T. C., Jr. & 13 others, eds. (1984). *Leg LXXIV*. Initial Reports of the Deep Sea Drilling Project, 74, 894 pp.

Moullade, M. & Nairn, A. E. M., eds. (1978a). *The Phanerozoic Geology of the World II, The Mesozoic, A*. Amsterdam: Elsevier, 529 pp.

Moullade, M. & Nairn, A. E. M., eds. (1978b). *The Phanerozoic Geology of the World II, The Mesozoic, B*. Amsterdam: Elsevier, 450 pp.

Nakanishi, M., Tamaki, K. & Kobayashi, K. (1989). Mesozoic magnetic lineations and seafloor spreading history of the northwestern Pacific. *Journal of Geophysical Research*, **94**, 15 437–62.

Nalivkin, D. V. (translated by S.I. Tomkeieff) (1960). *The Geology of the U.S.S.R. – A Short Outline: International Series of Monographs on Earth Sciences*. New York: Pergamon Press, 170 pp. plus geologic map of U.S.S.R.

Nalivkin, D. V. (translated by N. Rast) (1973). *Geology of the USSR*. Edingurgh: Oliver & Boyd, 855 pp.

Norton, I. O. & Sclater, J. G. (1979). A model for the evolution of the Indian Ocean and the breakup of Gondwanaland. *Journal of Geophysical Research*, **84**, 6803–30.

Olausson, E., ed. (1988). The Southern Ocean – The Antarctic: Present and past: Special Issue of *Palaeogeography, Palaeoclimatology, Palaeoecology*, **67**, 1–179.

Orth, C. J., Gilmore, J. S., Knight, J. D., Pillmore, C. L., Tschudy, R. H. & Fassett, J. E. (1981). An iridium abundance anomaly at the palynological Cretaceous–Tertiary boundary in northern New Mexico. *Science*, **214**, 1341–3.

Paull, C. K., Ussler, W., III & Dillon, W. P. (1991). Is the extent of glaciation limited by marine gas-hydrates? *Geophysical Research Letters*, **18**, 433–4.

Pedersen, T. E. & Calvert, S. E. (1990). Anoxia vs. productivity: what controls the formation of organic-carbon-rich sediments and sedimentary rocks. *American Association of Petroleum Geologists Bulletin*, **74**, 454–66.

Petters, S. W. (1981). Stratigraphy of Chad and Iullemmeden basins. *Eclogae Geologicae Helvetiae*, **74**, 139–59.

Piper, D. Z. & Codispoti, L. A. (1975). Marine phosphorite deposits and the nitrogen cycle. *Science*, **188**, 15–8.

Pollastro, R. M. & Pillmore, C. L. (1987). Mineralogy and petrology of the Cretaceous–Tertiary boundary clay bed and adjacent clay-rich rocks, Raton basin, New Mexico and Colorado. *Journal of Sedimentary Petrology*, **57**, 456–66.

Pomerol, Ch. & Premoli-Silva, I., eds. (1986). *Terminal Eocene Events: Developments in Palaeontology and Stratigraphy, v. 9*. Amsterdam: Elsevier, 414 pp.

Porter, S. C. (1975). Equilibrium-line altitudes of Late Quaternary glaciers in the Southern Alps, New Zealand. *Quaternary Research*, **5**, 27–47.

Powell, C. McA., Roots, S. R. & Veevers, J. J. (1988). Pre-breakup continental extension in East Gondwanaland and the early opening of the eastern Indian Ocean. *Tectonophysics*, **155**, 261–83.

Rampino, M. R. & Stothers, R. B. (1988). Flood basalt volcanism during the past 250 million years. *Science*, **241**, 663–8.

Renne, P. R. & Basu, A. R. (1991). Rapid eruption of the Siberian traps flood basalts at the Permo-Triassic boundary. *Science*, **253**, 176–8.

Riggs, S. R. (1984). Paleoceanographic model of Neogene phosphorite deposition, U. S. Atlantic continental margin. *Nature* **223**, 123–31.

Robinson, J. M. (1990). Lignin, land plants, and fungi: biological evolution affecting Phanerozoic oxygen balance. *Geology*, **15**, 607–10.

Rowley, D. B. & Lottes, A. L. (1988). Plate kinematic reconstructions of the North Atlantic and Arctic: late Jurassic to present. *Tectonophysics*, **155**, 73–120.

Ruddiman, W. F., Raymo, M. E., Martinson, D. G., Clement, B. M. & Backman, J. (1989). Pleistocene evolution: northern hemisphere ice sheets and North Atlantic Ocean. *Paleoceanography*, **4**, 353–412.

Rudkevich, M. Ya. & Maksimov, Ye. M. (1988). Cyclicity in the geologic development of the West Siberian platform during the cratonal stage. *Petroleum Geology*, **24**, 265–9.

Schlanger, S. O., Jenkyns, H. C. & Premoli-Silva, I. (1981). Volcanism and vertical tectonics in the Pacific basin related to global Cretaceous transgressions. *Earth and Planetary Science Letters*, **52**, 435–49.

Schmitz, B. (1988). Origin of microlayering in worldwide distributed Ir-rich marine Cretaceous/Tertiary boundary clays. *Geology*, **16**, 1068–72.

Schubert, G. & Sandwell, D. (1989). Crustal volumes of the continents and of oceanic and continental submarine plateaus. *Earth and Planetary Science Letters*, **92**, 234–46.

Schwan, W. (1985). The worldwide active middle/late Eocene geodynamic episode with peaks at +/- 45 and +/- 37 m.y. B.P., and problems of orogeny and sea-floor spreading. *Tectonophysics*, **115**, 197–234.

Scotese, C. R. & Sager, W. W., eds. (1988). Mesozoic and Cenozoic plate reconstructions. Special Issue of *Tectonophysics*, **155**, 1–399.

Sharpton, V. L. & Ward, P. D., eds. (1990). *Global Catastrophes in Earth History*. Geological Society of America Special Paper 247, 631 pp.

Sheridan, R. E. & Grow, J. A., eds. (1987). *The Atlantic Continental Margin: U.S.: The Geology of North America, vol. I-2*. Boulder, Colorado: Geological Society of America, 610 pp.

Sigurdsson, H., D'Hondt, S., Arthur, M. A., Bralower, T. J., Zachos, J. C., van Fossen, M. & Channell, J. E. T. (1991). Glass from Cretaceous–Tertiary boundary in Haiti. *Nature*, **349**, 482–7.

Silver, L. T. & Schultz, P. H., eds. (1982). *Geological Implications of Impacts of Large Asteroids and Comets on the Earth*. Geological Society of America Special Paper 190, 528 pp.

Sloss, L. L. ed. (1988). *Sedimentary Cover – North American Craton:*

U.S.: *The Geology of North America, vol. D-2*. Geological Society of America, 506 pp.

Smit, J. (1990). Meteorite impact, extinctions, and the Cretaceous/Tertiary boundary. *Geologie en Mijnbouw*, **69**, 187–204.

Stearn, C. & Carroll, R. (1989). *Paleontology: The Record of Life*. New York: John Wiley and Sons, 453 pp.

Stein, R., Rullkotter, J. & Welte, D. H. (1986). Accumulation of organic-carbon-rich sediments in the Late Jurassic and Cretaceous Atlantic Ocean – a synthesis. *Chemical Geology*, **56**, 1–32.

Sundquist, E. T. & Broecker, W. S., eds. (1985). *The Carbon Cycle and Atmospheric CO₂: Natural Variations Archean to Present*. American Geophysical Union Monograph 32, 627 pp.

Tankard, A. J. & Balkwill, H. R. (1989). *Extensional Tectonics and Stratigraphy of the North Atlantic Margins*. American Association of Petroleum Geologists Memoir 46, 641 pp.

Thierstein, H. R. & Berger, W. H. (1978). Injection events in ocean history. *Nature*, **276**, 461–6.

Thompson, G. A. & Morgan, J., eds. (1984). S. Thomas Crough Memorial. Special Section of *Journal of Geophysical Research*, **89**, 9867–10 108.

Vandamme, D., Courtillot, V., Besse, J. & Montigny, R. (1991). Paleomagnetism and age determinations of the Deccan traps (India): results of a Nagpur–Bombay traverse and review of earlier work. *Reviews of Geophysics*, **29**, 159–90.

Veevers, J. J., ed. (1984). *Phanerozoic Earth History of Australia*. Oxford: Oxford University Press, 418 pp.

Veevers, J. J. (1989). Middle/Late Triassic (230 +/- 5 Ma) singularity in the stratigraphic and magmatic history of the Pangean heat anomaly. *Geology*, **17**, 784–7.

Vogt, P. R. (1991). Bermuda and Appalachian–Labrador rises: common non-hotspot processes. *Geology*, **19**, 41–4.

Walliser, O. H., ed. (1986). *Global Bio-Events: Lecture Notes in Earth Sciences 8*. Berlin: Springer Verlag, 442 pp.

Weins, D. A., DeMets, G., Gordon, R .G., Stein, S., Argus, D., Engeln, J. F., Lundgren, P., Quible, D., Stein, C., Weinstein, S. & Woods, D. F. (1985). A diffuse plate boundary model for Indian Ocean tectonics. *Geophysical Research Letters*, **12**, 429–32.

Whitman, J. M., Harrison, C. G. A. & Brass, G. W. (1983). Tectonic evolution of the Pacific Ocean since 74 Ma. In Convergence and Subduction, ed. T. W. C. Hilde and S. Uyeda. Special issue of *Tectonophysics*, **99**, nos 2–4, 241–9.

Woods, M. T. & Davies, G.F. (1982). Late Cretaceous genesis of the Kula plate. *Earth and Planetary Science Letters*, **58**, 161–6.

Young, T. P. & Taylor, W. E. G., eds. (1989). *Phanerozoic Ironstones*. Geological Society of London Special Publication 46, 251 pp.

Zachos, J. C., Arthur, M. A. & Dean, W. E. (1989). Geochemical and paleoenvironmental variations across the Cretaceous/Tertiary boundary at Braggs, Alabama. *Palaeogeography, Palaeoclimatology, and Palaeoecology*, **69**, 245–66.

Zhou, L., Kyte, F. T. & Bohor, B. F. (1991). Cretaceous/Tertiary boundary of DSDP Site 596, South Pacific. *Geology*, **19**, 694–7.

Pyroclastics at Volcan Irazu, Costa Rica. The volcano is one of the consequences of subduction around the rim of the Pacific Ocean. □

9

THE MESOZOIC AND CENOZOIC – PART II. CONSUMPTION, COLLISION AND THE DEVELOPMENT OF SMALL OCEAN BASINS

9.0 Introduction

THE SEAFLOOR spreading described in Chapter 8 for the Mesozoic and Cenozoic was responsible for the consumption of oceanic (locally continental) lithosphere along two broad belts roughly perpendicular to each other. One was the southern margin of Eurasia, including the Alps, eastern Europe and western Asia, the Himalayas, and mountainous regions in western China and southeastern Asia. The second was the circum-Pacific region, including the Andes, the North American Cordillera, the island arcs and rifted fragments of eastern Asia and Australia, and New Zealand. We consider this Mesozoic and Cenozoic orogenic activity in one chapter because it appears to have been continuous across the paleontologic system boundary generated by the catastrophic events that separated Mesozoic and Cenozoic biotic activity.

Before geologists began to think of earth processes in terms of the movement of rigid lithospheric plates, the perpendicularity of the two mountain belts was regarded as evidence of a shrinking earth. A sphere that shrinks (under compression) maintains a spherical shape only if all diameters undergo equal contraction, hence requiring shortening along two perpendicular great circles. Although the concept never really worked very well and has long since been discarded, we mention it here as an historical note because several generations of geologists were raised with it.

The compressive belts along the Pacific rim are the result of spreading in the Atlantic and Indian Oceans and western Pacific marginal seas as well as the Pacific Ocean. Development of these ocean basins pushed continents and continental fragments toward the Pacific, adding their velocity vectors to those caused by Pacific spreading. In consequence, velocities of consumption of oceanic lithosphere in the Pacific-margin subduction zones may have been as high as 20 cm/yr (possibly more) in local areas.

The relative youthfulness and the complexity of Mesozoic/Cenozoic orogenic belts provide us with the best available opportunity to study orogenic process caused by plate interactions. In order to keep the discussion at a manageable length, we will emphasize somewhat different processes in each of the areas discussed (Fig. 9.1). Also, because of the intimate relationship between orogenic belts and the development of modern back-arc basins, the two topics are discussed together. The major events are as follows.

- Continental fragments began accreting to the southern margin of Asia at least during the late Paleozoic (possibly earlier) and continued episodically throughout the Mesozoic and Cenozoic. The most recent increments

259

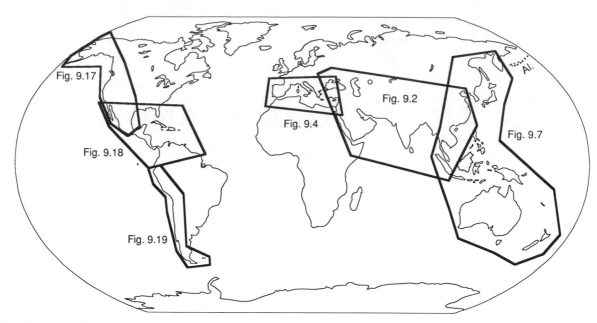

Fig. 9.1.   Locations of areas discussed in Chapter 9. (Robinson projection centered on Greenwich meridian.)

were India and Arabia, resulting from the opening of the Indian Ocean and Red Sea, respectively. The most impressive result of the suturing is the Himalayas, but several generations of orogenic belts lie nearly on top of each other along the entire Asian margin.

- The Alpine orogenic belt is the westernmost continuation of the Himalayan and other mountain chains along the southern margin of Eurasia (Section 9.2). The Alps and related belts formed in the exceptionally complex area of interaction between Africa, Europe, and possibly various fragments from Africa; it involved several stages of seafloor spreading in the area between all of these continental blocks. The Mediterranean is a relic of these interactions; perhaps its most interesting historical episode was the isolation that developed a deep evaporative basin, essentially a drying up of a body of water nearly identical to the present Mediterranean.
- Consumption of oceanic lithosphere in the western Pacific and southeastern Asia created deformed magmatic/sedimentary complexes associated with the opening of 'back-arc' ocean basins (Section 9.3). We investigate problems concerning: 1) the relationship between the age of orogenic activity and the age of separation of islands from the mainland; 2) the origin of the marginal basins behind the subduction zones; and 3) the origin of 'median tectonic lines' (see below) that separate adjacent orogenic belts metamorphosed under different T/P conditions.
- The Cordilleran region of western North America has undergone some type of compressive deformation nearly continuously throughout the Phanerozoic. In

consequence, the western continental margin is now significantly farther west than the Precambrian margin, partly caused by marginal growth and partly by accretion of exotic terranes (Section 9.4). Part of the complex history results from the interaction between the irregular coastline of the continent and the spreading ridge–transform boundary that was consumed against the continental margin. This interaction caused strike-slip faulting to be particularly important in the development of the Cordillera and also led to the development of the broad extensional area known as the Basin and Range.

- The Gulf of Mexico and the Caribbean are adjacent but have totally different histories (Section 9.5). The Gulf of Mexico developed by seafloor spreading as a result of the movement of North and South America away from each other. The principal issue in its formation is the manner in which continental terranes moved into part of the gap left behind (present Gulf of Mexico). The Caribbean also developed in response to the movement of South America and possibly North America; it is floored by lithosphere of uncertain origin and bounded through time by a series of subduction zones and transform faults. The differences between the two basins provide an opportunity to investigate the process of development of small oceans.
- The Andes are the 'ideal' continental-margin orogenic belt, with their highest elevations only a short distance from the coastline (Section 9.6). Furthermore, outcrop of pre-Andean continental crust near, or at, the coast indicates that this range was constructed mostly by simple subduction of oceanic lithosphere. Our discussion

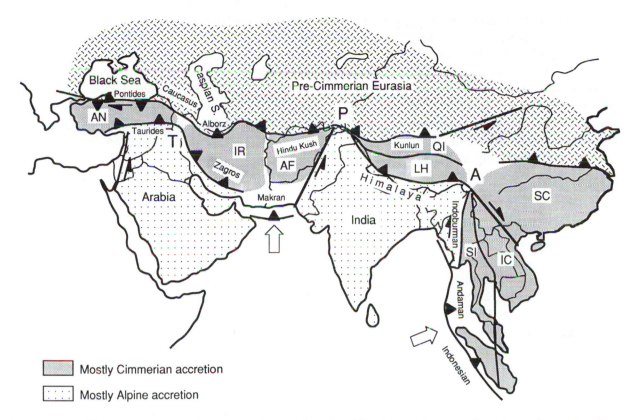

Mostly Cimmerian accretion

Mostly Alpine accretion

Fig. 9.2. Mesozoic/Cenozoic accretion of terranes against southern Asia. The diagram shows two broad ages of accretion: 1) Cimmerian, in the Mesozoic; and 2) Alpine–Himalayan, in the Cenozoic. Much of the Cimmerian record is obscured by the tectonic overprint of Cenozoic accretion. The two principal blocks accreted in the Cenozoic are India and Arabia (see Fig. 8.2). Ages of accretion, particularly of Cimmerian blocks, are controversial, and this map should be compared with the equivalent map of possible Paleozoic accretion (Fig. 7.5).

Unpatterned areas within Asia represent zones between recognizable blocks; they include uncertain borders between the Anatolian, Iranian, and Afghan blocks, the Makran accretionary prism, and a large flysch basin in southwestern China (northwest of the South China block). Thrust vergences are shown by conventional saw-tooth symbols. Inferred direction of movement of the downgoing slab in subduction zones is shown by the large white arrows. Mountain ranges, and other subduction zones, are labeled. The stable blocks are: SC, South China; IC, Indochina; SI, Sibumasu (Shantai); LH, Lhasa; QI, Qiangtang; AF, Afghanistan; IR, Iran; AN, Anatolia. (Mercator projection.)

Areas where late-Cenozoic collision of the Indian and Arabian continental blocks has caused extreme crushing of orogenic belts to form 'syntaxes' are labeled: A, Assam; P, Pamir; T, Turkish.   ☐

emphasizes the problems of developing an orogenic belt without recognizable collision of other terranes.

9.1 Accretion along the southern margin of Asia

ACCRETION of terranes against the southern margin of Asia was such a continuous process throughout the Phanerozoic that it is difficult to subdivide the accretionary process into distinct pulses, and the various defined 'orogenies' tend to blur into each other.

Additional complications arise from several factors, including: age(s) of rocks in accreted blocks are commonly different (both older and younger) from the age of suturing; diachronous closure occurred along many sutures; younger orogenic belts overprinted older ones; and many blocks were transported laterally along the continental margin after suturing.

Different workers apply numerous names to the various accreted terranes, intervening oceans, orogenic belts, and time periods; in this section we use only a very generalized terminology. The term 'Cimmerian' (Kimmerian) broadly refers to activity from approximately the Middle Permian until some time in the Cretaceous; i.e. roughly the Mesozoic. The terms 'Alpine' and 'Himalayan' (or 'Alpine–Himalayan') are applied to Cenozoic activity, although they include many Cretaceous events. The Alpine–Himalayan orogeny overprinted nearly all of the Cimmerian deformation zones, thus complicating the description of the older orogeny. Suturing of Cimmerian blocks to southern Asia resulted from the closure (by subduction) of an early Tethys Ocean, which we refer to as 'Paleotethys' ('Tethys I'). The destruction of Paleotethys was caused by spreading between the Cimmerian fragments and the northern edge of Gondwanaland, and the new ocean basin is referred to as 'Neotethys' ('Tethys II'). Movement of India and Arabia closed Neotethys and opened the Red Sea and Indian Ocean. A simplified map of the accretionary area is shown in Fig. 9.2. Several of the terranes on this map are also shown on Fig. 7.5 as possible Paleozoic accretionary blocks, and at least some suturing occurred during the Mesozoic in the area marked as 'Pre-Cimmerian Eurasia'.

Blocks accreted during the Cimmerian tend to be small, and the ones shown in Fig. 9.2 are mostly amalgams of a bewildering array of even smaller, sutured, fragments. The various blocks include several types of crust. Some are microcontinental fragments and continental-margin accretionary prisms formed on the passive northern margins of the microcontinents; despite their continental nature, exposed Precambrian crust is minor. Some blocks consist of whole and dismembered intra-oceanic arcs, most of which developed above southward-directed subduction zones within the disappearing Paleotethys. Oceanic lithosphere is preserved as ophiolite bodies and other fragments, both along the major suture zones and within the mapped blocks. Despite this fragmental nature within the terranes, some workers have proposed that the Cimmerian blocks were all welded together before suturing to Asia, thus producing a 'Cimmerian continent' that accreted nearly as a coherent unit.

As in the Paleozoic, destruction of Paleotethys and docking of the Cimmerian terranes was accomplished largely by northward subduction under Asia. Many investigators interpret the Black Sea and Caspian Sea as the result of back-arc spreading above the northward-descending slabs. The subduction was probably oblique in many areas, and reversal of strike-slip movement has been proposed for some zones. For example, the North Anatolian fault zone of the Pontide region of Turkey is now right lateral but may have been left lateral throughout much of the later Mesozoic.

The two major plates that docked in the Cenozoic are the Arabian and Indian blocks. Both of these terranes consist largely of exposed Precambrian crust and overlying platform sediments, mostly along the northern margins. These covering sediments range in age throughout much of the Phanerozoic, indicating a long exposure of the blocks along the northern margin of Gondwanaland. The Arabian plate has moved only a short distance as the Red Sea opened (Section 4.2), and has been partly subducted under the Tauride and Zagros mountains. The Indian plate moved very rapidly over much of the distance between the east coast of Africa and its present position south of the Himalayas (see Section 8.1 on the Indian Ocean).

Both the Arabian and Indian plates moved along large strike-slip faults to their present positions. The western edge of the Arabian plate is the Aqaba/Dead Sea lineament, with approximately 100 km of displacement (Sections 4.2 and 8.1). Lateral movement has been even greater along the two edges of the Indian plate, but exact estimates are difficult to obtain. In addition to lateral movements, both edges of the Indian plate have undergone compression, forming the complex Indo-Burman ranges on the east and a deformation belt along the Pakistan–Afghanistan border on the west.

The ramming of the Indian and Arabian continental blocks into a resistant Asia produced three areas in which older orogenic belts were crushed together and overprinted by intense younger deformation (Assam, Pamir and Turkey; Fig. 9.2). These areas are referred to as 'syntaxes', implying a gathering of many disparate tectonic elements. The collision of India with Asia also produced stresses that required oblique movements of parts of eastern Asia away from the intruding block. These movements are shown diagrammatically along two major fault zones extending eastward from the Lhasa and Qiangtang areas (Fig. 9.2), but similar motion has occurred along numerous other zones not shown in the diagram.

Himalayas

The Himalayas contain the highest peaks, deepest gorges and steepest slopes of any mountain range in

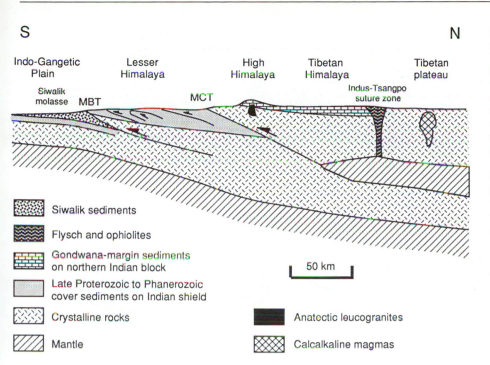

S

Indo-Gangetic
Plain

Siwalik
molasse   MBT

Lesser
Himalaya

MCT

High
Himalaya

Tibetan
Himalaya

Indus-Tsangpo
suture zone

Tibetan
plateau

N

Siwalik sediments

Flysch and ophiolites

**Gondwana-margin sediments
on northern Indian block**

**Late Proterozoic to Phanerozoic
cover sediments on Indian shield**

Crystalline rocks

Mantle

50 km

Anatectic leucogranites

Calcalkaline magmas

Fig. 9.3. Diagrammatic cross section of the Himalayas from India to Tibet. The elevation of the Himalayas results from the collision of India with Asia within the past ~25 m.y. The amount of underthrusting of Asia by continental rocks of the Indian plate is controversial. A wedge of mantle is shown in the Tibetan region to indicate the interspersing of mantle and crust along deep thrusts, and the actual relationships between mantle and crust are much more complicated than depicted here. The greatest thickness of 'pure' continental crust is shown under the high part of the Himalayas. (Vertical exaggeration 5×.)

the world (Fig. 9.3). Many areas are either inaccessible or can be reached only with great exertion (and with bottled oxygen). Geologic research commonly is based on traverses along river valleys or other areas of comparatively low elevation, thus yielding maps with only widely spaced control lines.

The Himalayas were produced as India and various fragments of continents and island arcs to the north of India collided with Asia, probably beginning in the Eocene (Fig. 8.3). The sweeping together of these small terranes caused only limited collision before the Miocene (at ~20 Ma) and resulted in the incorporation of entire volcanic arcs, plus smaller fragments, in the Himalayan zone. Northward-directed subduction is shown by geophysical data (see below) and by the occurrence of late-Cenozoic magmatism in southern Tibet, above the downgoing slab.

The extent of underthrusting of Tibet is controversial. The Himalayan crust is about 70 km thick, and early interpretations attributed this thickening to doubling of the crust by stacking Tibet above continental India. This stacking is mechanically difficult, however, and does not account for the roughly 50- to 60-km thickness throughout much of the Tibetan plateau. The crustal thickness is probably caused by overlapping of small slices of crust and crust/mantle, and seismic evidence shows local areas in which displaced Moho

overlies another layer of Moho. Despite the evidence for subduction under the Himalayas, it is unclear whether the slab extends under all of Tibet, possibly as far north as the Kunlun range. The intense compression has led to extremely rapid uplift rates in parts of the Himalayas (up to 700 m/m.y.) and rapid late-Cenozoic sedimentation in the northern Indian Ocean. The Himalayas are also sufficiently uplifted that north-directed normal faulting has occurred (Section 4.2).

As exposed on the surface, the join between India and Asia is the Indus–Tsangpo suture zone (Fig. 9.3). This zone, mostly in southern Tibet, contains ophiolites and flysch sequences in amphibolite to greenschist facies separating relatively undeformed rocks of the Tibetan plateau from the orogenic belt of the Himalayas. South of the suture zone, the Himalayas consist of three principal belts. The northernmost one is the High (Great) Himalayas and Tibetan (Tethyan) Himalayas, consisting of highly metamorphosed crystalline rocks of the Indian shield overlain by undeformed sediments originally deposited along the northern Indian (Tethyan) passive margin. The southern edge of the High Himalayas is a major thrust zone, referred to as the Main central thrust (MCT) in India. South of the MCT are the Lesser Himalayas, consisting mostly of lower-grade metasediments also deposited on the northern Indian shield. The Lesser Himalayas

A

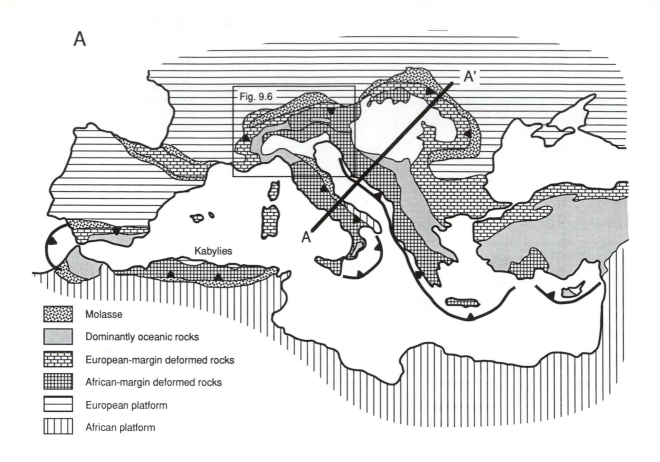

Molasse

Dominantly oceanic rocks

European-margin deformed rocks

African-margin deformed rocks

European platform

African platform

are bordered on the south by the Main boundary thrust (MBT), which overthrusts late-Cenozoic debris (Siwalik Group) shed from the uplifting mountain range into the Indo-Gangetic foreland basin.

All rocks in the Himalayas are derived from the Indian plate. Ages of the parental sediments that now constitute the metamorphic rocks of the Himalayas are as old as late Proterozoic (Vindhyan; Section 5.5) and as young as early Cenozoic. Locally the sediments have undergone melting to form syntectonic to post-tectonic granites. A major evaporite basin developed on the Gondwana margin of India in the latest Proterozoic and earliest Paleozoic, and Himalayan thrusts used the zone as a lubricant for southward thrusting in areas where the evaporites are particularly thick.

Although the Himalayas are commonly regarded as a model for mountain building, they show significant differences from many other ranges. One is the lack of involvement of the plate over-riding the subduction zone (the Tibetan plateau). In many mountain belts, the upper plate crosses the suture zone in a series of nappes or

thrust complexes, but this type of structure has not developed in the Himalayas. A second difference is the absence of a blueschist-facies suite, possibly because the Himalayas did not start to form until after all cold oceanic lithosphere had already been consumed and collision of two continental plates had begun. A third difference is the extremely high rate of subduction in the past 20 million years, probably caused by buildup of stress (and elevation?) in the Indian plate during collision prior to 20 Ma and rapid stress release after frictional resistance to subduction of the Indian continent was overcome.

[**References** – General information on the tectonics of Asia is provided by volumes edited by Spencer (1974) and McKerrow and Scotese (1990). Discussions of Tethys are in volumes edited by Aubouin, Le Pichon and Monin (1986), Audley-Charles and Hallam (1988), and Societe de la Geologique de France (1990). Sengor (1984, 1987) and papers in the Societe de la Geologique de France (1990) provide reviews of the evolution of Tethys. Southern and southeastern Asia are discussed in books by Valdiya (1984) and Hutchison (1989). Information for Pakistan is from Farah and DeJong (1979) and for the Himalayan mountain belt from Hirn et al. (1984a, b), Butler and Prior (1988), and Hodges, LeFort and Pecher (1988).]

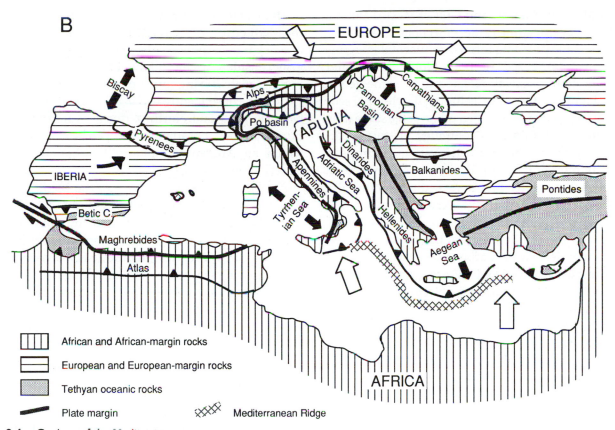

Fig. 9.4. Geology of the Mediterranean area.

(A) Distribution of major rock suites and some major tectonic zones. Rock suites of the area are grouped into undeformed platform assemblages of the European and African plates, rocks deformed along the margins of both plates, dominantly oceanic suites deposited between the two plates, and a foredeep molasse deposited in front of the growing Alpine orogen. The location of a cross section (A–A'; Fig. 9.5) and a map of the Alps (Fig. 9.6) are also shown

(B) Interpretation of Fig. 9.4A, showing plate margins and intra- and inter-plate orogenic features. The area is shown with three plates: Europe, Africa, and Apulia, which may have been a promontory of Africa or an independent continental plate. Much of the tectonic complexity of the region is caused by the crushing of Apulia between Europe and Africa. Europe is shown as the downgoing slab in the major subduction zones (white arrows) along the northern edge of the Alpine system, and Africa is shown as the downgoing slab in the subduction zones along its northern margin. Opening of major extensional basins is shown by opposing black arrows (e.g. Aegean Sea). Rotation of the Iberian block is shown by the curved arrow. Other symbols as in Fig. 9.2. ☐

9.2 Development of the Mediterranean Sea and Alpine orogenic belts

THE PRESENT Mediterranean Sea (Fig. 9.4A and B) largely occupies an area that was rifted apart in the early Mesozoic following the closure of Africa and Baltica (Section 7.1). The northern rim of the Mediterranean and the western part of the southern margin are the sites of Mesozoic/Cenozoic compressive orogeny broadly referred to as 'Alpine'. These mountain belts extend eastward through Asia Minor into the generally 'Himalayan' orogens of the southern margin of Asia. We discuss the history of the Mediterranean region in three categories: 1) an overview of the history of the Mediterranean Sea itself, including the significance of its extensive evaporite deposits; 2) the general evolution of orogenic belts around the Mediterranean; and 3) a closer inspection of the Alps.

Mediterranean Sea

The Triassic opening of the area now occupied by the Mediterranean was synchronous with the opening of the central North Atlantic and Gulf of Mexico along a complex set of spreading centers (Sections 8.1 and 9.5). The rifting and spreading formed continental margins, now poorly preserved in both North Africa and Europe, around a seaway referred to by some geologists as 'Liguride'. Oceanic (including flysch-type) sediments accumulated in the open ocean at the same time (approximately Jurassic) as continued platform sedimentation on both the North African and European shelves. In present Italy (Apulian plate; North African margin) shallow-water limestone sedimentation on a highly block-faulted continental crust continued into the Jurassic in the north and Cenozoic in the south. Drowning of the northern part of the platform caused deposition of deep-water carbonates in the north, and the entire area was covered by orogenically derived clastic sediments beginning in the Neogene.

After rifting, the configuration of the central and eastern Mediterranean was dominated by the closure of Africa and Europe, partly shown by modern north-directed subduction (Fig. 9.4B). The eastern Mediterranean may be completely closed and underlain by continental crust, although some workers propose that the Mediterranean Ridge (and associated ridges) are remnant oceanic crust partly thrust onto the African continental margin. In the western Mediterranean, a complex series of magnetic lineaments indicates multiple directions and periods of seafloor spreading since approximately the Late Oligocene. The complexity is particularly shown by the separation of Corsica and Sardinia from Europe and the emplacement of two small fragments of European crust (the 'Kabylies') against the North African coast (Fig. 9.4A). Beginning in the Cretaceous, spreading in the Bay of Biscay (Section 8.1) caused counterclockwise rotation of the Iberian peninsula and right-lateral movement across the Straits of Gibraltar, reducing the area of exposed oceanic lithosphere in the Mediterranean.

Two major parts of the Mediterranean (the Aegean and Tyrrhenian Seas) were developed by back-arc spreading beginning in the Miocene. Structural trends connecting the Tauride belt of Turkey (Section 9.1) with the Hellenides and remnants of other east–west structures can be traced geophysically across the Aegean, causing the Aegean to be regarded as a type example of lithospheric separation by pure stretching (Section 4.2).

Many of the uncertainties regarding the evolution of the Mediterranean are related to the possible existence of a separate 'Apulian' plate (Fig. 9.4B). Some geologists consider Apulia to have been an independently moving plate during much of the later Mesozoic and Cenozoic, consequently leaving an oceanic lithosphere between itself and North Africa as it moved toward Europe. Other investigators propose that the plate is simply an 'African promontory' that has always been closely attached to Africa.

A major event in the history of the Mediterranean region was the development of a suite of Messinian (latest Miocene) evaporites exposed by uplift around the present basin margin and sampled by numerous deep-sea cores. The Messinian suite is the youngest of the evaporite sequences that developed in the area throughout much of the Phanerozoic. Most of the older sequences, however, were associated with some episode of rifting and shallow-sea development in small basins, whereas the Messinian event was apparently caused by drying up of the entire present Mediterranean region. The isolation of the Mediterranean that permitted this drying was partly caused by closure of its eastern end as a result of northward movement of the Arabian plate (Section 8.1).

At the peak of evaporation, the Mediterranean may have been a topographically deep 'desert' (Hsu, 1983). Evidence supporting this mode of formation of the evaporites includes: 1) abundance of shallow-water to subaerial features (e.g. desiccation cracks) in latest Miocene rocks, indicating deposition in evaporating pans; 2) concentration of late-crystallizing evaporites (mainly halite) in the center of broader areas of sulfate deposition – this 'bull's-eye' pattern demonstrates progressive drying and shrinking of an evaporating pan; 3) alternation of evaporites with deep-water sediments over a time interval too short to be explained by uplift and downdrop of the Mediterranean basin; 4) deep incision of valleys, including the Rhone River, around the Mediterranean basin during Messinian time, presumably the result of a deep basin into which the rivers flowed; and 5) occurrence of enormous buried blocks, presumably of surrounding continental material, that appear to have fallen off a highland near the Straits of Gibraltar into the Messinian sedimentary suite.

The Mediterranean desert may have been filled repeatedly by breaching of the Straits of Gibraltar,

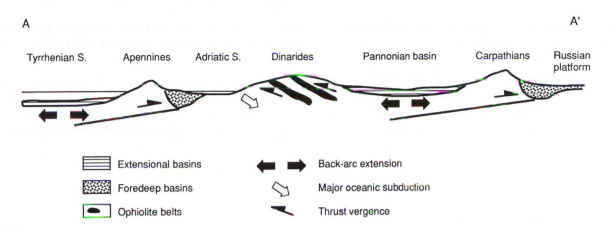

Fig. 9.5. Cross section from the Tyrrhenian Sea to the Russian platform (line A–A' on Fig. 9.4A). Continental crust is subducted beneath the Carpathians, and the Pannonian basin is shown as a 'back-arc' extensional feature. The Dinarides contain two major ophiolite belts, indicating at least two episodes of consumption of oceanic crust beneath them. The Adriatic Sea is floored by continental crust, and at least its western part is a foredeep basin created by the load of the Apennines. Extension in the Tyrrhenian Sea accompanies the thrusting of the Apennines toward the east. Vertical exaggeration is 10×.　□

which brought sealevel in the Mediterranean to the world norm, renewed abyssal sedimentation, and provided more sea water for the next episode of evaporation. This process provides us with an astonishing image – that of a repeated series of gigantic waterfalls with a drop of approximately four kilometers cascading from the Atlantic Ocean over the Straits of Gibraltar and down into the dry bottom of the Mediterranean. Some workers have estimated that the alternate evaporating and filling occurred 15 to 20 times.

Orogenic belts around the Mediterranean

The general geologic and tectonic characteristics of the Mediterranean region are shown in Fig. 9.4A and B. The major plates involved are Africa, Europe (Eurasia), and possibly Apulia (Adria). As discussed above, the Apulian 'plate' can be considered either as a promontory of Africa or as a detached plate. Regardless of the situation, the north–south closure of Europe and Africa/Apulia, plus the rotation of the Iberian peninsula, caused the development of a series of east–west-trending mountain belts.

The plate boundaries (Fig. 9.4B) are drawn between rocks of the African platform and their deformed marginal equivalents and rocks of the European platform and their deformed marginal equivalents. Thus, the boundaries pass mostly through oceanic (Tethyan)

rocks that consist of deep-water sediments, ophiolites, flysch and some arc-related sedimentary and volcanic deposits. In the Apennines and the central Alps (see below), the plate margins have been overstepped by rocks of the African continental margin, which are in direct contact with the foredeep molasse deposits in front of the thrust and nappe stacks. The northeastern margin of the Apulian plate is covered by young sediments of the back-arc Pannonian basin.

The tectonics of the Mediterranean area can be clarified by reference to the cross section shown in Fig. 9.5 (line A–A' on Fig. 9.4A). This section extends from the Russian platform, where the undeformed sediments are underlain by the Ukrainian shield, through three mountain chains (Carpathians, Dinarides, and Apennines), two back-arc basins that have undergone crustal extension (Pannonian and Tyrrhenian), and a broad foredeep (Adriatic). The only area of major subduction of oceanic crust is shown along the northeastern margin of the Apulian plate (Adriatic and westward). The crust from the Dinarides through the Carpathians is not really European basement but apparently consists of a mixture of old continental fragments, island arcs and oceanic material, including major ophiolite belts in the Dinarides and lesser ones in the Carpathians. Back-arc extension to form the Pannonian and Tyrrhenian basins began in the middle Cenozoic but was most active in the late Cenozoic, synchronous with the development of the shallow-level thrusts and minor volcanism that characterize both adjoining mountain belts.

None of the subsiding areas shown in Fig. 9.5 can be explained completely in terms of thermal downdrop of extended terranes (Pannonian and Tyrrhenian basins) or foredeep subsidence by crustal loading (Carpathian and Apennine foredeeps). Rates of thermal subsidence cannot be higher than subsidence rates of asthenosphere exposed at oceanic ridge crests (Section 4.2), but subsidence rates in the Tyrrhenian Sea may be as high as 1 km/m.y., higher than over mid-oceanic ridges. Similarly, subsidence rates in the Pannonian basin are almost as high as along ridge crests, which is unrealistic for a basin floored by thin continental crust. In the two foredeeps, the loads provided by the adjacent mountain belts cannot cause the observed extent of basement downwarp. Thus, the subsidence in both the extensional basins and foredeeps is accentuated by some process, not clearly understood, related to the complex collisional regime in the area.

The deformed area that rings much of the Mediterranean has been subdivided into a large number of individual mountain chains, each with a somewhat separate geologic history. In the next few paragraphs, we record only a few observations on these various belts.

- In the eastern Mediterranean, the Dinaride and Hellenide orogenic belts correlate with the Tauride belt of southern Turkey, and the Balkanides and Pontides represent the deformed southern margin of Eurasia (Fig. 9.4B). Two distinct ophiolite belts along the Hellenides and Dinarides (Fig. 9.5) imply closure of multiple small ocean basins. Compressional deformation in the Hellenide/Dinaride area probably extends from the Jurassic (Cimmerian; Section 9.1) into the early Cenozoic.
- The Carpathians contain minor oceanic (flysch and ophiolite) suites but consist primarily of platformal sediments and underlying (mostly Paleozoic) basement thrust northeastward over the Eurasian foreland. Thus, oceanic closure in the Carpathians is probably minor. The Carpathian foreland is the edge of the Ukrainian shield along the Tornquist line that marked the Paleozoic southwestern margin of stable Asia (Section 7.3). In both the Carpathians and the Alps (see below), compression was associated with southward subduction of Europe.
- The Apennines were formed by compression resulting from seafloor spreading in the western Mediterranean. This spreading caused counterclockwise rotation of Corsica and Sardinia in the middle to late Cenozoic, leading to eastward thrusting of flysch and a further thrust covering of these materials by platformal carbonates of the African (Apulian) plate.

- In the western Mediterranean, compressive orogeny was enhanced by the counterclockwise rotation of the Iberian peninsula (see above). This rotation created a two-sided, essentially 'intracratonic' mountain chain in the Pyrenees, with compression from near the end of the Cretaceous to the middle Cenozoic. The rotation also enhanced the compression in the Betic Cordillera and the western extension of the 'Maghrebides', mostly during the middle Cenozoic. The Atlas Mountains can also be regarded as essentially intracratonic, with deformation from the Cretaceous through much of the Cenozoic.

The Alps

For the tectonic addict, few sights can be more satisfying than an aerial view of the northern face of the Alps. The relatively undeformed Jura Mountains form a gentle northern margin for the Molasse basin, the industrial and agricultural heartland of Switzerland. The basin terminates abruptly to the south against the thrust complexes that have been glacially sculpted to form the horns and deep valleys characteristic of Alpine scenery. The combination of geology and scenery (perhaps also good restaurants) has invited intense geologic scrutiny of the Alps.

General relationships in the Alpine belt(s) are shown in Fig. 9.6 (see area on Fig. 9.4A). The boundary between the European and African plates is probably best mapped within the Penninic nappe sequence, including the area where the Penninic suites are overridden by the Austro-Alpine nappe sequence. The European plate shown in Fig. 9.6 is largely the area stabilized during the Variscan (Hercynian) orogeny of the late Paleozoic (Section 7.3).

Much of the history of the Alps must be explained in terms of the development of different nappe sequences, which moved at different times and carried suites of rocks from diverse locations (Section 4.1; Fig. 9.6). Nappes are large, recumbent, commonly crystalline-cored, folds that move toward the orogenic foreland (in the Alps, the European platform) and cause the crustal thickening and complex structures that characterize so many deeply exposed mountain belts. Rocks of the European margin occur in the slightly deformed Jura Mountains and Maritime Alps and are more complexly folded and thrusted in the *Helvetic* nappes of the Alps. These Helvetic nappes are overthrust by the *Penninic* nappes, which consist largely of rocks deposited on Tethyan oceanic lithosphere but include some microcontinental fragments.

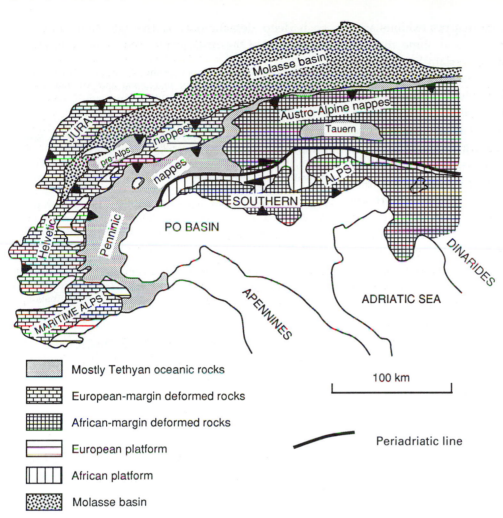

Fig. 9.6. Diagrammatic geology of the Alps. Rock suites are the same as in Fig. 9.4A (with no patterns on the European, African and Apulian platforms). Tethyan oceanic suites (flysch, ophiolites, etc.) form a partly obscured suture, mostly in the Penninic nappes and Tauern window, between Europe and colliding southern plates. Deformed rocks of the African margin have over-ridden this suture in the Austro-Alpine nappes. The Helvetic nappes and Jura contain deformed rocks of the European margin, and the Southern Alps contain deformed rocks of the African/Apulian margin. ☐

Legend:
- Mostly Tethyan oceanic rocks
- European-margin deformed rocks
- African-margin deformed rocks
- European platform
- African platform
- Molasse basin

100 km

Periadriatic line

Alpine ophiolites occur largely in the Pennine area. The small area of *Pre-Alpine* nappes also appears to be related to the Penninic suite. The *Austro–Alpine* nappes are derived from the southern margin of Tethys and, thus, represent platformal sediments from the African plate that have crossed the Alpine suture zone by thrusting toward the foreland. The large Tauern window exposes Tethyan and European rocks, some of which show evidence of Variscan deformation. The Southern Alps is an area of southward thrusting of African-margin rocks onto the Apulian (African?) continental plate. The border between the areas mapped as Penninic and Austro–Alpine is very complex and not well understood.

The Periadriatic line (Fig. 9.6) is a zone of major right-lateral movement. It is composed of numerous, smaller, fault segments, of which the best known is the Insubric line along the southern edge of the highest part of the Alpine chain. The combination of right-lateral movement and the southward subduction of Europe under Africa/Apulia is responsible for the complexity of strike-slip and thrust faulting in the Alpine region. Although the Periadriatic line has been considered by some geologists as the suture between Europe and Africa, most recent investigators regard it as lying within the African/Apulian plate, thus representing the latest stages of the counterclockwise rotation of Apulia. The Apulian plate just south of the Periadriatic line has been faulted upward during late Cenozoic movements, exposing a section deep into the crust (Ivrea zone) and possibly to the mantle.

The timing of the events in the Alps has been reasonably well established. The basement beneath the nappes (partly exposed in windows) and moved in the cores of some of the crystalline nappes exhibits ages through much of the Paleozoic, and some may be Precambrian. This basement is overlain by subaerial Pennsylvanian and Permian sediments, some of which were affected by late Variscan deformation. Marine platform sedimentation in the present Alpine region began in the Triassic and extended into the Jurassic, when major fragmentation, rifting and oceanic opening occurred. The rifting established small (Liguride or Penninic) basins, partly bounded by faults that were later reactivated during Alpine thrusting (forming 'inversion structures'). Thick oceanic sediments particularly accumulated during the (Early) Cretaceous.

Compression in the Alps probably began in the Middle Cretaceous. This time is the age of the youngest ophiolite, implying elimination of oceanic lithosphere, and the beginning of thick flysch accumulation in the former oceanic area and along rapidly subsiding continental margins. Some uplift in the Jura and development of the Molasse basin (Fig. 9.6) began in the Late Cretaceous, and areas south of the European margin were the sites of thick flysch sedimentation during the early Cenozoic. The main phase of Alpine deformation probably began in the Late Cretaceous, when Africa/Apulia collided with the suite of microcontinents and remnants of oceanic basins and began shoving them toward Europe. The first thrust movements were the emplacement of the Austro–Alpine nappes over this complex, largely Penninic, collage. The Penninic nappes moved largely during the Eocene and Oligocene, the Helvetic during the Miocene, and the Jura during the Pliocene. Thus, in general, the age of movement of the thrust complexes becomes younger to the north (toward the foreland).

Alpine metamorphic rocks record peak metamorphism mostly during the major ages of thrust movement. Earlier metamorphism in blueschist-facies series is partly overprinted by higher T/P conditions. In particular, high-pressure metamorphism affected some deep-seated rocks before their detachment from the continental margins on which they had formed prior to Alpine collision.

The basal detachment for the Alpine orogen is presumed to dip southward from near-surface in the Jura to very deep (several tens of kilometers) under the high Alps. Roughly comparable detachment surfaces may also exist at greater depths, particularly under the surface of 'stable' Europe; an example of the effect of such deep detachments is the late Cenozoic uplift (inversion) of Mesozoic graben sediments in Europe.

[**References** – The evolution of the Mediterranean basin is discussed in volumes edited by Biju-Duval and Montadert (1977), Berckhemer and Hsu (1982), Hsu (1983), Dixon and Robertson (1984) and Stanley and Wezel (1985). Information concerning orogeny in the Mediterranean area is from a review article by Channell, D'Argenio and Horvath (1979) and volumes edited by Spencer (1974), Biju-Duval and Montadert (1977), Berckhemer and Hsu (1982), Dixon and Robertson (1984), Stanley and Wezel (1985) and Aubouin *et al.* (1986). Specific information is from: Burchfiel (1976), on Romania; Sengor (1984), on the western part of the Cimmerian belt; Jacobshagen (1986), on Greece; Ziegler (1988), on Europe; and Royden and Karner (1984), on lithospheric processes in the Mediterranean region. The volume edited by Scotese and Sager (1988) also provides general information on the region. The Alps are discussed in volumes edited by Rybach and Lambert (1980) and Coward, Dietrich and Park (1989), a guidebook by Trumpy (1980) and a paper by Ernst (1975); the volume on inversion tectonics edited by Cooper and Williams (1989) also provides information.]

9.3 Subduction and back-arc spreading in the western Pacific

THE WESTERN MARGIN of the Pacific basin was the site of consumption of oceanic crust throughout the entire Mesozoic and Cenozoic (Fig. 9.7 and Section 8.1). This subduction created numerous continental-margin orogenic belts, many of which are now exposed in islands separated from Asia by later back-arc spreading, caused further deformation in the eastward-moving continental fragments, and developed intra-oceanic island arcs. Evidence of Paleozoic activity in these belts is largely obscured by later overprinting. Fig. 9.8 is a diagrammatic representation of the area during the (presumably late) Mesozoic before the initiation of back-arc spreading. Island groups, such as Japan and the Philippines, are shown in their present configurations although internal deformation has clearly occurred in them. All of the Mesozoic orogenic belts preserved on western-Pacific islands were formed while they were still attached to the mainland.

The location of the 'continental margin' depicted in Fig. 9.8 is arbitrarily chosen as the apparent seaward limit of 'old' (probably Paleozoic and older) continental crust. Mesozoic orogenic belts landward of this margin are deformed volcano–sedimentary materials deposited on marginal continental crust and/or allochthonous materials deposited on oceanic crust

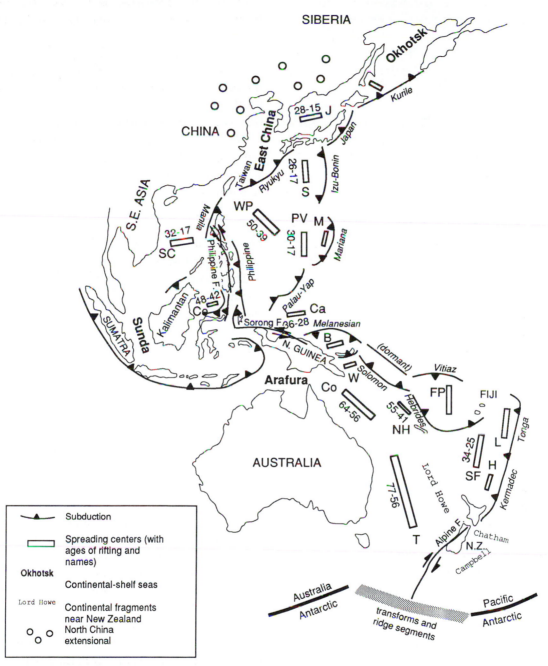

Fig. 9.7. Major tectonic features of the margins of the southwestern Pacific and northeastern Indian Oceans. The area contains an extraordinary complexity of subduction zones and small (back-arc?) basins. The orientation of back-arc spreading centers is approximately perpendicular to the main direction of extension, but many of the basins do not show definitive magnetic stripes that indicate the presence of a linear spreading center. Ages of opening in the basins are taken from Miyashiro (1986). The back-arc basins are: J, Japan; S, Shikoku; WP, West Philippine; SC, South China; PV, Parece Vela; M, Mariana; Ce, Celebes; Ca, Carolina; B, Bismarck; W, Woodlark; Co, Coral; NH, New Hebrides; FP, Fiji plateau; L, Lau; SF, South Fiji; H, Havre; T, Tasman. In addition to the information shown in the legend, the southern part of the map shows the spreading ridges between the Antarctic plate and the Pacific and Australian plates, together with a long zone of small ridges and transforms that connects the two major ridges. (Zenithal equidistant projection centered at 160°E.)

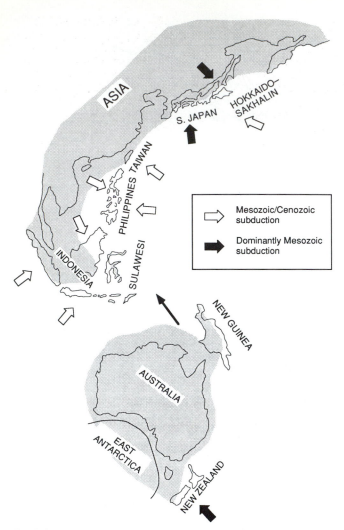

Fig. 9.8. Diagram of orogenic activity in the western Pacific region. The shaded area represents pre-Mesozoic continental crust; the Japanese islands are shown in position against Asia before the opening of the Sea of Japan (Fig. 9.7). The short arrows show the direction of movement of downgoing plates at various times. The long arrow indicates the counterclockwise rotation of the Australian plate, whose collision with the Indonesian area has created great structural complexity in the region (see Fig. 9.7). Most of the major orogenic belts are illustrated in Figs 9.9 to 9.14.   □

and transported onto the continent. The position of this line is made particularly uncertain by the difficulty of recognizing pre-Mesozoic crust within the deformational belts. In at least two areas (Taiwan and Sumatra) the most recent orogenic belts are constructed on the continental margin, and back-arc spreading has not caused separation of the islands from a continental mainland.

During the Mesozoic and Cenozoic, the western Pacific was affected by at least two, and possibly four, independent sources of stress. The major one was spreading in the Pacific Ocean. Throughout much of the Mesozoic and early Cenozoic, the Pacific basin contained several spreading ridges, of which only the East Pacific rise remains (Section 8.1). Westward subduction caused by spreading on this ridge compressed eastern Asia throughout the Mesozoic and Cenozoic but was greatly enhanced at about 40 Ma (Eocene), when the Pacific plate changed course from a northwesterly direction to a more direct westerly one (Section 8.1). This subduction down toward the west requires more stress than down toward the east because of resistance in the mantle caused by the earth's west-to-east rotation.

The second major source of stress in the western Pacific was provided by spreading on the Southeast Indian Ocean ridge, which caused northeastward-directed subduction beneath the Sunda shelf and farther east and was primarily responsible for the development of the Indonesian region. The ridge also propagated eastward, causing separation of Antarctica and Australia, and became connected with the East Pacific rise, probably during the Oligocene. Spreading on this ridge system was responsible for rotation of the Australian–New Guinea continent northward and northwestward, with resultant collision and deformation in eastern Indonesia.

These two major sources of stress may have been supplemented by two other independent sources. One may have been a mantle thermal anomaly, which accentuated the extensional stresses associated with subducting slabs and contributed to (or largely caused?) much of the opening of marginal basins in the western Pacific (see below). A second independent stress may have been translational movement of the Pacific basin not caused solely by spreading on ridges in the Pacific or eastern Indian Oceans. Strike-slip faults are abundant throughout the western Pacific, including: the large area of transform shear between the Australia–Antarctic ridge and the East Pacific ridge; the Alpine fault of New Zealand; the Sorong and Philippine faults, and the median tectonic line of Japan. Much of the present configuration of the Philippines and Sulawesi is the result of strike-slip accretion of crustal slivers, including fragments of Australian continental crust in eastern Sulawesi.

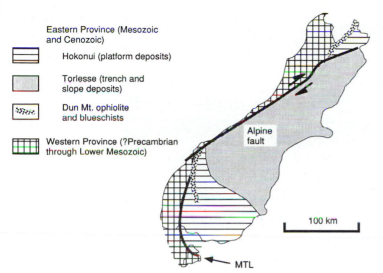

Eastern Province (Mesozoic and Cenozoic)

Hokonui (platform deposits)

Torlesse (trench and slope deposits)

Dun Mt. ophiolite and blueschists

Western Province (?Precambrian through Lower Mesozoic)

Alpine fault

100 km

MTL

Fig. 9.9. Diagrammatic geology of the South Island of New Zealand. The major Alpine transform fault juxtaposes old crust against Mesozoic/Cenozoic deposits (Hokonui and Torlesse) in the northern part of the island. The Dun Mountain ophiolite suite shows complex relationships with old crust and younger sedimentary suites in the southern part of the island. (Modified from Landis and Coombs, 1967; Suggate, 1978; Korsch and Wellman, 1988.)

Translational movement also was responsible for the northward migration of the triple junction presently in Japan from an initial position near the Philippines.

We discuss the evolution of the western Pacific margin under two headings: 1) marginal orogenic belts and median tectonic lines; and 2) opening of the marginal seas. A discussion of subduction-related magmatism, such as characterizes the western Pacific orogenic belts, is in Section 4.1.

Marginal orogenic belts and median tectonic lines

The islands of the western Pacific all have slightly different geologic histories, but a consistent feature of many of them is the 'median tectonic line' (MTL). The term 'MTL' has a variety of definitions. Commonly it is the join between high-T/P (greenschist facies series) metamorphism on the landward side of the MTL and low-T/P (blueschist facies series) metamorphism on the seaward side. The juxtaposition of greenschist-series belts and blueschist-series belts is referred to as a 'paired metamorphic belt' although the adjacent terranes are commonly separated by an MTL or other strike-slip fault zone. The MTL may also be the edge of the old continental craton, separating a landward shelf facies from a seaward trench facies. Syntectonic calcalkaline batholiths are commonly restricted to the landward side of the MTL. The MTL is invariably the site of faulting, possibly following the plane of weakness along the old continental margin. The direction of fault

movement, however, is variable along the MTL from place to place at any one time and from time to time at any one place. In order to demonstrate the variety of orogenic histories (and their relationships to MTLs) in the western Pacific, we now present capsule summaries of the South Island of New Zealand, New Guinea, Sumatra, Taiwan and southern Japan, plus brief statements about Sulawesi and the Philippines.

The South Island of New Zealand (Fig. 9.9) currently straddles the transform join (partly Alpine fault; Fig. 9.10) between the Tonga–Kermadec subduction zone, which borders the volcanically active North island, and the complicated zone of transforms and ridge segments that joins the Australia–Antarctic ridge with the East Pacific rise. Because of its curvature, part of the transform is the site of minor thrusting, and a fossil spreading center (Macquarie) lies just to the east of the oceanic part of the fault. The age of movement on the Alpine fault is uncertain, with estimates ranging from wholly Mesozoic to wholly Cenozoic and Recent. The amount of movement is at least 500 km and possibly as much as 1000 km. The relationship between the Alpine fault and the oroclinal bend in the metamorphic rocks to the east of the fault is also unclear. Complicated stress relationships along the Alpine fault yield uplift rates in the Southern Alps just east of the fault in the range of 1 km/m.y., the highest known anywhere on earth.

Most of the metamorphism recorded on the South Island is the result of the mid-Cretaceous Rangitata orogeny, which occurred before the opening of the

Fig. 9.10. Alpine fault zone of the South Island of New Zealand, seen from the coast. The fault is one of the major transforms on the margin of the Pacific Ocean. ☐

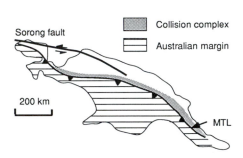

Fig. 9.11. Diagrammatic geology of New Guinea. The southern part of the island is old crust of the Australian platform. It is separated from accreted volcano–sedimentary suites along a median tectonic line (MTL). The Sorong fault is a major transform in the southwestern Pacific (Fig. 9.7). (Modified from Davies and Smith, 1971; Brown, Pigram and Skwarko, 1980; Mason and Heaslip, 1980; Milsom, 1985.) ☐

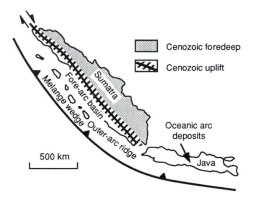

Fig. 9.12. Diagrammatic geology of the Sumatra and Java area of Indonesia. Subduction of lithosphere of the Indian Ocean has created a typical sequence of melange, outer-arc ridge, and fore-arc basin. The principal (Barisan) uplift caused development of a foredeep on the northern part of the island. (Modified from Hayes and Taylor, 1978; Hamilton, 1979; Katili and Reinemund, 1982.) ☐

Tasman Sea. The MTL, as the term is used in New Zealand, separates a Paleozoic (Australian?; Antarctic?) western crust, dominantly in greenschist and amphibolite facies, from younger rocks in blueschist, lower greenschist, and zeolite facies. The Eastern Province contains at least one major suture, the Dun Mountain ophiolite and blueschist terrane (type area for dunite). In New Zealand, the MTL does not lie along the major ophiolite belt or the join between platform (Hokonui) and trench (Torlesse) sediments.

New Guinea is primarily a continental-margin orogenic belt for the Australian landmass (Fig. 9.11). The MTL separates an arc-oceanic complex containing

ophiolites and flysch–argillite sequences metamorphosed to blueschist facies from unmetamorphosed clastic/carbonate deposits on the Australian shelf. The MTL is also the site of underthrusting of the continent by the collision complex. Orogenic activity began in the middle Cenozoic and has continued to the present. Much of the orogenic belt has undergone sinistral strike-slip faulting, consistent with the movement on the major Sorong fault (Fig. 9.7).

Sumatra is the southwestern margin of the Sunda shelf, a drowned extension of the Paleozoic and older continental crust of Southeast Asia (Fig. 9.12). *Java* is an

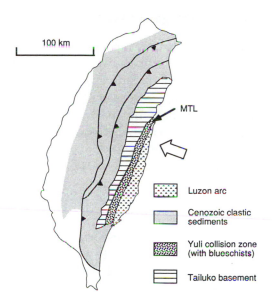

Fig. 9.13. Diagrammatic geology of Taiwan. Subduction of Pacific lithosphere developed several west-vergent thrust complexes. Basement is exposed between the thrust complexes and a blueschist/ophiolite zone bordered by a median tectonic line (MTL). The Luzon arc is a fragment of a typical volcanic arc. (Modified from Ernst, Ho and Liou, 1985.)   □

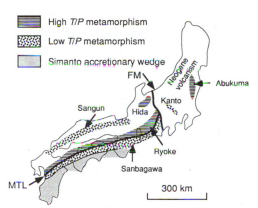

Fig. 9.14. Diagrammatic geology of Japan (except Hokkaido). The southern part of the area (south of the Fossa Magna, FM) is dominated by at least two series of paired metamorphic belts – the older Sangun and Hida, and the younger Sanbagawa and Ryoke. The Simanto accretionary wedge is currently active. Neogene volcanism north of the Fossa Magna results from subduction of Pacific lithosphere beneath Japan. (Modified from Miyashiro, 1973; Uyeda and Miyashiro, 1974; Kanaori, 1990; Ukawa, 1991.)   □

intra-oceanic arc east of the Sunda shelf. The Indonesian subduction zone (Java–Sumatra trench) is considerably offshore from the Sunda shelf, forming a broad slope and a fore-arc basin probably developed on thin continental crust (Section 4.1). Islands southwest of Sumatra are the result of uplift of the fore-arc basin. Possibly because flysch/trench deposits are not involved in the presently exposed orogenic area, no MTL has been proposed for Sumatra. The lithosphere of the southeastern Indian Ocean is moving obliquely toward Sumatra, causing dextral shear along the major orogenic belt.

Orogenic deformation and granitic magmatism began in the late Paleozoic in Sumatra and moved progressively southwestward. Activity since the middle Cenozoic is in the Barisan belt, which is not a thrust zone but is the primary locus of modern calcalkaline volcanism. Prior to the middle Cenozoic uplift, sediments in the foredeep basin of northeastern Sumatra were supplied largely by the Southeast Asian mainland, but Neogene strata were derived largely from the Barisan highland.

The island of *Taiwan* is an uplifted part of the east Asian continental margin (Fig. 9.13). Deformation of a

suite of Mesozoic/Cenozoic clastic sedimentary and volcanic rocks began in the Late Cretaceous and continued into the Neogene. All thrusting was west vergent as the subducted oceanic lithosphere disappeared under the continental margin. The oldest rocks are uplifted along the landward side of the MTL, which is bordered on the east by an ophiolite/blueschist terrane. The ophiolite terrane is a suture zone for an island arc referred to as the 'Luzon arc' because it probably formed along the northward extension of a former subduction zone along the northeastern side of Luzon, in the Philippines.

Southern *Japan*, south of the island of Hokkaido, displays a steady progression of marginal outbuilding and orogeny throughout much of the period from the late Paleozoic to the present (Fig. 9.14). The earliest evidence of this process is in western Japan, where the Sangun and Hida metamorphic terranes may constitute a typical paired metamorphic belt (see above) developed by early and middle Mesozoic deformation of upper Paleozoic and lower Mesozoic volcano–sedimentary suites. Farther east, the Ryoke and Abukuma (greenschist to granulite facies) and Sanbagawa (blueschist facies) belts are the type example of paired metamorphic belts and were used by A. Miyashiro to establish the concept of facies series. The Ryoke/Sanbagawa metamorphism occurred in the late Mesozoic to early Cenozoic and affected mostly Mesozoic clastic sediments deposited seaward of the older orogenic suites. The youngest example of seaward progression is the Simanto clastic wedge, which has undergone only low-grade blueschist-facies (prehnite–pumpellyite) and zeolite-facies metamorphism.

The MTL is a sharp demarcation between the Ryoke and Sanbagawa metamorphic suites. It presumably does not represent an old cratonic edge but is an abrupt transition between high thermal gradients landward and low gradients seaward. The movement pattern on the MTL is very complicated, probably involving a series of thrust, normal fault, and strike-slip motions at various times through much of the late Mesozoic and Cenozoic. A large proportion of the movement prior to the Neogene was left-lateral, and the entire Japanese island area may have been part of a broad sinistral shear system, bordered on the east by the MTL and on the west by a fault largely obscured by spreading in the Japan Sea. Neogene movement on the MTL, associated with the opening of the Japan Sea, was apparently mostly right lateral.

Neogene activity shows considerable differences across the Fossa Magna (FM) and adjacent structures (Fig. 9.14). Most of the active volcanism is to the north, beginning with massive outpourings of silicic pyroclastic rocks and continuing as a dominantly basaltic suite in which H. Kuno established his type example of variation in volcanic compositions with depth to the subduction zone (Section 4.1). This volcanism was initiated by the change of movement of the western Pacific plate from northwest to westerly at 40 Ma, thus providing a more perpendicular collision with Japan. Because of spreading in the Japan Sea, rapid spreading on the East Pacific rise, and approximate parallelism of the two spreading vectors, present subduction under Japan is estimated to be a high rate of nearly 10 cm/yr.

The *Philippines* and *Sulawesi* are almost entirely composed of upper Mesozoic to Cenozoic oceanic and subduction-zone complexes. The Philippines were assembled largely by strike-slip movements. Sulawesi owes its irregular shape to the accretion of several terranes to form the different 'arms' of the island. The join between western and eastern units in Sulawesi is regarded as an MTL. Eastern and southeastern Sulawesi contain some fragments of Australian continental crust brought westward by strike-slip movement.

Opening of the marginal seas

The numerous subduction zones that rim the western margin of the Pacific basin (the 'Andesite line'; Section 8.1) separate a typical spreading ocean from a set of highly variable small ocean basins (Fig. 9.7). Many of these basins are floored by oceanic lithosphere and show magnetic stripe patterns caused by ridge spreading, in some cases around rotational poles close to the basins themselves (e.g. Tasman Sea). Some apparent single basins contain two or more spreading centers separated by intra-basinal ridges of uncertain origin (e.g. the area east of the Philippines and Taiwan). Continental fragments, and possibly plateaus of other origins, occur in some basins (e.g. three apparently continental plateaus connected with New Zealand). Several seas are clearly underlain by subsided continental crust (Okhotsk, East China, Sunda, and Arafura), and these areas have abrupt borders with oceanic lithosphere (e.g South China Sea and the southern part of the Okhotsk Sea).

The process of formation of oceanic lithosphere in these small basins has attracted considerable research effort. Pertinent observations include the difficulty of recognizing organized magnetic anomaly patterns, the average elevation higher than that expected for oceanic lithosphere of the apparent age of the basins (Section 4.2), and the relatively high heat flow. The elevation and heat-flow data have been used as arguments against early geological thought that the basins were simply old Pacific Ocean lithosphere trapped by the development of intra-oceanic subduction zones, and the mapping of at least poorly defined spreading ridges has dispelled that concept.

A present question regarding the small ocean basins is the extent to which their origin can be explained purely as the result of extensional stresses established above deep parts of subducting slabs. Many of the smallest and youngest basins, which have opened within the past 10 Ma, probably have such an origin (e.g. Bismarck, Woodlark, Lau, and Havre basins and the basin on the Fiji plateau). Opening of larger and older basins, however, probably cannot be attributed solely to stresses generated by the downgoing slab. One possibility is that the purely 'back-arc' stretching has been enhanced by an additional mantle thermal anomaly that passed through the area of the western Pacific from the late Mesozoic to the present. This thermal anomaly has been termed a 'hot region' by A. Miyashiro, who distinguished it from a mantle plume that retains a relatively fixed position with respect to a mantle reference frame. Evidence for such a mantle anomaly is seen in the approximate decrease in age of the basins northward from the Tasman Sea to the Japan Sea, and the northern extension of the anomaly may be responsible for the area of Neogene volcanism and extension in northeastern China (Fig. 9.7).

The opening of the marginal basins at different times has permitted complex interactions of subduction directions under western Pacific island chains. The Philippines, Solomons and Hebrides (Vanuata) have all experienced reversal (flip-flop) of the subduction zones from one side to the other. Furthermore, the subduction of hot, young lithosphere covered by thick sequences of turbidites derived from adjacent arc and continental terranes has provided an abundant supply of fertile source material for the volcanic products.

[**References** – General information on the Pacific orogenic belts is provided by volumes edited by Uyeda, Murphy and Kobayashi (1979), Howell (1985), Nairn, Stehli and Uyeda (1985, 1988), Leitch and Scheibner (1987) and Ben-Avraham (1989). Edited volumes on island arcs include Talwani and Pitman (1977), Wezel (1986) and Hamilton (1988). Southeastern Asia is discussed by Hayes and Taylor (1978), Katili and Reinemund (1982) and papers in the volume edited by Hayes (1980). Other general papers include: Uyeda and Miyashiro (1974), on the Pacific and Japan; Kamp (1986), on the southwest Pacific; Charvis and Pelletier (1989), on the New Hebrides; Xu *et al.* (1989), on eastern Asia; Jolivet, Huchon and Rangin (1989), on the western Pacific; and Grady, James and Parker (1990) on Australasia. Paired metamorphic belts and other aspects of metamorphism are discussed by Miyashiro (1973) and Ernst (1975). Western Pacific basins are discussed by Miyashiro (1986). Doglioni (1991) discusses westward-directed subduction.

Information for specific areas is from the following.
New Zealand – Landis and Coombs (1967), Suggate (1978), and Korsch and Wellman (1988);
New Guinea – Davies and Smith (1971), Brown, Pigram and Skwarko (1980), Mason and Heaslip (1980), and Milsom (1985);
Philippines – Roeder (1977), Karig, Sarewitz and Haeck (1986), Barrier, Huchon and Aurelia (1991), and the volume edited by Flower and Hawkins (1989);
Indonesia – Katili (1978), Hayes and Taylor (1978), Hamilton (1979), and Katili and Reinemund (1982);
Taiwan – Ernst, Ho and Liou (1985);
Japan – Miyashiro (1973), Uyeda and Miyashiro (1974), Tatsumi *et al.* (1989), Kanaori (1990), and Ukawa (1991).]

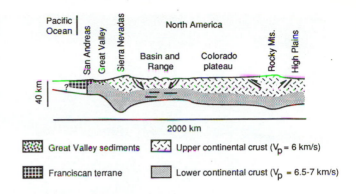

Fig. 9.15. Cross section from the High Plains of the United States to the Pacific Ocean. Upper and lower continental crust are shown with different P-wave velocities. The thicknesses of both the upper crust and entire crust are greatest under the Rocky Mountains and Sierra Nevadas and thinnest in the Basin and Range. The Colorado plateau is an area that has undergone little compressive deformation throughout the Phanerozoic. The San Andreas transform fault is approximately along the border between continental and oceanic crust, although the actual transition is uncertain. The Great Valley is a foredeep basin west of the Sierra Nevadas. (Vertical exaggeration 10×.)   ☐

9.4 Marginal growth and terrane movement in the North American Cordillera

THE CORDILLERAN region of North America, extending from Mexico to Alaska, is probably the most intensely studied part of the earth. The wealth of investigations stems partly from the excellent exposure throughout most of the area, including vegetation-free deserts, steep mountains (but not too high to climb), and river valleys incised more than 1 km into the countryside. The area is also home to a large number of geologists, particularly in the western US, who can reach their field areas in a few hours (provided they can obtain funding to do so).

Evolution of the western margin of North America in the Mesozoic and Cenozoic was dominated by the interaction of the continent with the East Pacific rise and, in Alaska, with the Kula–Pacific spreading center (Section 8.1). By some method, whose nature is highly controversial, these interactions transmitted compressional stresses up to 1500 km into the continental interior. The entire area was also affected by translational movements, with dextral shear in much of the late Mesozoic

Fig. 9.16. Eastern edge of Sierra Nevadas, California. Some of the normal faults that form the present scarp, with a relief of 3 to 4 km, have a total vertical movement of 20 km. The Sierra block is the western edge of the Basin and Range province.   □

and all of the Cenozoic and sinistral shear likely in the earlier Mesozoic. The large Basin and Range province represents continental extension in the Neogene.

As an introduction to the area, Fig. 9.15 shows a cross section in the western United States, extending from the High Plains east of the deformed area to the Pacific Ocean. Upper and lower continental crust (separated by a Conrad discontinuity?) are somewhat arbitrarily distinguished. Mantle at the Moho is recognized by P-wave velocities of 8.0 to 8.1 km/s except under the Basin and Range, where velocities are commonly slightly <8 km/s. The '?' in the crust under the Pacific Ocean indicates a complex, and little understood, relationship among accreted Franciscan (oceanic) melange, possible thin continental or transitional crust, and oceanic crust.

In Fig. 9.15, the Southern Rocky Mountains are diagrammatically depicted as 'basement-cored' uplifts along reverse faults with dips in the range of 30° to 45°. To the west of the Rockies is the broad Colorado plateau, which has a crustal thickness of 45 to 50 km and a nearly complete Phanerozoic section that, for reasons still uncertain, has escaped the deformational activity that has surrounded it for more than one hundred million years. The Basin and Range (Great Basin) is diagrammatically shown by two bounding normal faults, although the entire area is cut by listric faults that have formed a conventional horst-and-graben topography (Section 4.2). The lower crust under the Basin and Range shows numerous horizontal seismic reflectors, possibly the result of ductile shearing during the crustal extension. Extension above a rising asthenosphere is shown

by the thin crust (25 to 30 km), low upper-mantle P-wave velocities, and high heat flow.

The western margin of the Basin and Range extensional area is the extraordinary scarp of the Sierra Nevadas, formed by normal faults with 20 km (or more) of vertical movement (Fig. 9.16). The thick crust of the Sierras was developed during Mesozoic batholithic magmatism, whose products now constitute most of the high parts of the range. To the west, the Great Valley sequence occupies a down-dropped block of uncertain tectonic evolution that was filled by sediments shed from the rising Sierras. The California Coast range, west of the Great Valley, consists of oceanic and continental-margin sedimentary and magmatic rocks, including the Franciscan blueschist/ophiolite/melange complex. At the latitude of the cross section, the Coast range is cut by the San Andreas fault.

In the following sections, we discuss the evolution of the Cordillera during four different time periods (early and middle Mesozoic, Cretaceous, Paleogene and Neogene). The diagrams that depict this history (Fig. 9.17A–D) do not show one instant of time but, rather, attempt to show events that happened throughout the time period indicated. That is, they are not snapshots but time exposures. Furthermore, in these diagrams the configuration of the Cordillera has not been restored by reversing the movements on faults. The reason for this lack of palinspastic reconstruction is simply the desire to avoid controversies of what moved where, how far, and when. For example, estimates for extension in the Basin and Range are from 20% to 100% (and possibly more); regardless of the exact estimates, it is clear that the western margin of North America would have been much farther east at the latitude of the Basin and Range if the diagrams had been corrected for this extension.

Many of the controversies over the history of the Cordillera are between geologists who describe the seaward outgrowth of the continental margin in terms of sedimentation and magmatism on the margin and geologists who attribute the outgrowth to accretion of exotic terranes. Both processes have undoubtedly operated, but their relative importance is unclear. A further controversy concerns the nature of recognized (or proposed) terranes. Those that consist of island arcs formed very near the continental margin or parts of the margin that have been translated by strike-slip faulting along it are essentially part of the process of marginal outgrowth. Conversely, some terranes have been proposed to have formed in very different parts of the earth and to have been brought to the Cordilleran margin by seafloor spreading (Section 8.1). Some of these proposed 'far-travelled' terranes contain sedimentary rocks (mostly limestones) with a 'Tethyan' (generally more tropical) fauna than is found in rocks that are indigenous to the Cordillera.

Early Mesozoic

During the latest stages of the Paleozoic and beginning of the Mesozoic, the Cordilleran continental margin in the western United States was greatly modified by two processes. One was the docking of a large terrane, presumably mostly an island arc, during the Permo-Triassic Sonoma orogeny (Section 7.4). The other was left-lateral truncation of the Paleozoic miogeocline, which formerly extended an unknown distance southwestward from its present boundary (Fig. 9.17A). This major faulting was accompanied by other left-lateral shearing along the entire Cordilleran margin but appears to have preceded the left-lateral movement that rotated Mexico and Central America counterclockwise into the opening left as South America drifted away from the present Gulf of Mexico (Section 9.5). Opening of the Canadian basin in the Arctic Ocean also occurred at some poorly defined time during the Mesozoic, moving Alaska counterclockwise to its present position.

At a time in the early Mesozoic before the movements of Alaska and Central America, and after the docking and shearing in the western United States, the Cordilleran western margin must have been a fairly straight edge that separated continental platform and miogeocline from oceanic lithosphere (Fig. 9.17A). Whereas Paleozoic compression in the area had resulted largely in subduction under offshore arcs, rather than under North America, during the early and middle Mesozoic the Pacific oceanic lithosphere began active subduction under North America and created a large magmatic arc. This subduction approximately coincided with the early-Mesozoic inception of seafloor spreading and drifting in the North Atlantic (Section 8.1).

Fig. 9.17A shows an area along the western margin of the North American craton in which Paleozoic and Mesozoic sedimentary sequences are thicker than in the continental interior. Parts of this area are recognizably a Paleozoic miogeocline, particularly in the United States, with a zone (Wasatch line) separating the area of greater subsidence to the west from the area of lesser sedimentary accumulation to the east. Later Mesozoic (Laramide) thrusting appears to have involved mostly the thicker sedimentary sequences, and the Wasatch line localized the east-vergent sole thrusts for some of the Laramide orogenic movements.

Cretaceous

By Cretaceous time, magmatism and compressive deformation were widespread throughout the entire Cordillera (Fig. 9.17B). In Mexico and, more particularly, in the United States, thrusting developed far inland, possibly because of shallowness of the subducting slab. The eastward limit of major thrusting was largely along the Wasatch line and other predeformational tectonic borders, but thrusting also occurred to the east of that line, and some thrusts involved crystalline basement. Major batholithic suites (Coast Range, Sierra Nevada and Peninsular) developed mostly in the Jurassic and Cretaceous above underthrust oceanic lithosphere. Translational movements had shifted from left-lateral in the early Mesozoic to right lateral. Thus, the entire west coast of North America was under dextral transpression, but the geometry of oceanic spreading made translational movements more significant opposite the Kula plate than opposite the Farallon plate. The Kula plate and its bounding ridges moved northward during the Cretaceous and was subducted under the Aleutians in the Paleogene (Section 8.1).

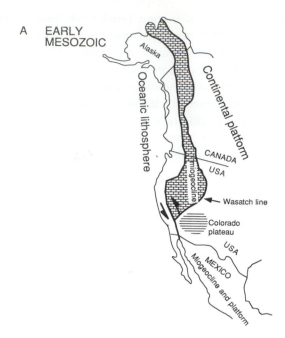

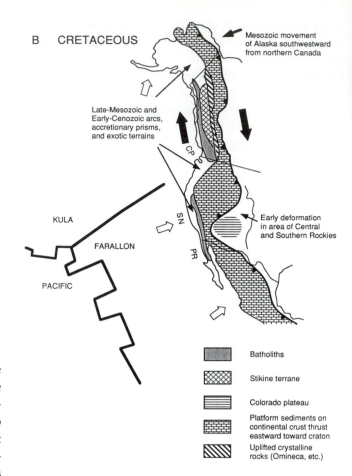

▨	Batholiths
⬚	Stikine terrane
▤	Colorado plateau
▦	Platform sediments on continental crust thrust eastward toward craton
▧	Uplifted crystalline rocks (Omineca, etc.)

Mesozoic deformation in the Canadian part of the Cordillera can be described in terms of five north–south-trending belts (Fig. 9.17B). The eastern-most belt (Eastern Cordillera) is autochthonous to North America and consists of a series of east-vergent thrusts that displace platform and miogeoclinal sediments and some basement. Compressive forces for this deformation have been ascribed to the docking of exotic terranes to the west. West of the Eastern Cordillera is a belt of crystalline rocks (Omineca) that consists largely of highly metamorphosed and intruded middle-Proterozoic to middle-Paleozoic mio-geoclinal sediments, some late-Paleozoic to early-Mesozoic volcanogenic material, and minor exposures of older-Precambrian crystalline basement. The Omineca belt is either the western margin of autochthonous North America or a continental frag-ment accreted before the Mesozoic.

To the west of the Omineca belt is the Intermontane part of the Canadian Rockies, a complex series of indi-vidual structural blocks consisting largely of island-arc volcanic rocks, volcaniclastic sediments, minor lime-stone, and ophiolites; whether they formed directly offshore from North America or at greater distances is unclear. Many of the structural blocks appear to have coalesced (e.g. Stikine) by the middle of the Mesozoic, and they may have docked against North America as large units rather than in separate episodes. This domi-

nantly oceanic assemblage is partly thrust under the Omineca crystalline belt.

The Cretaceous and early-Cenozoic Coast Plutonic complex intrudes the allochthonous and oceanic assemblages and separates the Intermontane belt from the Insular belt. The batholiths of the complex are typi-cally calcalkaline, and their emplacement was probably assisted by the right-lateral shearing that left areas of low lateral stress in the crust. The Insular belt may con-sist of separate terranes that had fused in the middle to late Mesozoic before docking in the Cretaceous. Some plutons, partly of the Coast Plutonic complex, 'stitch' the sutures between the various terranes and indicate completion of the amalgamation by the early Cenozoic.

The broad area of Alaska south of the autochthonous crust shown in Fig. 9.17B represents almost continual outbuilding of the continental margin during the later Mesozoic and the entire Cenozoic. The area is domi-nated by south-vergent thrusts in flysch (accretionary wedge) volcano–sedimentary sequences with inter-spersed ophiolites and melange zones. Although major,

C PALEOGENE

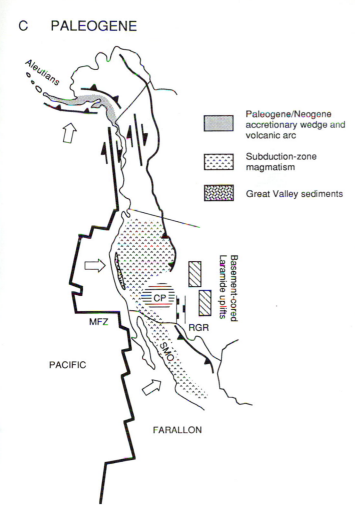

Fig. 9.17 **Fig. 9.17** Series of sketch maps indicating activity in the Cordillera of North America (Mexico to Alaska) within various time ranges through the Mesozoic and Cenozoic. Each map depicts events through a long period of time and does not imply synchroneity of all events shown. Directions of downgoing slabs are shown by the white arrows. The legend is on part B.

(A) Early Mesozoic. The miogeocline consists of rocks deposited along the western margin of cratonic North America; it is shown as wider in the area of the present Basin and Range only because of extension after deposition. The Wasatch line is the eastern limit of significant subsidence.

(B) Cretaceous. The offshore spreading–transform system is shown very diagrammatically; in particular, the position of the ridge between the Kula and Farallon plates is highly controversial. Northward movement of the Pacific lithosphere relative to stable North America is well established and will continue until the present. Platform sediments are being thrust eastward over the craton (Sevier orogenic belt). Several definable tectonic belts are shown in western Canada (see text). Major batholiths formed by subduction of oceanic lithosphere are: CP, Coast Plutonic complex; SN, Sierra Nevada; PR, Peninsular Range.

(C) Paleogene. Subduction-generated magmatism is widespread, and a very broad accretionary wedge causes outgrowth along the southern margin of Alaska. The southern Rocky Mountains is created largely by basement uplifts (diagonal lines) along reverse faults on both sides of the uplift. RGR is the Rio Grande rift. Other symbols are: SMO, Sierra Madre Occidental; CP, Colorado plateau; MFZ, Mendocino fracture zone.

(D) Neogene. Extension is shown by opposing black arrows in the Basin and Range (B&R) and Gulf of California (GC; extension of the spreading ridge between the Pacific and Cocos plates). Subduction occurs beneath southern Alaska, the Cascade Range (Cas) and western Mexico. The Yellowstone hotspot (Y) may be the terminus of a plume track along the Snake River plain (SRP). The San Andreas fault (SA) represents the dominant effect of northward movement of the Pacific plate relative to North America. Other symbols are: GJ, Gorda and Juan de Fuca plates; CP, Colorado plateau; MFZ, Mendocino fracture zone; RGR, Rio Grande rift; TMVB, Trans-Mexican volcanic belt. □

separately identifiable, thrusts have been mapped, the whole region can also be described as a broad thrust zone with pervasive shearing of the underplated oceanic crust and accretionary sediments over a cross-strike distance of several hundred kilometers.

Paleogene

The Cenozoic history of the Cordillera is separable into two periods by a major reorganization that began in the western US at ~20 Ma, approximately the Paleogene/Neogene boundary (Fig. 9.17C and D). This transition was caused by the irregular intersection of the Pacific–Farallon plate boundary with the western edge of North America. During the Paleogene, the Farallon plate became a thin sliver against the western edge of the continent, and compression ceased in the Canadian Cordillera. The Kula, and later the Pacific,

plates continued subduction and underplating of the Alaskan margin. Volcanism occurred along the entire Aleutian Island chain until ~40 Ma, when a shift of plate motion from northward to northwestward (Section 8.1) converted the subduction zone along the western Aleutians into a strike-slip fault and terminated magmatism. Subduction in Mexico led to the development of the basaltic to (mostly) silicic volcanic rocks of the Sierra Madre Occidental.

In the western US, subduction formed a broad area of calcalkaline volcanism with complexly shifting patterns of location and time. In the early Paleogene, thrusting continued to the east of the magmatic arc, continuing the Laramide orogeny that had begun in the Late Cretaceous. The area east of the Laramide

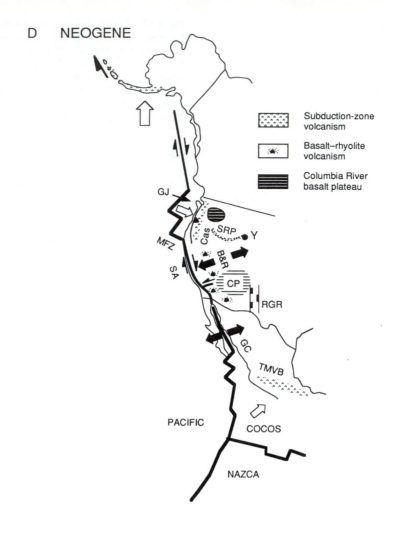

For legend see page 281.

thrust belts was also affected by compression, but the result was the uplift of 'basement-cored' blocks, probably along reverse faults of moderate dip. Accretion (of exotic terranes and/or marginal-basin deposits) continued on the western margin of the continent, and debris was shed from the Sierra Nevadas to form the Great Valley sequence, overlying some of the accretionary complexes. The oldest extension in the Cordillera occurred along the Rio Grande rift, between the Colorado plateau and the Laramide uplifts.

Neogene

Near the beginning of the Neogene, the western edge of North America intersected the triple junction formed by the East Pacific rise and the very long Mendocino fracture zone (MFZ; Fig. 9.17C and D).

This intersection converted the triple junction into a junction between the Pacific plate, the remains of the Farallon plate (Gorda and Juan de Fuca plates) and North America. As the East Pacific rise entered the subduction zone, it either was subducted and continued spreading under the continent, or according to more recent geologic concepts, was terminated as a spreading center. This termination left a 'slab window' between the San Andreas fault and the onland projection of the MFZ in which the subducted slab does not exist.

The Pacific–North American plate boundary is now the San Andreas fault, which extends from the Mendocino fracture zone to a complex intersection with a series of spreading ridges and transforms that has caused the Neogene opening of the Gulf of California (Fig. 9.17D). The fault partly localizes right-lateral motion between the two plates and has continually lengthened as the MFZ moved northward, with estimates of displacement up to 500 km. Andesite-dominated volcanism has continued only in those areas where active subduction can still occur: i.e. in southern Alaska; in the Cascade Range, opposite the Gorda and Juan de Fuca plates; and along the Trans-Mexican volcanic belt, opposite the Cocos plate (Section 8.1).

Two features shown in Fig. 9.17D are discussed elsewhere: the evolution of the Basin and Range is discussed in more detail in Section 4.2; the possibility that the Columbia River plateau and Snake River plain form a plume track from Yellowstone is mentioned in Section 8.2.

[References – General information on the North American Cordillera is provided by: volumes edited by R. Smith and Eaton (1978), Howell (1985) and Pakiser and Mooney (1989); and review papers by Burchfiel (1979), Saleeby (1983) and Oldow *et al.* (1989). The Basin and Range is discussed by Glazner and Ussler (1989) and D. Smith and Miller (1990). Evolution of the Canadian Cordillera is discussed by Monger and Price (1979), Monger, Price and Tempelman-Kluit (1982), Lambert and Chamberlain (1988) and Struik (1988). Information for Mexico is from De Cserna (1989). Information on Alaska is from Coney and Jones (1985) and papers in the volumes edited by Page (1989) and Patton and Box (1989). Interactions between Pacific plates and North America are discussed by Debiche, Cox and Engebretson (1987) and Atwater (1989). Kay and Rapela (1990) provide an edited volume on Cordilleran magmatism.]

9.5 Formation of two contrasting small ocean basins: the Gulf of Mexico and the Caribbean

EVEN a casual glance at the margins of the Gulf of Mexico and the Caribbean will suggest profound differences in their origins. The Gulf is virtually a land-locked basin that receives a high percentage of the debris eroded from North America. This influx is partly distributed along the shorelines, forming barrier islands, long sandy beaches and coastal swamps. The surrounding continental margins are rifted and passive.

In contrast, the Caribbean is surrounded by active margins and only partly isolated from the world oceans. Its waters lap against the rocky shores of vol-canic islands with small beaches tucked into protected coves. Only locally is it bordered by the reefs and swamps of passive margins, such as the 'back side' of the Central American volcanic arc. In this section, we discuss the geologic histories that have developed two such different basins.

Gulf of Mexico

A major problem in the reassembly of Pangea is the fit between North and South America. Simple northwest-ward movement of South America fills the Gulf of Mexico but causes overlap of Mexico and most of Central America onto the South American continental crust (Fig. 9.18A). A similar, though more tractable, problem involves the closure of the Atlantic, which requires overlap of Africa onto the present Bahamas. For these reasons, most reconstructions of this part of Pangea assume that Mexico and northern Central America, plus parts of the Bahamas, were rotated counterclockwise (left laterally) into their present positions during the Mesozoic (Fig. 9.18A and B).

A possible pre-rift configuration of Africa and the Americas is shown in Fig. 9.18A. This reconstruction places all of Mexico and those parts of Central America (northern areas) that are underlain by pre-Mesozoic crust to the west of South America. An alternative reconstruction, which has been proposed more recently, places the Yucatan area of Mexico, Guatemala, and Belize in the northwestern part of the present Gulf of Mexico, implying that Yucatan is separate from the rest of the Maya block and moved to near its present position, mostly during the Jurassic. Interpretation is made difficult by the thick covering of Cretaceous and lower Cenozoic carbonate and evaporitic rocks in Yucatan. Some investigators also place a rift along the site of the present Mississippi River, where Cretaceous and Cenozoic marine rocks extend far inland.

Following the development of rift basins now buried along the North American coast, marine waters were able to enter the stretched area between the Americas at some time during the Triassic, although the direction (Atlantic or Pacific) from which the marine transgression occurred is not known. This shallow ocean became the site of widespread accumulation of evaporites, including thick halite deposits (Fig. 9.18B). Evaporite/carbonate sequences, generally less

A   EARLY MESOZOIC

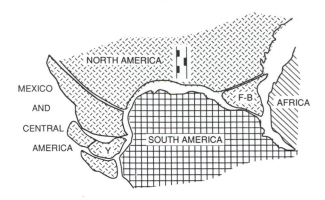

B   LATE MESOZOIC

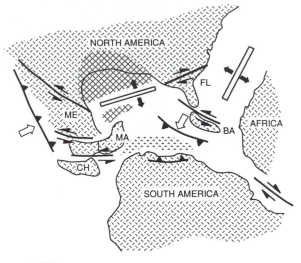

▨▨▨ Cretaceous basalt of Caribbean basin

▨▨▨ Jurassic salt covered by younger sediment

▨▨▨ Pre-Mesozoic continental crust

rich in halite, underlie Cenozoic carbonate accumulations in Yucatan and in the (now-separated) areas of Florida and the Bahama banks.

Ultimately, probably in the Late Jurassic to Early Cretaceous, the stretching between North and South America permitted the beginning of seafloor spreading and the creation of oceanic lithosphere. This lithosphere now underlies the central Gulf of Mexico, but little is known about it because it is covered by more than 10 km of clastic sediments shed from North America and Mexico. P-wave velocities in the central Gulf are as high as 8.3 to 8.4 km/s, significantly higher than the 8.0 to 8.1 km/s characteristic of major oceans. The high density of the upper mantle is presumably related to the very slow opening of the Gulf and the great thickness of sediment.

The present shape of the Gulf of Mexico was partly controlled by movement of continental plates into the gap left between North and South America (Fig. 9.18B). Although most of the movement was left lateral, right-lateral motion of the Florida–Bahama block may have occurred along a buried line of faulting in Georgia (Georgia, or Sewanee, embayment). The Bahamas probably separated from Florida by left-lateral movement on one or more faults and became a northeastern buttress against which subduction-zone assemblages from the Caribbean area collided. The northwestern part of the Bahamas has apparently accumulated as much as 10 km of carbonate rocks overlying a stretched continental crust (Fig. 9.18C), but the southeastern part of the Bahama platform may have formed on oceanic lithosphere. The Florida platform did not undergo as much thinning as the Bahamas but accumulated a thick sequence of Mesozoic/Cenozoic carbonates (intermingled with

evaporites during the Jurassic); the submarine western margin of this platform has apparently become extremely steep because of solution by groundwater.

Left-lateral movement of micro-continental blocks in Mexico and northern Central America is required by continental fitting and is partly confirmed by sparse paleomagnetic data. Unfortunately, the major zones of movement are difficult to locate. The northernmost one (the Mojave–Sonora megashear) has been proposed to show a displacement of 800 km, but it does not have a surface expression, and correlation of suites across it is problematic. Based on paleomagnetic data, left-lateral movement of 300 km of the Maya block relative to northern Mexico may have occurred along what is now the Trans-Mexican volcanic belt, but the only field evidence that it was a shear zone is the present volcanic activity, implying a zone of weakness. The Chortis block and the Maya block are separated by the combined Motagua–Polochic fault zones, along which both strike-slip and compressional movement can be proven to have occurred. Complete reconstruction of a pre-rift North America, however, requires approximately 1300 km of left lateral movement along the Motagua–Polochic and related faults, for which evidence is unavailable.

Sedimentation in the open Gulf of Mexico was con-

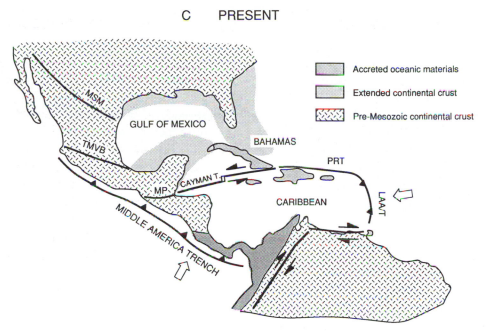

Fig. 9.18.   Sequential development of the Gulf of Mexico and Caribbean. Directions of downgoing slabs are shown by the white arrows.

(A) Closed configuration at start of Mesozoic. F-B, Florida–Bahamas block; Y, Yucatan block. The rift symbol indicates possible Mesozoic rifting along the Mississippi embayment.

(B) Approximately late Mesozoic. Spreading centers are shown in the Atlantic Ocean and in the developing Gulf of Mexico between North America and the Maya block; the Maya block may be separated into two parts, with spreading affecting only the Yucatan part (see Fig. 9.18A). Except for possible right-lateral movement of the Florida block (FL), left-lateral movement is shown for all other blocks (ME, Mexico; MA, Maya; CH, Chortis; BA, Bahamas). This lateral movement fills part of the Gulf of Mexico left by the separation of North and South America.

(C) Present configuration. Left-lateral movement along several shear zones in Mexico and Central America has essentially rotated Central America into part of the formerly larger Gulf of Mexico; symbols are: MSM, Mojave–Sonora megashear; TMVB, Trans-Mexican volcanic belt; MP, Motagua–Polochic fault zones. The Caribbean plate is outlined by: strike-slip faults along its northern margin (with one small spreading center in the Cayman area); strike-slip faults along its southern margin; and subduction beneath the Caribbean plate along its eastern and western margins; LAA/T is the Lesser Antilles trench. The Puerto Rico trench (PRT) is a former subduction zone that now shows only strike-slip movement.   □

trolled by three processes. One was the pouring of enormous amounts of clastic sediment into the nearly closed basin by rivers draining North America. The major river system (the Mississippi) has built a series of sediment fans that have advanced the continental shelf seaward and provided debris for the deep gulf. A second process, concurrent with sedimentation, was seaward movement of a 'hingeline', generally coincident with the shoreline at any time, which separated a landward area of uplift from a seaward area of extreme downwarp. A third process was movement of salt beneath the continental margins. This movement created salt domes throughout most of the marginal area, and flowage of salt toward the deep gulf caused deformation of the overlying clastic pile.

Caribbean

Although the Caribbean is now a 'plate' and presumably has been one throughout at least the later Mesozoic and the Cenozoic, even the present borders are not completely determined (Fig. 9.18A–C). At present, the Caribbean is effectively a 'tongue' of oceanic lithosphere projecting into the Atlantic as North and South America slide past it on bordering strike-slip faults. Strike-slip faulting on the northern margin partly follows the former Puerto Rican subduction zone and the Motagua–Polochic fault zone of Central America. The southern strike-slip margin extends westward into a very complex zone of dextral strike-slip faulting in the northern ranges of the Andes. A major subduction zone, the Lesser Antilles arc, has formed where spreading in the Atlantic sends oceanic lithosphere directly beneath the Caribbean lithosphere. The western side of the Caribbean plate is the Middle America trench, caused by eastward movement of the Cocos plate away from the East Pacific rise and Galapagos spreading centers (Section 8.1).

North and South America have moved independently away from the Mid-Atlantic ridge from the Jurassic to the present, showing general separation of the two continents punctuated by periods during which they came slightly closer together. During their approach, the Caribbean plate was squeezed in a north–south direction at the same time as subduction occurred on both the eastern and western edges, leading to deformation within the Caribbean and compressional orogenic belts around its margins. Compression and subduction of the Caribbean beneath the northern margin of South America caused the construction of the Venezuelan Coast ranges during the Late Cretaceous and possibly early Cenozoic. Subduction of the Atlantic lithosphere beneath the northern margin of the Caribbean occurred from the Cretaceous into the Paleogene, forming the Puerto Rico trench under the eastern part of the Greater Antilles and creating sparse examples of blueschist-facies metamorphism. Subduction may have been in opposite directions under both the Greater Antilles and the Lesser Antilles before flipping to the orientations now demonstrated by the trenches.

The history of the accretionary complexes around the Caribbean is roughly similar in different areas. The oldest rocks are probably middle-Mesozoic ophiolite suites. Organized southward (or southwestward) subduction under the Greater Antilles began in the Cretaceous and caused the formation of a 'primitive island-arc' volcanic suite, characterized by subaqueous volcanism, low LIL abundances, relatively nonradiogenic isotopes, and a bimodal SiO_2 distribution (Section 4.1). Younger volcanism changed to calcalkaline and largely subaerial, and the arc migrated northeastward toward the Bahamas in the early part of the Tertiary, after which time subduction ceased. Early-Mesozoic subduction at the eastern edge of the Caribbean changed to westward-directed subduction in the Lesser Antilles in the middle Cenozoic. The accretionary area in southern Central America is mostly the product of Neogene volcanism. This vol-

canism separated the Caribbean and Pacific oceanic waters (Section 8.1).

With the exception of a small part of northern South America and the Chortis block of Central America (Fig. 9.18B), the Caribbean plate consists of oceanic and arc magmatic rocks and sediments derived from them. The absence of terrigenous debris has characterized the entire history of the area, with the result that sediments within the deep Caribbean oceanic basin are dominantly biogenic. In the basinal areas, typical mantle P-wave velocities of >8 km/s are not reached until depths of 15 to 20 km, significantly deeper than the 10 to 12 km typical of major oceans. Deep drilling through several hundred meters of Caribbean biogenic sediments has penetrated to a layer of Middle-Cretaceous (~100 Ma) tholeiitic basalt that is very similar to basalts preserved in older suites on some of the surrounding land masses. This basalt may represent the top of Caribbean oceanic lithosphere or may lie within a dominantly sedimentary sequence formed on deeper oceanic crust.

Three major proposals have been made for the formation of the Caribbean-wide Cretaceous basalt. One is that it is a plateau-type 'flood basalt' with approximately MORB compositions. Another proposal is that the basalt originated at the Galapagos hotspot as the eastern Pacific lithosphere moved eastward and was isolated by the westward-moving American continents. This origin might also generate a basalt with MORB-like compositions, but there is no evidence that the Galapagos hotspot was active before ~2 Ma (Section 8.2). A third possibility is that the Caribbean represents a former spreading center, possibly a northeastward extension of the Galapagos spreading ridge, with typical development of MORB along it. An origin as a spreading center should yield magnetic stripes in the Caribbean, and evidence for their occurrence is doubtful.

Regardless of its nature or mode of origin, the Caribbean now acts as a small, independent plate. In size, shape, and general location (protruding eastward into the Atlantic), it is similar to the Scotia arc between South America and the Antarctic Peninsula (Section 8.1). The basin behind the Scotia arc, however, contains well-developed magnetic stripes, which are absent in the Caribbean.

[**References** – General references on the Gulf of Mexico and the Caribbean include volumes edited by Nairn and Stehli (1975) and

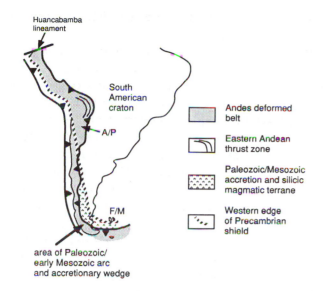

Huancabamba
lineament

South
American
craton

A/P

F/M

area of Paleozoic/
early Mesozoic arc
and accretionary wedge

Andes deformed
belt

Eastern Andean
thrust zone

Paleozoic/Mesozoic
accretion and silicic
magmatic terrane

Western edge
of Precambrian
shield

Fig. 9.19. Andes south of the Huancabamba lineament (deflection). The western edge of cratonic South America is very near the coast in the north but eastward of a zone of Paleozoic outgrowth in the south. The Andean deformed belt is situated between subducted oceanic lithosphere on the west and underthrust continental lithosphere on the east. The Altiplano–Puna (A/P) is a high plateau where the Andean belt is wide. F/M is the Falkland/Malvinas Islands. □

Scotese and Sager (1988) and articles by Anderson and Schmidt (1983), Pindell (1985) and Donnelly (1989a). Specific information on the Gulf of Mexico is from Pilger (1978) and a volume edited by Pilger (1980). Specific information on the Caribbean is from review articles by Donnelly (1989b) and Dengo and Case (1990) and a volume edited by Bonini, Hargraves and Shagam (1984). Mexico is discussed by Young (1983) and Urrutia-Fucugauchi and Bohnel (1988). Paull and Neumann (1987) discuss steepening of carbonate bank margins.]

9.6 Formation of continental-margin arcs in the Andes and Antarctic Peninsula

THE ANDES rise almost directly from the crashing surf of the Pacific Ocean along the entire western coast of South America, the product of subduction of Pacific Ocean lithosphere over a distance of nearly 7000 km (Fig. 9.19). The peaks of recently active volcanoes project above the mostly arid landscape of intermittent river courses and hardy people accustomed to breathing thin air. Although access to many areas is poor,

outcrops are good, and geologists in increasing numbers have seized the opportunity to discover what really happens when an oceanic plate is subducted directly beneath a continent. Because the evolution of the Antarctic Peninsula (Figs 7.1 and 9.1) appears to have been caused by similar processes during the same time period, we discuss the Antarctic Peninsula briefly in this section.

The present configuration of the eastern Pacific places three distinct plates of oceanic lithosphere against the western edge of South America. In the north, the Cocos plate impinges on the relatively low-lying, multiple, Andean ranges of Ecuador and Colombia, which were constructed mostly on a basement of oceanic lithosphere by a combination of subduction and strike-slip movement (Section 9.5). Current subduction in this area is mostly northeastward, building the Central American isthmus and causing only minor, but commonly catastrophic, volcanic activity in northwestern South America. The southern boundary of the Cocos plate is the Galapagos spreading ridge (Section 8.1), which intersects South America slightly north of the major bend of the Andes (Huancabamba deflection, or lineament; Fig. 9.19) and separates the Nazca and Cocos plates.

The major lithosphere currently being subducted under South America is the Nazca plate (Section 8.1). This subduction, and the consumption of earlier plates in the same area, has constructed the high part of the Andes, perhaps most typified by the Altiplano/Puna region of modern volcanic suites covering upper Mesozoic and Cenozoic deformed sediments. The Nazca plate is separated from the Antarctic plate by a series of ridge segments and transform faults (Section 8.1). In southernmost South America, the Antarctic plate intersects South America at a low (glancing?) angle and has caused only a small amount of modern compressional and magmatic activity. Thus, the southernmost part of the Andes is topographically lower than farther north and retains a Mesozoic deformational and magmatic history relatively unmodified by younger events.

The Antarctic Peninsula is now separated from the southern tip of South America by the Drake passage, the site of interchange of water between the Atlantic and Pacific and a major factor in the evolution of Cenozoic climates (Section 8.4). During the Mesozoic, before separation, the Antarctic Peninsula and the southern Andes may have formed a nearly linear magmatic arc facing the Pacific Ocean.

The western margins of South and North America have strikingly different geologic histories despite the fact that they have been bordered by oceanic lithosphere throughout the Phanerozoic and possibly earlier. By some combination of terrane accretion, marginal sedimentation, extension, and magmatism, the western margin of North America is now significantly farther to the west than it was at the beginning of the Phanerozoic (Sections 7.4 and 9.4). The Andes, however, are built almost wholly on the edge of the Precambrian basement of South America and exhibit little accretion of exotic terranes or outbuilding caused by accretionary wedges. A further distinction is that subduction under the western (in present orientation) margin of North America did not begin until the Mesozoic, with earlier accretion caused by west-vergent subduction; under South America, in contrast, subduction has been directed under the continental margin throughout the Phanerozoic.

Andes

The long geologic history of the Andean region is difficult to decipher because of the tendency of succeeding orogenic pulses to affect the same area as earlier ones. The pre-Mesozoic record is very fragmentary except in an eastward bulge of the central part of the Andes, where exposures of Paleozoic (probably miogeoclinal) sediments have undergone Mesozoic/Cenozoic east-vergent thrusting (Fig. 9.19; Section 7.5), and in areas of Paleozoic magmatism and deformation east of the southern part of the Andes.

The most decipherable record of the Andes starts in the early Mesozoic and is best displayed toward the southern tip of South America, where the greatest width of land area occurs between the western margin of the continent and the western limit of known Precambrian shield (Fig. 9.19). Here, Paleozoic volcano–sedimentary terranes on the shield margin were intruded and covered by a lower Paleozoic to upper Mesozoic granite–rhyolite suite similar to the Proterozoic 'granite–rhyolite province' of southern North America (Section 5.2). The new early-Mesozoic margin became the site of a volcanic arc, sedimentary wedge and ocean-floored back-arc basin that were active until approximately the middle Cretaceous and

built the continental margin westward to its present position.

Farther north along the Andes, the early and middle Mesozoic were times of development of marginal magmatic arcs and back-arc basins between the arc and the platformal areas of the South American craton. Throughout much of the length of the Andes, the basins were characterized by a shallow-water depositional environment of most of the sediments and an abundance of ignimbritic and other pyroclastic deposits. These basins apparently were floored by continental crust, indicating very little stretching of the continental margin and no creation of separate offshore 'terranes' that might later be regarded as 'exotic'.

The arc and basin complexes along western South America were compressed against the craton with increasing intensity toward the latter part of the Mesozoic. The major event in the orogenic history was in the middle Cretaceous, when accelerated seafloor spreading in the South Atlantic caused more rapid clockwise rotation of South America toward the Pacific (Section 8.1). This enhanced compression crushed the magmatic and sedimentary belts into, and onto, the crystalline basement and was primarily responsible for developing the present Andean orogenic belt. The Cretaceous compression has continued, with intervals of abatement, to the present.

Much of the near-coastal area of the Andes is occupied by large composite batholiths. The two major ones are the Coastal batholith, mostly in Peru, and the Patagonian batholith, in Argentina and Chile. Intrusion ages range generally from Jurassic to early Cenozoic, with some indication that magmatic suites become younger toward the west. Intrusion level is quite shallow for many bodies, and some plutons appear to intrude the volcanic pile formed by eruption from their own magma chambers.

Following closure of the marginal basins, volcanism has occurred throughout most of the Andes and in some areas east of the deformed belt. Because the volcanic activity has continued until the present, much of the older history of the Andes is buried. An early classification of Andean volcanic suites consisted of a 'Rhyolite Formation' that formed the surface of the high plateaus (principally Altiplano/Puna) and an 'Andesite Formation' that formed modern volcanic cones (where the original andesites were described). This classification no longer has chronological significance because of the contemporaneity of rhyodacitic and andesitic volcanism, but the ratio of andesitic to rhyodacitic volcanism appears to have increased toward the later Cenozoic. The magmatic activity in the Andes occupies the same area as the compressed sediments from marginal basins and the older magmatic arcs, thus forming a very narrow orogenic/magmatic belt very different from the broad deformational area of North America.

Because of the exceptional concentration of calcalkaline intrusive and extrusive suites in the Andes, and the covering of older metamorphic rocks by the volcanic blanket, many geologists have regarded the Andes as consisting largely of igneous material newly extracted from the mantle. This 'juvenile' igneous rock could cause the increase in volume of the Andes required by the deep root (to 70 km under the high plateaus). An alternative explanation is derivation of the magmas at least partly by remelting of crust of the South American craton, which would not generate any new continental material and would require thickening of the Andean belt only by structural rearrangements of existing material.

One solution to the preceding problem is to attribute the thickening of the Andes largely to tectonic shortening by stacking of thrust sheets. Restoration of pre-deformation widths of Paleozoic sedimentary sections along the eastern edge of the Andes (Fig. 9.19) suggests that much of the geophysically determined volume of the Andes could be accounted for by pre-Andean sedimentary and basement rocks. This origin would greatly limit the amount of igneous material added from the mantle and is consistent with crustal remelting as a source for much of the Andean magmas.

Antarctic Peninsula

The ice cover, the remoteness and the climate all combine to make the Antarctic the least-understood continental area of the earth. The continent can, however, be divided into a Precambrian craton (East Antarctica; Section 3.0) and a probable collage of disparate continental and arc fragments (West Antarctica). The two regions are separated by the Transantarctic Mountains, a compressive orogenic belt active during both Paleozoic and somewhat younger time (Section 7.5). Because of its complexity and the lack of information, West Antarctica has been called the 'problem

child of Gondwanaland' (Dalziel and Elliot, 1982). The Antarctic Peninsula is the best exposed and most accessible part of West Antarctica (Fig. 7.1A).

The geologic history of the Antarctic Peninsula appears to be very similar to that of the southern Andes. Many reconstructions of Gondwana place the peninsula alongside of, or the southern continuation of, the southern tip of South America (Fig. 7.1). Thus, rocks in the Antarctic Peninsula record subduction from the Paleozoic (and late Proterozoic?) into at least the early Cenozoic and possibly younger. Separation of the Antarctic Peninsula from South America was caused by the development of the Scotia Sea, which contains a complex pattern of two small spreading ridges and several transform faults (Section 8.1). Both the northern and southern margins of the Scotia plate are strike-slip faults that have moved the Scotia plate relatively eastward past the South American and Antarctic plates. The plate configuration and strike-slip movement are virtually identical to those in the Caribbean, the difference being the clear evidence that the Scotia plate is oceanic lithosphere formed by spreading of organized ridges, in contrast to the uncertain origin of the Caribbean crust (Section 9.5).

[**References** – General information is provided by the volumes edited by Nairn *et al.* (1985) and Monger and Francheteau (1987) and a review by Mpodozis and Forsythe (1983). Specific information on the tectonics of the Andes and pre-Andean activity is from a book by Zeil (1979) and papers by Dalziel (1986), Jordan and Allmendinger (1986), Ramos (1988), Soler, Carlier and Marocco (1989), Sheffels (1990) and Wilson (1991). Andean magmatism is discussed in volumes edited by Linares, Cordani and Munizaga (1982) and Kay and Rapela (1990), a book by Pitcher *et al.* (1985), a volume published by Congreso Geologico Argentino (1987) and papers by Baker and Francis (1978) and Kay *et al.* (1989). Dalziel and Elliot (1982) discuss West Antarctica, and general information on the Antarctic is provided by volumes edited by Oliver, James and Jago (1983) and Thomson, Crame and Thomson (1991).]

References

Anderson, T. H. & Schmidt, V. A. (1983). The evolution of Middle America and the Gulf of Mexico–Caribbean Sea region during Mesozoic time. *Geological Society of America Bulletin*, **94**, 941–66.

Atwater, T. (1989). Plate tectonic history of the northeast Pacific and western North America. In *The Eastern Pacific Ocean and Hawaii, The Geology of North America, Volume N*, ed. E. I. Winterer, D. M. Hussong & R. W. Decker, pp. 21–72. Boulder, Colorado: Geological Society of America.

Aubouin, J., Le Pichon X. & Monin, A. S., eds (1986). Evolution of the Tethys. Special Issue of *Tectonophysics*, **123**, 315 pp.

Audley-Charles, M. G. & Hallam, A., eds (1988). *Gondwana and Tethys*. Geological Society of London Special Publication 37, 317 pp.

Baker, M. C. W. & Francis, P. W. (1978). Upper Cenozoic volcanism in the Central Andes – ages and volumes. *Earth and Planetary Science Letters*, **6**, 175–87.

Barrier, E., Huchon, P. & Aurelia, M. (1991). Philippine fault: a key for Philippine kinematics. *Geology*, **19**, 32–5.

Ben-Avraham, Z., ed. (1989). *The Evolution of the Pacific Ocean Margins*, Oxford Monographs on Geology and Geophysics No. 8. New York: Oxford University Press, 234 pp.

Berckhemer, H. & Hsu, K., eds (1982). *Alpine–Mediterranean Geodynamics*: American Geophysical Union Geodynamics Series, vol. 7, 216 pp.

Biju-Duval, B. & Montadert, L., eds (1977). *Structural History of the Mediterranean Basins*. Paris: Editions Technip, 448 pp.

Bonini, W. E., Hargraves, R. B. & Shagam, R., eds. (1984). *The Caribbean–South American Plate Boundary and Regional Tectonics*. Geological Society of America Memoir 162, 421 pp.

Brown, C. M., Pigram, C. J. & Skwarko, S. K. (1980). Mesozoic stratigraphy and geological history of Papua New Guinea. *Palaeogeography, Palaeoclimatology, and Palaeoecology*, **29**, 301–22.

Burchfiel, B. C. (1976). *Geology of Romania*. Geological Society of America Special Paper 158, 82 pp.

Burchfiel, B. C. (1979). Geologic history of the central western United States. *Nevada Bureau of Mines and Geology Report* 33, pp. 1–11.

Butler, R. W. H. & Prior, D. J. (1988). Anatomy of a continental subduction zone: the main mantle thrust in Northern Pakistan. *Geologische Rundschau*, **77**, 239–55.

Channell, J. E. T., D'Argenio, B. & Horvath, F. (1979). Adria, the African promontory, in Mesozoic Mediterranean palaeogeography. *Earth Science Reviews*, **15**, 213–92.

Charvis, P. & Pelletier, B. (1989). The northern New Hebrides back-arc troughs: history and relation with the North Fiji basin. *Tectonophysics*, **170**, 259–77.

Coney, P. J. & Jones, D. L. (1985). Accretion tectonics and crustal structure in Alaska. *Tectonophysics*, **119**, 265–83.

Congreso Geologico Argentino (1987). *Andean magmatism and its tectonic setting*. X Congreso Geologico Argentino, vol. 4, 379 pp.

Cooper, M. A. & Williams, G. D., eds (1989). *Inversion Tectonics*. Geological Society of London Special Publication 44, 375 pp.

Coward, M. P., Dietrich, D. & Park, R.G., eds (1989). *Alpine Tectonics*. Geological Society of London Special Publication 45, 450 pp.

Dalziel, I. W. D. (1986). Collision and Cordilleran orogenesis: an Andean perspective. In *Collision Tectonics*, ed. M. P. Coward & A. C. Ries, pp. 389–404. Geological Society of London Special Publication 19.

Dalziel, I. W. D. & Elliot, D. H. (1982). West Antarctica: problem child of Gondwanaland. *Tectonics*, **1**, 3–19.

Davies, H. J. & Smith, I. E. (1971). Geology of eastern Papua. *Geological Society of America Bulletin*, **82**, 3299–312.

Debiche, M. G., Cox, A. & Engebretson, D. C. (1987). *The Motion of Allochthonous Terranes across the North Pacific Basin*. Geological Society of America Special Paper 207, 49 pp.

De Cserna, Z. (1989). An outline of the geology of Mexico. In *The Geology of North America; An Overview: The Geology of North America, Volume A*, ed. A. W. Bally & A. R. Palmer, A.R., pp. 233–64. Boulder, Colorado: Geological Society of America.

Dengo, G. & Case, J.E., eds. (1990). *The Caribbean Region: The Geology of North America*, vol. H. Boulder, Colorado: Geological Society of America, 528 pp.

Dixon, J. E. & Robertson, A. H. F., eds. (1984). *The Geological Evolution of the Eastern Mediterranean*. London: Geological Society of London, 824 pp.

Doglioni, C. (1991). A proposal for the kinematic modelling of W-dipping subductions - possible applications to the Tyrrhenian–Apennines system. *Terra Nova*, **3**, 423–34.

Donnelly, T. W. (1989a). History of marine barriers and terrestrial connections: Caribbean paleogeographic inference from pelagic sediment analysis. In *Biogeography of the West Indies*, ed. C. Woods, pp. 103–118. Leiden: E.J. Brill.

Donnelly, T. W. (1989b). Geologic history of the Caribbean and Central America, In *The Geology of North America – An Overview: The Geology of North America, vol. A*, ed. A. W. Bally & A. R. Palmer, pp. 299–321. Boulder, Colorado: Geological Society of America.

Ernst, W. G. (1975). Systematics of large-scale tectonics and age progressions in Alpine and circum-Pacific blueschist belts. *Tectonophysics*, **26**, 229–46.

Ernst, W. G., Ho, C. S. & Liou, J. G. (1985). Rifting, drifting, and crustal accretion in the Taiwan sector of the Asiatic continental margin. In *Tectonostratigraphic Terranes of the Circum-Pacific Region*, ed. D. G. Howell., pp. 375–89. Houston, Texas: Circum-Pacific Council for Energy and Mineral Resources.

Farah, A. & DeJong, K. A., eds (1979). *Geodynamics of Pakistan*. Quetta, Pakistan: Geological Survey of Pakistan, 361 pp.

Flower, M. F. J. & Hawkins, J. W., eds. (1989). Ophiolites and crustal genesis in the Philippines. Special Issue of *Tectonophysics*, **168**, 1–237.

Glazner, A. F. & Ussler, W., III (1989). Crustal extension, crustal density, and the evolution of Cenozoic magmatism in the Basin and Range of the western United States. *Journal of Geophysical Research*, **94**, 7952–60.

Grady, A. E., James, P. R. & Parker, A. J. (1990). Australasian Tectonics. Special Issue of *Journal of Structural Geology*, **12**, 519–803.

Hamilton, W. B. (1979). *Tectonics of the Indonesian Region*. U.S. Geological Survey Professional Paper 1078, 345 pp.

Hamilton, W. B. (1988). Plate tectonics and island arcs. *Geological Society of America Bulletin*, **100**, 1503–27.

Hayes, D. E., ed. (1980). *The Tectonic and Geologic Evolution of Southeast Asian Seas and Islands*. American Geophysical Union, Geophysical Monograph 23, Part 1, 326 pp; Part 2, 395 pp.

Hayes, D. E. & Taylor, B. (1978). *A Geophysical Atlas of the East and Southeast Asian Seas*. Geological Society of America Map Series MC-25.

Hirn, A., Lepine, J.-C., Jobert, G., Sapin, M., Wittlinger, G., Xin, X. Z., Yuan, G. E., Jing, W. X., Wen, T. J., Bai, X. S., Pandey, M. R. & Tater, J. M. (1984a). Crustal structure and variability of the Himalayan border of Tibet. *Nature*, **307**, 23–5.

Hirn, A., Nercessian, A., Sapin, M., Jobert, G., Xin, X. Z., Yuan, G. E., Yuan, L. D. & Wen, T. J. (1984b). Lhasa block and bordering sutures – a continuation of a 500-km Moho traverse through Tibet. *Nature*, **307**, 25–7.

Hodges, K. V., LeFort, P. & Pecher, A. (1988). Possible thermal buffering by crustal anatexis in collisional orogens: thermobarometric evidence from the Nepalese Himalaya. *Geology*, **16**, 707–10.

Howell, D. G., ed. (1985). *Tectonostratigraphic Terranes of the Circum-Pacific Region*. Earth Science Series No. 1. Houston, Texas: Circum-Pacific Council for Energy and Mineral Resources, 581 pp.

Hsu, K. J. (1983). *The Mediterranean Was a Desert*. Princeton: Princeton University Press, 197 pp.

Hutchison, C. R. (1989). *Geological Evolution of South-East Asia*. Oxford: Oxford University Press, 368 pp.

Jacobshagen, V. (1986). *Geologie von Griechenland*. Berlin: Gebruder Borntrager, 363 pp.

Jolivet, L., Huchon, P. & Rangin, C. (1989). Tectonic setting of western Pacific marginal basins. *Tectonophysics*, **160**, 23–47.

Jordan, T. E. & Allmendinger, R. W. (1986). The Sierras Pampeanas of Argentina: a modern analogue of Rocky Mountain foreland deformation. *American Journal of Science*, **286**, 737–64.

Kamp, P. J. J. (1986). The Cretaceous–Cenozoic tectonic development of the Southwest Pacific region. *Tectonophysics*, **121**, 225–51.

Kanaori, K. (1990). Late Mesozoic–Cenozoic strike-slip and block rotation in the inner belt of Southwest Japan. *Tectonophysics*, **177**, 381–99.

Karig, D. E., Sarewitz, D. R. & Haeck, G. D. (1986). Role of strike-slip faulting in the evolution of allochthonous terranes in the Philippines. *Geology*, **14**, 852–5.

Katili, J. A. (1978). Past and present geotectonic position of Sulawesi, Indonesia. *Tectonophysics*, **45**, 289–322.

Katili, J. A. & Reinemund, J. A. (1982). *Southeast Asia: Tectonic Framework, Earth Resources and Regional Geological Programs*. International Union of Geological Sciences Publication No. 13, 68 pp.

Kay, S. M., Ramos, V .A., Mpodozis, C. & Sruoga, P. (1989). Late Paleozoic to Jurassic silicic magmatism at the Gondwana margin: analogy to the middle Proterozoic in North America. *Geology*, **17**, 324–8.

Kay, S. M. & Rapela, C. W., eds. (1990). *Plutonism from Antarctica to Alaska*. Geological Society of America Special Paper 241, 263 pp.

Korsch, R. J. & Wellman, H. W. (1988). The Geological Evolution of New Zealand and the New Zealand region. In *The Ocean Basins and Margins, vol. 7B, The Pacific Ocean*, ed. A. E. M. Nairn, F. G. Stehli & S. Uyeda, pp. 411–82. New York: Plenum Press.

Lambert, R. St.J. & Chamberlain, V. E. (1988). Cordillera revisited, with a three-dimensional model for Cretaceous tectonics in British Columbia. *Journal of Geology*, **96**, 47–60.

Landis, C. A. & Coombs, D. S. (1967). Metamorphic belts and orogenesis in southern New Zealand. *Tectonophysics*, **4**, 501–18.

Leitch, E. C. & Scheibner, E., eds. (1987). *Terrane Accretion and Orogenic Belts*. American Geophysical Union Geodynamics Series, 19, 343 pp.

Linares, E., Cordani, U. & Munizaga, F., eds. (1982). Magmatic Evolution of the Andes: Special Issue of *Earth-Science Reviews*, **18**, 199–443.

Mason, D. R. & Heaslip, J. H. (1980). Tectonic setting and origin of intrusive rocks and related porphyry copper deposits in the western highlands of Papua New Guinea. *Tectonophysics*, **63**, 125–37.

McKerrow, W. S. & Scotese, C. R., eds. (1990). *Paleozoic Palaeogeography and Biogeography*. Geological Society of London Memoir 12, 435 pp.

Milsom, J. (1985). New Guinea and the western Melanesian arcs. In *The Ocean Basins and Margins, vol. 7A, The Pacific Ocean*, ed. A. E. M. Nairn, F. G. Stehli & S. Uyeda, pp. 551–605. New York: Plenum Press.

Miyashiro, A. (1973). *Metamorphism and Metamorphic Belts*. New York: Halsted Press, 492 pp.

Miyashiro, A. (1986). Hot regions and the origin of marginal basins in the western Pacific. *Tectonophysics*, **122**, 195–216.

Monger, J. W. & Francheteau, J., eds. (1987). *Circum-Pacific Orogenic Belts and Evolution of the Pacific Ocean Basin, Geodynamics Series, vol. 18*. American Geophysical Union, 165 pp.

Monger, J. W. H. & Price, R. A. (1979). Geodynamic evolution of the Canadian Cordillera – progress and problems. *Canadian Journal of Earth Sciences*, **16**, 770–90.

Monger, J. W. H., Price, R. A. & Tempelman-Kluit, D. J. (1982). Tectonic accretion and origin of the two major metamorphic and plutonic welts in the Canadian Cordillera. *Geology*, **10**, 70–5.

Mpodozis, C. & Forsythe, R. (1983). Stratigraphy and geochemistry of accreted fragments of the ancestral Pacific floor in southern South America. *Palaeogeography, Palaeoclimatology, and Palaeoecology*, **41**, 103–24.

Nairn, A. E. M. & Stehli, F. G., eds. (1975). *The Ocean Basins and Margins, vol. 3, The Gulf of Mexico and the Caribbean*. New York: Plenum Press, 706 pp.

Nairn, A. E. M., Stehli, F. G. & Uyeda, S., eds. (1985). *The Ocean Basins and Margins, vol. 7A, The Pacific Ocean*. New York: Plenum Press, 783 pp.

Nairn, A. E. M., Stehli, F. G. & Uyeda, S., eds. (1988). *The Ocean Basins and Margins, vol. 7B, The Pacific Ocean*. New York: Plenum Press, 642 pp.

Oldow, J. S., Bally, A. W., Ave Lallemant, H. G. & Leeman, W. P. (1989). Phanerozoic evolution of the North American Cordillera: United States and Canada. In *The Geology of North America; An Overview: The Geology of North America, Volume A*, ed. A. W. Bally & A. R. Palmer, pp 139–232. Boulder, Colorado: Geological Society of America.

Oliver, R. L., James, P. R. & Jago, J. B., eds. (1983). *Antarctic Earth Science*. Cambridge: Cambridge University Press, 697 pp.

Page, R. A., ed. (1989). Special Section: Northern Chugach Mountains-Southern Copper River basin segment of the Alaskan Transect. Part 1, *Journal of Geophysical Research*, **94**, 4253–466; Part 2: *Journal of Geophysical Research*, **94**, 16 021–82.

Pakiser, L. C. & Mooney, W. D., eds. (1989). *Geophysical Framework of the Continental United States*. Geological Society of America Memoir 172, 826 pp.

Patton, W. W., Jr. & Box, S. E., eds (1989). Yukon–Koyukuk basin and its borderlands, western Alaska Special Section of *Journal of Geophysical Research*, **94**, 15 805–16 020.

Paull, C. K. & Neumann, A. C. (1987). Continental margin brine seeps: Their geological consequences. *Geology*, **15**, 545–8.

Pilger, R. H., Jr. (1978). A closed Gulf of Mexico, pre-Atlantic ocean plate reconstruction and the early rift history of the Gulf and North Atlantic. *Gulf Coast Association of Geological Societies Transactions*, **28**, 385–93.

Pilger, R. H., Jr., ed. (1980). *The Origin of the Gulf of Mexico and the Early Opening of the Central North Atlantic Ocean*. Baton Rouge, Louisiana: School of Geoscience, Louisiana State University, 103 pp.

Pindell, J. L. (1985). Alleghenian reconstruction and subsequent evolution of the Gulf of Mexico, Bahamas, and proto-Caribbean. *Tectonics*, **4**, 1–39.

Pitcher, W. S., Atherton, M. P., Cobbing, E. J. & Beckinsale, R. D. (1985). *Magmatism at a Plate Edge – The Peruvian Andes*. Glasgow: Blackie and Son, 328 pp.

Ramos, V. A. (1988). The tectonics of the central Andes; 30° to 33° S latitude. In *Processes in Continental Lithospheric Deformation*, ed. S. P. Clark., Jr., B. C. Burchfiel & J. Suppe, pp. 31–54. Geological Society of America Special Paper 218.

Roeder, D. (1977). Philippine arc system – Collision or flipped subduction zones? *Geology*, **5**, 203–6.

Royden, L. & Karner, G. D. (1984). Flexure of lithosphere beneath Apennine and Carpathian foredeep basins: evidence for an insufficient topographic load. *American Association of Petroleum Geologists Bulletin*, **68**, 704–12.

Rybach, L. & Lambert, A., eds. (1980). Symposium Alpine Geotraverses. *Eclogae Geologicae Helvetiae*, **73**, 353–679.

Saleeby, J. R. (1983). Accretionary tectonics of the North American Cordillera. *Annual Review of Earth and Planetary Sciences*, **15**, 45–73.

Scotese, C. R. & Sager, W. W., eds. (1988). Mesozoic and Cenozoic plate reconstructions. Special Issue of *Tectonophysics*, **155**, 1–399.

Sengor, A. M. C. (1984). *The Cimmeride Orogenic System and the Tectonics of Eurasia*. Geological Society of America Special Paper 195, 82 pp.

Sengor, A. M. C. (1987). Tectonics of the Tethysides: orogenic collage development in a collisional setting. *Annual Reviews of Earth and Planetary Sciences*, **15**, 213–44.

Sheffels, B. M. (1990). Lower bound on the amount of crustal shortening in the central Bolivian Andes. *Geology*, **18**, 812–15.

Smith, D. L. & Miller, E. L. (1990). Late Paleozoic extension in the Great Basin,western United States. *Geology*, **18**, 712–15.

Smith, R. B. & Eaton, G. P., eds. (1978). *Cenozoic Tectonics and Regional Geophysics of the Western Cordillera*. Geological Society of America Memoir 152, 388 pp.

Societe de la Geologique de France (1990). Tethys: Special Issue of *Bulletin de la Societe Geologique de France*, ser. 8, vol. **6**, 867–1048.

Soler, P., Carlier, G. & Marocco, K. (1989). Evidence for the subduction and underplating of an oceanic plateau beneath the south Peruvian margin during the Late Cretaceous: structural implications. *Tectonophysics*, **163**, 13–24.

Spencer, A. M., ed. (1974). *Mesozoic–Cenozoic Orogenic Belts*. Geological Society of London Special Publication 4, 809 pp.

Stanley, D. J. & Wezel, F.-C., eds. (1985). *Geological Evolution of the Mediterranean Basin*. New York: Springer Verlag, 589 pp.

Struik, L. C. (1988). Crustal evolution of the eastern Canadian Cordillera. *Tectonics*, **7**, 727–47.

Suggate, R. P., ed. (1978). *The Geology of New Zealand (two volumes)*. Wellington: New Zealand Geological Survey, 820 pp.

Talwani, M. & Pitman, W.C., III, eds. (1977). *Island Arcs, Deep-Sea Trenches and Back-Arc Basins*. American Geophysical Union Maurice Ewing Series, vol. 1, 470 pp.

Tatsumi, Y., Otofuji, Y.-I., Matsuda, T. & Nohda, S. (1989). Opening of the Sea of Japan back-arc basin by asthenospheric injection. *Tectonophysics*, **166**, 317–29.

Thomson, M. R. A., Crame, J. A. & Thomson, J. W. (1991). *Geological Evolution of Antarctica*. Cambridge: Cambridge University Press, 722 pp.

Trumpy, R. (1980). *An Outline of the Geology of Switzerland: Part A of Geology of Switzerland, A Guide-Book*. Basel: Schweizerische Geologische Kommission: Wepf and Co., 102 pp.

Ukawa, M. (1991). Collision and fan-shaped compressional stress pattern in the Izu block at the northern edge of the Philippine Sea plate. *Journal of Geophysical Research*, **96**, 713–28.

Urrutia-Fucugauchi, J. & Bohnel, H. (1988). Tectonics along the Trans-Mexican volcanic belt according to paleomagnetic data. *Physics of the Earth and Planetary Interiors*, **52**, 320–9.

Uyeda, S. & Miyashiro, A. (1974). Plate tectonics and the Japanese islands: A synthesis. *Geological Society of America Bulletin*, **85**, 1159–70.

Uyeda, S., Murphy, R. W. & Kobayashi, K., eds. (1979). *Geodynamics of the Western Pacific: Advances in Earth and Planetary Sciences, vol. 6*. Japanese Center for Academic Publications, 392 pp.

Valdiya, K. S. (1984). *Aspects of Tectonics – Focus on South-Central Asia*. New Delhi, India: Tata McGraw-Hill, 319 pp.

Wezel, F.-C., ed. (1986). *The Origin of Arcs: Developments in Geotectonics, vol. 21*. Amsterdam: Elsevier, 597 pp.

Wilson, T. J. (1991). Transition from back-arc to foreland basin development in the southernmost Andes: stratigraphic record from the Ultima Esperanza District. *Geological Society of America Bulletin*, **103**, 99–111.

Xu, J., Tong, W., Zhu, G., Lin, S. & Ma, G. (1989). An outline of the pre-Jurassic tectonic framework in east Asia. *Journal of Southeast Asian Earth Sciences*, **3**, 29–45.

Young, K. (1983). Mexico. In *The Phanerozoic Geology of the World – The Mesozoic, B*, ed. M. Moullade & A. E. M. Nairn, pp. 61–88. Amsterdam: Elsevier.

Zeil, W. (1979). *The Andes – A Geological Review: Beitrage zur Regionalen Geologie der Erde, vol. 13*. Berlin: Gebruder Borntrager, 260 pp.

Ziegler, P. A. (1988). *Evolution of the Arctic–North Atlantic and the Western Tethys*. American Association of Petroleum Geologists Memoir 43, 198 pp.

AUTHOR INDEX

Page numbers in italic type refer to figure captions.

293

SUBJECT INDEX

Page numbers in italic type refer to figures and captions.